PROGRESS IN COLLOID & POLYMER SCIENCE

Editors: H.-G. Kilian (Ulm) and G. Lagaly (Kiel)

Volume 73 (1987)

New Trends in Colloid Science

Guest Editor: H. Hoffmann (Bayreuth)

Springer-Verlag
Berlin Heidelberg GmbH

ISBN 978-3-662-15931-6   ISBN 978-3-7985-1697-7 (eBook)
DOI 10.1007/978-3-7985-1697-7
ISSN 0340-255 X

© 1987 by Springer-Verlag Berlin Heidelberg
Originally published by Dr. Dietrich Steinkopff Verlag GmbH & Co. KG, Darmstadt in 1987
Softcover reprint of the hardcover 1st edition 1987

Chemistry editor: Heidrun Sauer; Copy editing: Deborah Marston; Production: Holger Frey.

Preface

From October 1–3 1986 the workshop "New Trends in Colloid Science" was held in Como, Italy. More than 100 scientists from 14 different countries attended the conference, representing most of the European groups working in this field. From their oral and poster presentations, 25 papers and a further 25 extended summaries have been selected for this volume.

During the conference the European Colloid and Interface Society was officially founded. Through a number of objectives, the society aims to strengthen colloid science by advancing information exchange and cooperation between scientists in Europe. Annual society conferences will now be held, the proceedings of which will be published in the series "Progress in Colloid and Polymer Science" by Steinkopff Verlag, Darmstadt.

It is a great pleasure for us to thank the organisers, Vittorio Degiorgio and Mario Corti, for a most inspiring and enjoyable workshop. It is the hope of the editors that the annual ECIS conference will be a stimulating event for colloid science, and that these books will help to achieve this.

H. Hoffmann
H.-G. Kilian
G. Lagaly

Contents

Abstracts

Progress in Colloid & Polymer Science   Progr Colloid & Polymer Sci 73:1–4 (1987)

Q-particles: Size quantization effects in colloidal semiconductors

A. Henglein

Hahn-Meitner-Institut Berlin, Bereich Strahlenchemie, Berlin, F.R.G.

Abstract: Quantum mechanical effects in colloidal particles of semiconductor materials are described (Q-particles). They are observed in the case of extremely small particles (1–10 nm). In these particles the eletronic energy levels experience a transition from semiconductor behavior to molecular behavior. The optical and photocatalytical properties of the particles change with their size in this transition range.

Key words: Q-particles, fluorescence, photo-catalysis, inorganic colloids, spectroscopy.

Colloid chemistry and quantum chemistry developed as separated sciences in the first decades of this century. The reason was that the interest of colloid chemists was focussed on the shape and crystal structure of rather big particles consisting of many thousands or even millions of atoms. The electronic structure of these particles was practically the same as for macropieces of the material. Figure 1 shows the front page of the famous book „Die Welt der vernachlässigten Dimensionen" (the world of the neglected dimensions) which Wolfgang Ostwald wrote at the beginning of this century to attract the interest of his fellow chemists to the new field of colloid science [1]. In this book he ranged the size of colloidal particles from 0.1 μ to 1 mμ. He then continued "it should always be emphasized that we are dealing with an arbitrary definition and that transition systems of optional degrees of dispersity exist between colloids and suspensions as well as between *colloids and solutions of molecular dispersity*".

The transition from colloids to molecules was forgotten in the further development of colloid science. If Ostwald were to write a book on colloid science today, he would perhaps choose the old attractive title but point to the 1 nm particles as the modern neglected dimension. This dimension has attracted the attention of the author's laboratory for a number of years, and methods have been developed for the preparation and investigation of extremely small colloidal particles of materials which are metals or semiconductors in the macrocrystalline state. Particles consisting of a few dozen atoms are often not stable and grow upon aging.

However, using the fast methods of modern chemical kinetics, such as stop-flow techniques, pulse radiolysis

DIE WELT
DER VERNACHLÄSSIGTEN
DIMENSIONEN

EINE EINFÜHRUNG
IN DIE MODERNE KOLLOIDCHEMIE
MIT BESONDERER BERÜCKSICHTIGUNG
IHRER ANWENDUNGEN

VON

DR. WOLFGANG OSTWALD
PRIVATDOZENT AN DER UNIVERSITÄT LEIPZIG

DRESDEN UND LEIPZIG
VERLAG VON THEODOR STEINKOPFF
1915

Fig. 1. Front page of a famous book on colloid science

and flash photolysis, one can study the properties of short lived particles. For example, the method of the "growing micro-electrode" enabled the observation of changes in the standard potential of the redox system $Ag_n \rightleftarrows Ag^+ + e^- + Ag_{n-1}$ with changing agglomeration number of the particles (starting with $n = 1$) [2]. In the case of semiconductor particles, the changes in the electronic structure can often be observed at much larger sizes than for metal particles. A few of these changes are described below.

The reader should be reminded that the absorption of a light quantum in a semiconductor occurs by promoting an electron from the occupied valence band into an unoccupied level of the conduction band. A positive hole in the valence band is created, and both charge carriers can rapidly move to the surface of the colloidal particle and initiate chemical reactions [3]. With decreasing size of the colloidal particle a gradual transition from semiconductor properties to molecular properties occurs [4, 5]. The levels of the valence band are moderately shifted to lower energies, while those of the conduction band are strongly shifted to higher energies. As a consequence, the optical properties of the material, as well as the photocatalytic properties, drastically change. It has been proposed to characterize materials consisting of small particles which have properties different to those of the macrocrystals by the prefix "Q" [5]. The Q stands for the quantum mechanical effect, the so-called size quantization, which is responsible for these changes.

Up to date, Q-particles of a dozen materials, i. e. the sulfides and selenides of cadmium, zinc and indium, the phosphides and arsenides of cadmium and zinc, and the oxides of zinc, cadmium and indium have been prepared in this laboratory [6–8]. In many cases, the particles could be recovered from the colloidal solutions by carefully removing the solvent; the powders obtained could be redissolved to yield the original colloidal solutions. There is no doubt that many other inorganic semiconductor materials could also be made in the Q-state, the difficulty being finding suitable methods of preparation. The sulfides, selenides, phosphides and arsenides can be made by fast precipitation techniques from dilute aqueous polyphosphate solutions or in organic solvents at lower temperatures. The oxides are obtained by precipitation of the metal ions with NaOH in alcoholic solution.

In Figure 2 the absorption spectra of a conventional CdS sol and of two Q-CdS sols are shown. It can be seen that the onset of absorption of the small particles occurs at a much shorter wavelength than for the big-

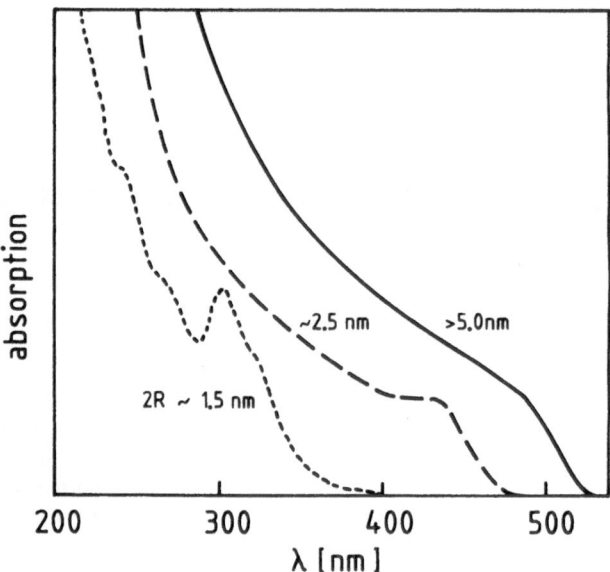

Fig. 2. Absorption spectra of a conventional CdS colloid (> 5 nm) and of two Q-CdS sols

ger ones. In the spectrum of Q-CdS a maximum about 40 nm below the onset is present. It is attributed to the optical transition to the "excition" state in which the electron and hole are not completely free but form a hydrogen-atom-like species in the CdS lattice. This optical transition cannot be resolved in the case of a macrocrystal unless the spectrum is taken at liquid helium temperature.

Figure 3 shows the wavelength of the onset of absorption of cadmium sulfide as a function of the particle diameter [9]. The absorption threshold of compact CdS lies at 515 nm corresponding to a band gap energy of 2.4 eV. As the particles become smaller than 4 nm the threshold is rapidly shifted to shorter wavelengths. The color of such colloidal solutions changes from yellow-orange to yellow-green and finally the solutions are colorless. Colorless CdS in the solid state has also been prepared.

Even stronger effects have been observed for cadmium phosphide, Cd_3P_2, and cadmium arsenide, Cd_3As_2. These materials are black in the macrocrystalline state. They start to absorb in the infrared, and have very low band gap energies (0.5 and 0.1 eV, respectively). Q-particles of both materials have been made having all possible colors in the visible spectrum depending on the particle size. In these particles the band gap (if one still can use this designation) is shifted from a

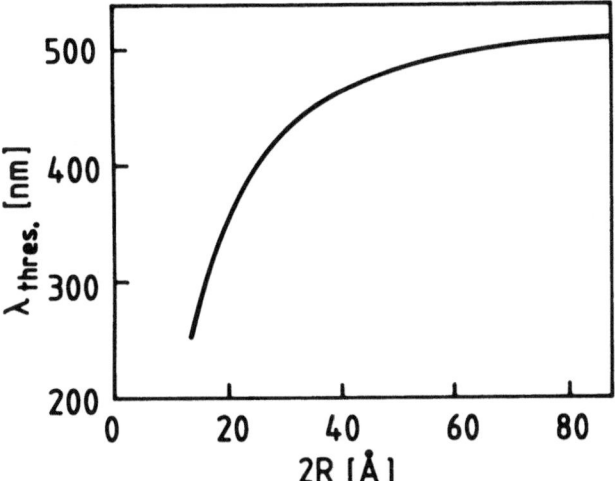

Fig. 3. Wavelength of the onset of absorption of CdS particles as a function of the diameter

few tenths of an eV up to about 3 eV for the smallest particles which are colorless.

Another property which changes with particle size is the color of the fluorescence light, which with decreasing size, is blue-shifted. Q-cadmium phosphide may fluoresce in the infrared, red, yellow or green depending on particle size. The very small particles luminesce with particularly high quantum yields. In Q-particles, the electron produced by light absorption is on a more negative potential and the hole generated is on a more positive potential than in the macrocrystalline materials. Chemical effects have been observed which can be explained by the stronger reducing properties of the electrons and the stronger oxidizing properties of the holes. For example, H_2 evolution was observed on illuminated PbSe and HgSe colloids and CO_2 reduction on CdSe when the particle size was less than 5 nm [10]. Q-CdS particles catalyse the formation of the hydroperoxide $(CH_3)_2C(OH)$-OOH when illuminated in aerated propanol-2 solution, while larger particles are not efficient in this respect [5].

The size effects are generally described by the well-known quantum mechanics of a "particle in a box" [4, 5, 11]. The electron and the positive hole are confined to potential wells of small dimension and this leads to a quantization of the energy levels (which in the bulk material constitute virtual continua in the conduction and valence bands, respectively). These phenomena arise when the size of the colloidal particle becomes comparable to the DeBroglie wavelength of the charge carriers. The quantization effects for an electron in an evacuated box become significant at box dimensions of some 0.1 nm. However, in the colloidal particles the effects can already be seen at a much larger particle size. The reason for this lies in the fact that the effective mass of a charge carrier, which moves in the periodic array of the constituents of the crystal lattice, is generally much lower than the mass of an electron in free space. This results in a larger DeBroglie wavelength. The smaller the effective mass of the charge carriers the more pronounced are the optical size effects.

As a final conclusion we may say that the discovery of size quantization effects has opened a new dimension in colloid science. A lot of studies on the optical and catalytic properties of Q-particles are expected to be performed in the near future. Finally we may mention that a few reports have recently appeared in which interesting non-linear optical properties of such semiconductor particles have been described [12, 13]. This aspect may also play an important role in future investigations.

References

1. Ostwald W (ed) (1915) Die Welt der vernachlässigten Dimension, 1 Aufl, Th Steinkopff, Dresden
2. Henglein A, Tausch-Treml R (1981) J Coll Interf Sci 80:84
3. Henglein A (1985) In: Eicke HF (ed) Modern trends of colloid science in chemistry and biology, Birkhäuser Verlag, Basel
4. Brus LE (1983) J Chem Phys 79:5566; (1984) 80:4403; (1986) J Phys Chem 90:2555
5. Fojtik A, Weller H, Koch U, Henglein A, Bunsenges B (1984) Phys Chem 88:969
6. Weller H, Fojtik A, Henglein A (1985) Chem Phys Lett 117:485
7. Fojtik A, Weller H, Henglein A (1985) Chem Phys Lett 120:552
8. Koch U, Fojtik A, Weller H, Henglein A (1985) Chem Phys Lett 122:507
9. Weller H, Schmidt HM, Koch U, Fojtik A, Baral S, Henglein A, Kunath W, Weiss K, Dieman E (1986) Chem Phys Lett 124:557
10. Nedeljković JM, Nenadović MT, Mićić OI, Nozik AJ (1986) J Phys Chem 90:12
11. Schmidt HM, Weller H (1986) Chem Phys Lett 129:615
12. Flytzanis C, Hache F, Ricard D, Roussignol Ph (1986) In: Kelly MJ, Weisbuch C (eds) The Physics and Fabrication of Microstructures and Microdevices, Springer-Verlag, Berlin
13. Henglein A, Kumar A, Janata E, Weller H (1986) Chem Phys Lett 132:133

Received November 26, 1986;
accepted January 23, 1987

Author's address:

Prof. Dr. A. Henglein
Hahn-Meitner-Institut für Kernforschung
Postfach 39 01 88, D-1000 Berlin 39, F.R.G.

A simple theory for the self-diffusion coefficients in binary mixtures of highly charged spherical macroions

G. Nägele, M. Medina-Noyola, J. L. Arauz-Lara[1]), and R. Klein

Fakultät für Physik, Universität Konstanz, Konstanz, F.R.G.
[1]) Department of Chemistry, Syracuse University, Syracuse, New York, U.S.A.

Abstract: The static and dynamic properties of a dilute binary macroion mixture, where a small amount of large and highly charged spheres is immersed in a system of small spheres, are theoretically discussed. From light scattering experiments on such systems, the self-diffusion of the large spheres within the system of small spheres is essentially determined. Such systems have recently been studied by Phillies, using suspensions of highly charged polystyrene spheres. Modelling the elements of the memory function matrix by single exponentials, relating the k-dependent amplitudes and relaxation rates to the moments of the one-particle propagators, compact analytic expressions are derived for the normalized mean squared displacements and long-time self-diffusion coefficients. The moments are calculated from a generalized Smoluchowski equation and are given by integrals over the static correlation functions. Using the so-called two-component macrofluid model to calculate the static correlation functions, analytic expressions of the static partial total correlation functions in k-space are derived within the mean-spherical approximation for the binary mixture. Because of the strong dilution of the mixture, a rescaling procedure has to be used. This rescaling procedure is a simple generalization of that given by Hansen and Hayter for the one-component macroion fluid. Comparison of the theoretically calculated long-time self-diffusion of the large spheres with the experimental results of Phillies was quite satisfactory. Also, calculations of the two self-diffusion coefficients as functions of the diameter ratio and salt concentration have been performed, as well as of the normalized mean squared displacements.

Key words: Colloidal suspension, self-diffusion, Brownian motion.

Introduction

We present a theory of the self-diffusion coefficients in binary mixtures of highly charged spherical Brownian particles. This is a direct extension of a theory [1] which describes the self-diffusion of a one-component Brownian system. Typical systems which have been investigated recently by dynamic light scattering experiments [2], are suspensions of polystyrene spheres in water. In dynamic light scattering experiments on binary mixtures of macroions, the measured structure factor

$$S^M(q,t) = \frac{1}{\bar{f}^2} [x_1 f_1^2 S_{11}(q,t) + x_2 f_2^2 S_{22}(q,t)$$

$$+ 2(x_1 x_2)^{1/2} f_1 f_2 S_{12}(q,t)] \qquad (1.1)$$

is obtained. The $S_{\alpha\beta}$ are the partial dynamic structure factors, $\bar{f}^2 = \sum_\alpha x_\alpha f_\alpha^2$, $x_\alpha = \varrho_\alpha/(\varrho_1 + \varrho_2)$ denotes the mole fraction and ϱ_α the number density of species $\alpha = 1, 2$. For the homogeneous spheres, the wave vector dependent form-amplitudes

$$f_\alpha \propto (n_\alpha - n_s)\, \sigma_\alpha^3 \qquad (1.2)$$

are proportional to the volume of the spheres with diameter σ_α and to the difference of the refractive indices n_α of particle of species α to the solvent refractive index n_s. If the mixture is composed in a way that $\varrho_1 \gg \varrho_2$, then

$$S_{22}(q,t) = G_s^{(2)}(q,t) \equiv \langle e^{i\underline{q}(\underline{r}^{(2)}(t) - \underline{r}^{(2)}(0))} \rangle, \qquad (1.3)$$

for the cross term in S_{22}, describing correlations between different particles of species 2, can be neglected with respect to the self-diffusion propagator $G_s^{(2)}$ of a tagged particle of species 2. Furthermore, if the condition $\varrho_1 f_1^2 \ll \varrho_2 f_2^2$ is fulfilled, only the few particles of species 2 contribute significantly to the scattered intensity. This can be achieved according to Equation (1.2) by choosing the diameter σ_2 appreciably larger than σ_1, or by matching the refractive index n_1 to the solvent index n_s. Then Equation (1.1) reduces to

$$S^M(q,t) \approx G_s^{(2)}(q,t) . \tag{1.4}$$

From the self-diffusion propagator, relevant properties like mean squared displacement $W^{(\alpha)}(t)$ and long-time self-diffusion coefficient $D_s^{(\alpha)}$ can be extracted. The self-diffusion coefficient $D_s^{(2)}$ has been recently measured by Phillies [3] for an asymmetric binary mixture of polystyrene spheres appreciably different in size. To our knowledge, the method of index-matching has so far only been applied to fairly high [4] or very high [5] concentrated binary mixtures of uncharged particles, the two species being almost equal in size.

2. The single exponential form of the memory function

In the time regime probed by dynamic light scattering, the diffusion of the Brownian particles can be described by the Smoluchowski equation [6]. Using the Mori-Zwanzig projection operator technique, the memory equations

$$\frac{\partial}{\partial t} G_s^{(\alpha)}(q,t) = - q^2 D_o^{(\alpha)} G_s^{(\alpha)}(q,t)$$
$$+ \int_0^t dt' M_s^{(\alpha)}(q,t-t') G_s^{(\alpha)}(q,t') \tag{2.1}$$

are derived neglecting hydrodynamic interactions. We intend to describe the self-diffusion properties of diluted, highly charged macroion mixtures; therefore, hydrodynamic interaction can be neglected, as has been justified by experiment. $D_o^{(\alpha)}$ is the diffusion coefficient of particles of species α at infinite dilution. For the memory function $M_s^{(\alpha)}$ we make the single exponential approximation (SEXP)

$$M_s^{(\alpha)}(q,t) = a_\alpha(q) \, e^{- b_\alpha(q) t} . \tag{2.2}$$

The amplitude $a_\alpha(q)$ and relaxation rate $b_\alpha(q)$ are expressed through the memory function and its time

derivative at $t = 0$. These two initial values are related to the moments of $G_s^{(\alpha)}$

$$m_j^{(\alpha)} = \left[\frac{\partial^j G_s^{(\alpha)}(q,t)}{\partial t^j} \right]_{t=0} \tag{2.3}$$

by

$$M_s^{(\alpha)}(q,0) = m_2^{(\alpha)}(q) - q^4 (D_o^{(\alpha)})^2$$

$$\frac{\partial}{\partial t} M_s^{(\alpha)}(q,0) = m_3^{(\alpha)}(q) + 2 q^2 D_o^{(\alpha)} m_2^{(\alpha)}(q)$$
$$- q^6 (D_o^{(\alpha)})^3 . \tag{2.4}$$

The first three moments of the self-diffusion propagator $G_s^{(\alpha)}$ have been calculated by J. L. Arauz-Lara [7]. The expressions for them will be given elsewhere.

With the definition

$$D^{(\alpha)}(t) = \frac{\langle (\underline{r}_1^{(\alpha)}(t) - \underline{r}_1^{(\alpha)}(0))^2 \rangle}{6t} \tag{2.5}$$

of the time-dependent self-diffusion coefficient of a tagged particle of species α, we obtain, by use of Equation (2.2), the following expressions

$$D_s^{(\alpha)} / D_o^{(\alpha)} = 1 - \frac{A_\alpha^2}{B_\alpha + C_\alpha} \tag{2.6}$$

$$D^{(\alpha)}(t) / D_o^{(\alpha)} = D_s^{(\alpha)} / D_o^{(\alpha)}$$
$$+ (1 - D_s^{(\alpha)} / D_o^{(\alpha)}) \frac{\tau_\alpha}{t} (1 - e^{-t/\tau_\alpha}) \tag{2.7}$$

$$\tau_\alpha = \frac{k_B T A_\alpha}{D_o^{(\alpha)} (B_\alpha + C_\alpha)} \tag{2.8}$$

for the long-time self-diffusion coefficient $D_s^{(\alpha)}$ and the time dependent self-diffusion coefficient $D^{(\alpha)}(t)$. The numbers A_α, B_α and C_α are expressed by the static pair distribution functions $g_{\alpha\beta}(r)$ and triple correlation functions $g_{\alpha\beta\gamma}(\underline{r}, \underline{r}')$ through

$$A_\alpha = \sum_{\gamma=1}^2 \varrho_\gamma \int d^3r \, g_{\alpha\gamma}(r) \, (\underline{q} \cdot \underline{\nabla})^2 \, u_{\alpha\gamma}(r)$$

$$B_\alpha = \sum_{\gamma=1}^2 \left(1 + \frac{D_o^{(\gamma)}}{D_o^{(\alpha)}} \right) \varrho_\gamma \int d^3r \, g_{\alpha\gamma}(r) \, [(\underline{q} \cdot \underline{\nabla}) \, \underline{\nabla} u_{\alpha\gamma}(r)]^2$$

$$C_\alpha = \sum_{\gamma,\delta=1}^2 \varrho_\gamma \varrho_\delta \iint d^3r \, d^3r' \, g_{\alpha\gamma\delta}(\underline{r}, \underline{r}') \, (\underline{q} \cdot \underline{\nabla})$$
$$\times (\underline{q} \cdot \underline{\nabla}') \, (\underline{\nabla} \cdot \underline{\nabla}') \, u_{\alpha\gamma}(r) \, u_{\alpha\delta}(r') . \tag{2.9}$$

$u_{\alpha\beta}$ is the pair potential between particles of species α and β. From our use of moments expansions it is clear that Equation (2.7) gives the correct short time behaviour. In the results below, however, we shall approximate $C_\alpha = \sigma$.

3. The two-component macrofluid model

In this simple model, the bare pair potential between two spherical macroions of species α and β is the sum of a hard-sphere potential plus a long-range screened Coulomb potential

$$u_{\alpha\beta}(r) = \frac{q_\alpha q_\beta}{\varepsilon(1 + \varkappa\sigma_\alpha/2)(1 + \varkappa\sigma_\beta/2)} \frac{e^{-\varkappa(r-\sigma_{\alpha\beta})}}{r}, r > \sigma_{\alpha\beta}$$
$$= \infty, r < \sigma_{\alpha\beta}. \qquad (3.1)$$

where $\sigma_{\alpha\beta} \equiv (\sigma_\alpha + \sigma_\beta)/2$, q_α is the charge of one macroion of species α and ε the relative dielectric constant of the solvent. The influence of the salt ions and the neutralizing counterions is only taken into account through the Debye screening length $\varkappa^{-1}$, where

$$\varkappa^2 = \frac{4\pi}{\varepsilon k_B T} \sum_j \varrho_j Q_j^2, \qquad (3.2)$$

the sum including all species of salt ions and counterions of charge Q_j and number density ϱ_j. With regard to the restriction $\varrho_2 \ll \varrho_1$ made in Section 1 to obtain the self-diffusion coefficient of species 2 by dynamic light scattering, we have approximately calculated the static pair correlation functions of the two-component macrofluid model in the limiting case $\varrho_2 = 0$. The approximation we used was the mean spherical approximation, making use of general results obtained by Blum et al. [8]. In addition, a simple extension of the rescaling procedure given by Hansen and Hayter [9] was applied. A detailed exposition of our calculations will be presented in a future publication. The rescaled expressions of the static pair correlation functions were used as input for the expressions of Equation (2.9).

4. Results

Throughout our calculations, we kept the temperature fixed to $T = 300$ K and used $\varepsilon = 80$ fo the dielectric constant of the solvent H_2O. In order to give an impression of the accuracy for the SEXP, Figure 1 shows a comparison of the time-dependent self-diffusion coefficient $D(t)/D_o$, calculated in the one-com-

ponent macrofluid model in SEXP and by Brownian dynamics computer simulation by Gaylor et al. [10]. The same parameters as in the computer simulation data were used, only the surface charge was increased to $q = 145$ e, in order to fit the height of the first peak of our rescaled $g(r)$ to that obtained by the computer simulations. The overall agreement is good.

In order to give suggestions to the experimentalist, several system parameters such as volume fraction $\phi_1 = \frac{1}{6}\pi\varrho_1\sigma_1^3$, diameter ratio $\lambda = \sigma_2/\sigma_1$ and $1-1$ electrolyte (e. g. added HCl) number concentration ϱ_s have been varied. The contribution of the added $1 - 1$ electrolyte to the screening is given by $(\varkappa\sigma_1)_{salt} = \left(\frac{8\pi e^2}{\varepsilon k_B T}\varrho_s\right)^{1/2}\sigma_1$, e being the elementary charge. Because of the condition $\varrho_2/\varrho_1 \ll 1$ used in our calculations of the static correlation functions, the static as well as the dynamic properties of species 1 are not affected by the presence of the few larger particles of species 2. The diameter of species 1 is fixed to $\sigma_1 = 380$ Å and $D_o^{(1)} = k_B T/(3\pi\eta\sigma_1)$ is set equal to $1.15 \cdot 10^6$ Å²/ms, where η is the shear viscosity of the solvent. If not varied, $\sigma_2 = 1500$ Å. The diffusion coefficient of species 2 at infinite dilution is related to that of species 1 by $D_o^{(2)} = D_o^{(1)}/\lambda$.

In Figure 2 we compare the long-time self-diffusion coefficient $D_s^{(2)}/D_o^{(2)}$, calculated with SEXP, with experimental data obtained by Phillies [3], in dependence on the volume fraction ϕ_1. For completeness,

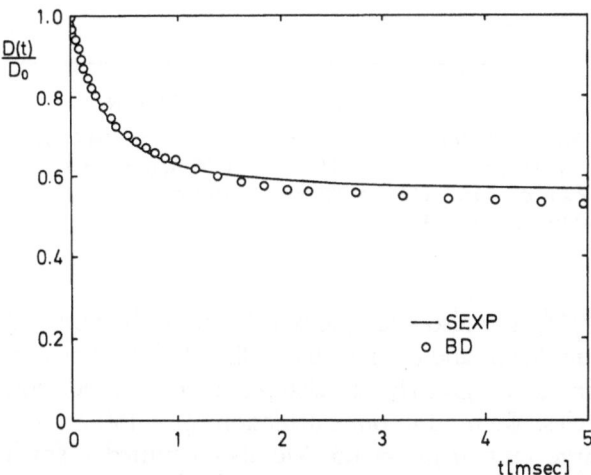

Fig. 1. Calculated (——) and computer simulated (O) [10] normalized time-dependent self-diffusion coefficient $D(t)/D_o$ of the one-component macrofluid model with volume fraction $\phi = 4.4 \cdot 10^{-4}$, charge $q = 145$ e, diameter $\sigma = 460$ Å and screening length $\varkappa^{-1} = 3080$ Å. $D_o = 9.5 \cdot 10^5$ Å²/ms

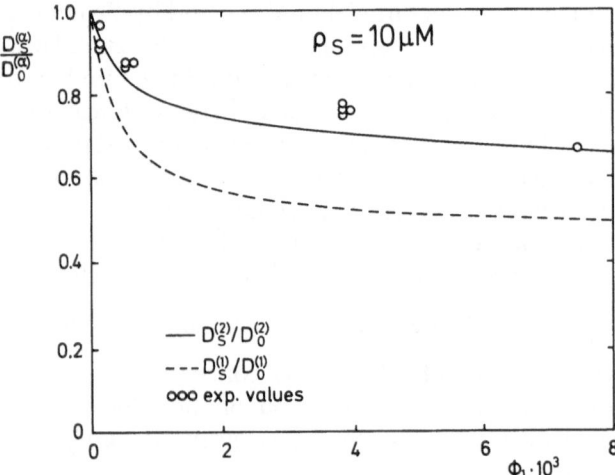

Fig. 2. In SEXP calculated (——) and by dynamic light scattering [3] measured long-time self-diffusion coefficient $D_s^{(2)}/D_0^{(2)}$ versus volume fraction $\phi_1 \cdot 10^3$ of species 1. The line (– – –) is the result for $D_s^{(1)}/D_0^{(1)}$. The used parameters are: $\sigma_1 = 380$ Å; $\sigma_2 = 1500$ Å; $q_1 = 150$ e; $q_2 = 225$ e and $(\varkappa\sigma_1)_{salt} = 0.395$, corresponding to 10 μM added 1 − 1 electrolyte

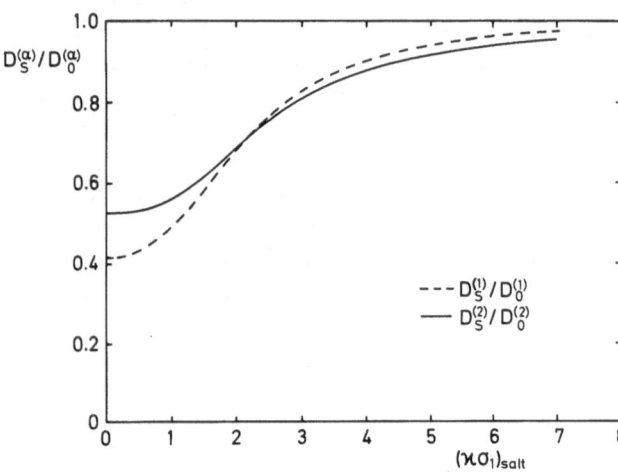

Fig. 3. $D_s^{(\alpha)}/D_0^{(\alpha)}$ for $\alpha = 2$ (——) and $\alpha = 1$ (– – –) versus $(\varkappa\sigma_1)_{salt}$ with system parameters $\sigma_1 = 380$ Å, $\sigma_2 = 1500$ Å, $q_1 = 200$ e, $q_2 = 300$ e and $\phi_1 = 0.02$. The contribution of counterions to the screening is $(\varkappa\sigma_1)_{C.I.} = 1.34$

$D_s^{(1)}/D_0^{(1)}$ has also been plotted. We used the nominal values for σ_1 and σ_2 given by Phillies. In his article, no estimate was given for the charges of the two macroion species. We used in our calculations $q_1 = 150$ e and a charge ratio $q_2/q_1 = 1.5$. We also assumed a small amount of 10 μM electrolyte to be left in the ion-exchanged suspension. Obviously, the self-diffusion coefficients decrease with increasing ϕ_1. The increase in $D_s^{(\alpha)}/D_0^{(\alpha)}$ due to the enlarged static screening by adding 1 − 1 electrolyte can be seen in Figure 3.

In Figure 4, we plot $D_s^{(2)}/D_0^{(1)}$ as a function of diameter ratio $\lambda = \sigma_2/\sigma_1$. The parameters of species 1 are kept fixed, $D_s^{(1)}/D_0^{(1)} = 0.41$. The dependence of the surface charge q_2 on the diameter σ_2 is assumed to vary in three different ways. Of course, $D_s^{(2)}$ decays much faster for $q_2 = \lambda^2 q_1$ than for q_2 kept fixed to q_1. The actual dependence of q_2 on σ_2 (and other parameters) is unknown, but experiments like that illustrated in Figure 4 might help to obtain further information.

In Figure 5 finally, the time-dependent self-diffusion coefficients $D^{(\alpha)}(t)/D_0^{(\alpha)}$ are plotted versus time.

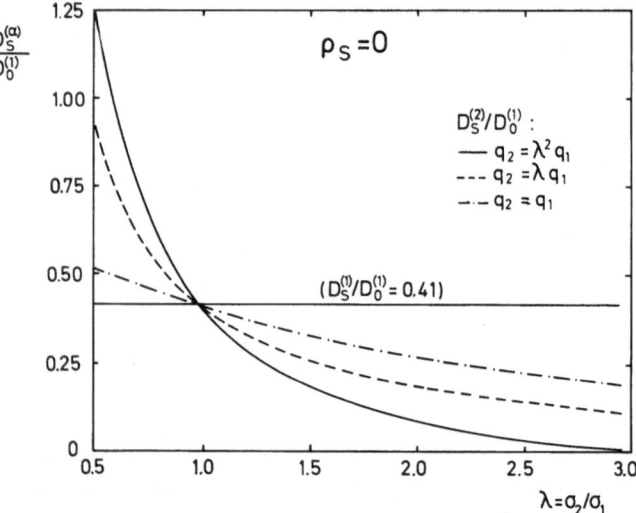

Fig. 4. $D_s^{(2)}/D_0^{(1)}$ as a function of the diameter ratio σ_2/σ_1. The surface charge q_2 of species 2 is related to the particle size in three different ways: $q_2 = \lambda^2 q_1$ (——), $q_2 = \lambda q_1$ (– – –) and $q_2 = q_1$ (–·–·–). $\sigma_1 = 380$ Å, $q_1 = 200$ e and $\phi_1 = 0.02$

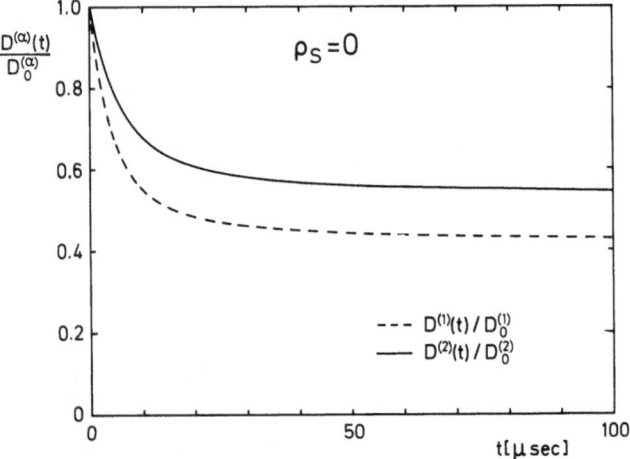

Fig. 5. Time-dependent self-diffusion coefficients $D^{(\alpha)}(t)/D_0^{(\alpha)}$ for $\alpha = 2$ (——) and $\alpha = 1$ (– – –) versus time (μs). $\sigma_1 = 380$ Å, $\sigma_2 = 1500$ Å, $q_1 = 200$ e, $q_2 = 300$ e, $\phi_1 = 0.02$ and $\varrho_s = 0$

Acknowledgements

This work is part of a joint research project supported by the Bundesministerium für Forschung und Technologie (F.R.G.), Conacyt (Mexico) and Cosnet-Sep (Mexico). M. Medina-Noyola acknoledges the Alexander-von-Humboldt Foundation for a research fellowship.

References

1. Arauz-Lara JL, Medina-Noyola M (1986) J Phys A19:L117–L121
2. Härtl W, Versmold H (1984) J Chem Phys 80:1387
3. Phillies GDJ (1984) J Chem Phys 81:1487
4. Kops-Werkhoven MM, Pathmamanoharan C, Vrij A, Fijnaut HM (1982) J Chem Phys 77:5913
5. van Megen W, Underwood SM, Snook IA (1986) J Chem Phys 85:4065
6. Murphy TJ, Aguirre JL (1972) J Chem Phys 57:2098
7. Arauz-Lara JL (1985) Ph D thesis, Cinvestav, Mexico
8. Blum L, Høye JS (1978) J Stat Phys 19:317; Blum L (1980) J Stat Phys 22:661
9. Hansen JP, Hayter JB (1982) Mol Phys 46:651
10. Gaylor KJ, Snook IK, van Megen WJ, Watts RO (1980) J Chem Soc Faraday Trans II 76:1067

Received December 24, 1986;
accepted January 29, 1987

Authors' address:

Prof. Dr. Rudolf Klein
Universität Konstanz
Fakultät für Physik
Postfach 55 60
D-7750 Konstanz 1, F.R.G.

Structural changes in anionic micelles induced by counterion complexation with a macrocyclic ligand: A neutron scattering study

K. A. Payne[1]), L. J. Magid[1])[2]), and D. F. Evans[2])

[1]) Department of Chemistry, University of Tennessee, Knoxville, Tennessee, U.S.A.
[2]) Department of Chemical Engineering and Materials Science, University of Minnesota, Minneapolis, Minnesota, U.S.A.

Abstract: The changes in micellar aggregation number which occur in SDS and C_7SS solutions when the sodium ions are complexed by the cryptand C 222 have been investigated at several concentrations of amphiphile and supporting electrolyte (NaCl). The observed micellar reorganizations are rationalized in terms of changes in head group repulsion and hydrocarbon-water contant.

Key words: Micelles, cryptands, sans.

Introduction

Recently Evans and coworkers [1] reported that the complexation by macrocyclic multidentate ligands of the bound counterions at the sodium dodecylsulfate (SDS) micellar surface in dilute SDS aqueous solutions causes a decrease in micellar size and an increase in micellar ionization. They proposed that the formation of the large sodium ion-ligand inclusion complexes draws the counterions away from the sulfate head groups, thus increasing the repulsion between head groups, inducing greater curvature at the micellar surface and a corresponding decrease in micellar size. Parallel behavior had been previously observed for cationic surfactants (alkyltrimethylammonium compounds) having highly hydrated counterions such as hydroxide or acetate [2].

We have been interested in how patterns of micellar aggregation differ for double-chain vs. single-chain amphiphiles; for sodium bis(n-alkyl)sulfosuccinates (C_nSS; [3–5]) in water, the first-formed spherical micelles have large areas per head group (> 100 Å^2) and low extents of counterion binding (fraction (β) of ca. 0.5), but with increasing surfactant concentration rapid micellar growth, deviation of the micellar shape from spherical and substantial increases in the fraction

of counterions bound are observed. We supposed that sodium ion complexation by macrocyclic ligands in dilute C_nSS solutions would produce minimum-sphere, highly ionized micelles (with β's even lower than observed with SDS). The addition of supporting electrolyte [1] would be expected to cause little, if any, increase in micellar size, a very attractive feature for a model system in which to explore, using small-angle scattering techniques, the evolution of micellar interactions from strongly repulsive to attractive under conditions where the micellar form factor is invariant.

We employed the macrocyclic ligand 4,7,13,16,21, 24-hexaoxa-1,10-diazabicyclo[8.8.8]hexacosane (C 222), and studied its effect, using small-angle neutron scattering (SANS), on micellar size for aqueous SDS and C_7SS solutions of varying amphiphile and supporting electrolyte concentrations. Examination of representative scattering curves, presented in Figure 1, shows immediately, without recourse to curve-fitting, that C 222 does indeed decrease the mean aggregation number of SDS micelles. In contrast, at some concentrations of C_7SS C 222 has the opposite effect. We proceed now to examine in more detail these differences in behavior.

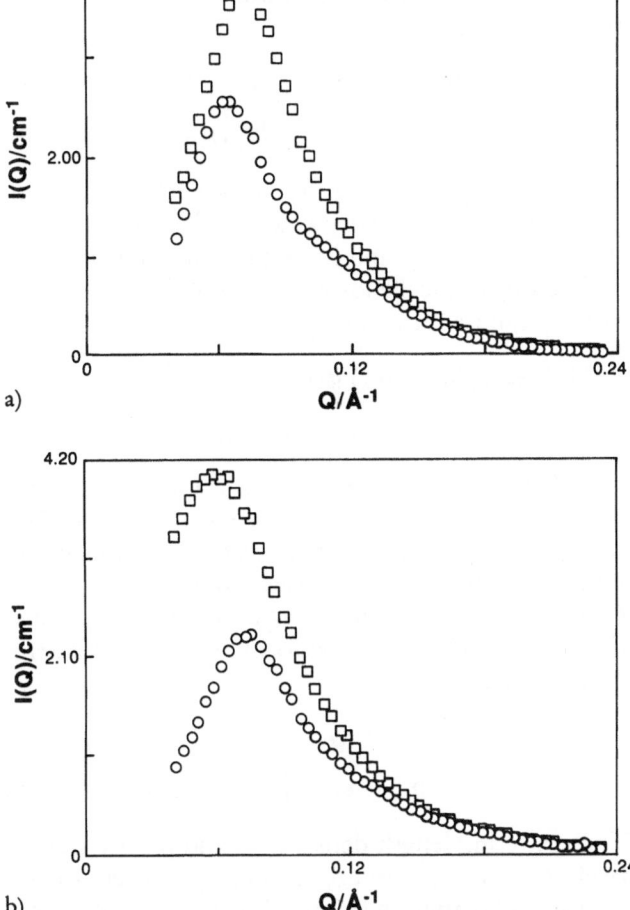

a)

b)

Fig. 1. SANS data for (a) 0.10 M SDS (at 25 °C) and (b) 0.10 M C_7SS (at 45 °C) micellar solutions without (O) and with (□) C 222

with Q being defined as $(4\pi/\lambda) \sin\theta$ (2θ is the scattering angle). The samples were contained in thermostatted quartz spectrophotometric cells with 0.2 cm path lengths.

Scattering from samples and solvent (D_2O) was corrected for detector background and sensitivity, empty cell scattering, calculated incoherent scattering and sample transmission. Solvent intensity was subtracted from that of the sample for each detector element, and the differences were converted to radially averaged intensities vs. Q by using programs provided by the Center. Absolute intensities were computed from calibration constants based on standards provided by the Center. For some of the solutions containing the larger micelles, measurements were made at two sample-to-detector distances and the data combined to form a single scattering curve.

Results

Conductimetric data for C_7SS solutions without and with C 222 (1:1 molar ratio) establish that the cryptate increases β, an effect opposite to that found for SDS. Table 1 summarizes the results; β was evaluated from the slopes of the specific conductivity vs. concentration below and just above the cmc, using the Evans equation [6]. It is evident from Figure 2 that in

Table 1. cmc's and degrees of counterions binding for SDS and C_7SS solutions containing C 222

Amphiphile	T, °C	cmc, mol/l	β	Reference
SDS	25	0.0080	0.82	[1]
SDS : C 222	25	0.0016	0.52	[1]
C_7SS	45	0.0040	0.52	this work
C_7SS : C 222	45	0.0025	0.64	this work

Experimental

Materials

Syntheses of the sulfosuccinates have been described previously [3]. SDS was obtained from BDH Chemicals, C 222 from Aldrich and D_2O from Norell. All solutions containing C 222 were prepared in a nitrogen atmosphere. SDS solutions were studied at 25 °C and C_7SS solutions at 45 °C, since all previous SANS data for the latter surfactant were collected at the higher temperature.

SANS

Neutron scattering measurements were performed on the 30-m (source to detector) SANS instrument of the National Center for Small-Angle Scattering Research (NCSASR) at the High Flux Isotope Reactor (HFIR) in Oak Ridge, Tennessee. The wavelength of the neutrons used was 4.75 Å; scattered intensities were recorded in the Q range form 0.011 (or 0.04 for the smaller micelles) to 0.24 Å⁻¹,

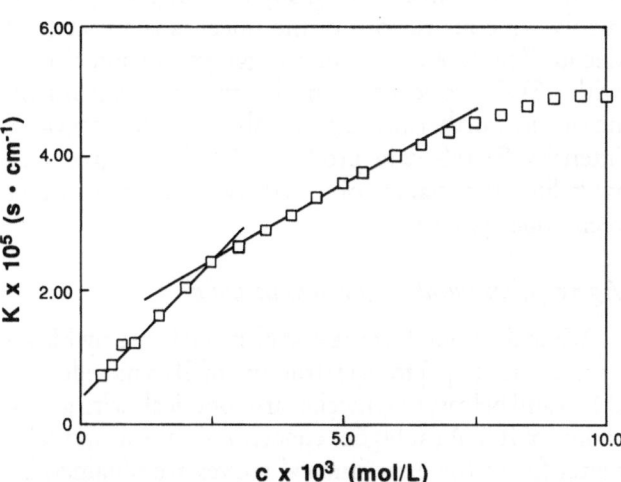

Fig. 2. Specific conductivity vs. concentration for solutions of 1:1 C_7SS : C 222 at 45 °C

the presence of C 222 at C_7SS concentrations above 9×10^{-3} M, β approaches a value of one.

The SANS curves arising from dispersions of charged globular micelles [7–9] are described by:

$$I(Q) = N_p \cdot [S(Q) \cdot \langle F(Q) \rangle_Q^2 + \Delta(Q)] \qquad (1)$$

where

$$\Delta(Q) = \langle |F(Q)|^2 \rangle_Q - \langle F(Q) \rangle_Q^2 \qquad (2)$$

N_p is the micelle number density, $S(Q)$ the intermicellar structure factor and $F(Q)$ the single-micelle form factor. $\Delta(Q)$ is zero if the micelles are monodispersed spheres, but differs from zero for anisometric particles or polydispersed spheres.

The micelles interact via a screened Coulombic potential; at high enough supporting electrolyte concentration, these electrostatic repulsions are screened, and the interaction peaks in the scattering curves disappear. $S(Q)$ is evaluated for a given micellar charge Z, hard sphere diameter and micellar volume fraction following Hayter and Penfold [10] and Hansen and Hayter [11]. Since we are neglecting the finite size of the counterions, the micellar charges may be overestimated by 10%–20% [12].

Practitioners of small-angle neutron scattering are sometimes accused of inferring details about micellar structure which cannot be supported by the resolution of their scattering data. Certainly there exist a plethora of SANS-derived micellar models for SDS in water [7, 13–17]. All are characterized by a large, dry hydrocarbon core having a radius, when the micelles are spherical, roughly equal to the length of an all-*trans* C_{12} chain; the number of methylene groups residing in the outer, highly aqueous regions of the micelles is a matter of debate. The evolution of mean aggregation number ($\bar{n}$) with SDS concentration is model-independent, however. Furthermore, as we shall see, our absolute intensity SANS data produce $\bar{n}$'s which agree well with literature values obtained by a variety of other techniques [18, 19].

Aggregation numbers for SDS in water

We shall adopt the core + shell model used by Hayter and Penfold [7] for the structure of SDS micelles. At 0.1 M and below, the micelles are spherical, with a core radius of 16.7 Å; at higher concentrations substantially better fits to the experimental curves are obtained by assuming they are prolate ellipsoids. The scattering length density of the core, ϱ_c, is computed to be -0.39

$\times 10^{10}$ cm^{-2}, using standard compilations of scattering lengths [20] and group volumes [21]. The volume of the shell is given by

$$V_{sh} = \bar{n} \cdot [V_{C_{12}H_{25}} + V_{OSO_3^-} + \beta \cdot V_{Na^+} \\ + V_{D_2O}(h_{OSO_3^-} + \beta \cdot h_{Na^+})] - V_c. \qquad (3)$$

The scattering length density of the shell, ϱ_{sh}, is computed by group additivity as well. Table 2 summarizes the group volumes and hydration numbers used.

The single-micelle form factor introduced in Equation (1) can be recast for spheres as

$$F(Q) = V_c(\varrho_c - \varrho_{sh}) \, \phi(QR_c) \\ + (V_c + V_{sh}) \, (\varrho_{sh} - \varrho_{D_2O}) \, \phi(QR_{tot}) \qquad (4)$$

with $\phi(x)$ given by $(\sin x - x \cos x)/(x^3)$, and R_{tot} equal to the overall radius of the hydrated micelle. For ellipsoids, the expression for the form factor is similar, but it depends on the micelles orientation with respect to Q, so the orientational averaging indicated in Equations (1) and (2) must be performed. At Q equals zero, $I(0)/S(0)$ becomes

$$I(0)/S(0) = N_p \cdot [V_c(\varrho_c - \varrho_{sh}) + (V_c + V_{sh}) \\ \cdot (\varrho_{sh} - \varrho_{D_2O})]^2. \qquad (5)$$

The scattering length densities ϱ_c and ϱ_{sh} are respectively $b_{mon,c}/V_{mon,c}$ and $b_{mon,sh}/V_{mon,sh}$, where the b's are sums of scattering lengths. We note that N_p is $(c - cmc)/\bar{n}$, V_c is $\bar{n} \cdot V_{mon,c}$ and V_{sh} is $\bar{n} \cdot V_{mon,sh}$. Then defining $b_{mon,t} = b_{mon,c} + b_{mon,sh}$ and $V_{mon,t} = V_{mon,c} + V_{mon,sh}$, Equation (5) can be rewritten as

$$I(0)/S(0) = (c - cmc) \cdot \bar{n} \cdot [b_{mon,t} - V_{mon,t} \cdot \varrho_{D_2O}]^2. \qquad (6)$$

Table 2. Group volumes and hydration numbers used in the data analysis

Group	Volume, Å³	Hydration number (h)
CH₃	54.3	
CH₂	26.9	
CH	20.6	
–CO₂–	37.0	
–OSO₃–	70.2	4
–SO₃–	60.6	4
Na⁺	– 6.0[a])	6
C 222 : Na⁺	567.0	0
D₂O	30.2	

[a]) Takes solvent electrostriction by the cation into account.

In this expression, the quantity in square brackets is independent of micellar hydration, since the inclusion of water molecules changes each term by the same amount. We will find the result that $I(0)/S(0)$ is proportional to the product of the number of micellized monomers per cc ($c - cmc$) and the mean aggregation number very useful in what follows.

The scattering curves for 0.025 M–0.50 M SDS in D_2O at 25 °C were fitted using Equations (1) and (4) in a weighted nonlinear least squares routine, with $\bar{n}$ and Z as adjustable parameters. β is, of course, $(\bar{n} - Z)/\bar{n}$. When the micelles are ellipsoids, the axial ratio is a third adjustable parameter; the overall and core axial ratios are assumed to be the same. Figure 3a shows a typical fit, and Table 3 summarizes the values obtained.

Aggregation numbers for SDS : C 222 in water

The complexation constant [22] for Na^+ by C 222 is 5×10^3, so that very little of the Na^+ present is uncomplexed. The volume of the C 222 : Na^+ complex is 567 $Å^3$ and its diameter is 10 Å; it is hydrogen-rich, with a ϱ of 0.74×10^{10} cm^{-2}. In our first attempt to fit the scattering curves for the SDS : C 222 micellar solutions, we again used group additivity, computing the contents of the micellar shells from the adjustable parameters $\bar{n}$ and Z, as was done for the SDS cases. Because of C 222's large volume, this gave Z a prominent role in determining the volume and ϱ of the micellar shells. The fits gave values of ϱ_{sh} of 2.2×10^{10} cm^{-2}, and therefore high contrast with the D_2O solvent. The mean aggregation numbers thus obtained were concentration-invariant (average of 51.5), but the fits were poor as a result of the experimental and calculated $I(Q)$'s differing by a proportionality constant.

We therefore took a different approach: $\bar{n}$ was fixed (actually a range of $\bar{n}$ was tested, and the values reported in Table 4 reflect this) and Z, V_{sh} and ϱ_{sh} were allowed to vary. The micellar core has a radius slightly less than 16.7 Å; it contains all the C_{12} chains. The fits are now very good; Figure 3b shows a typical result. Allowing V_{sh} to vary independently, thus decoupling it from the micellar charge, makes it evident that these SDS : C 222 micelles have a voluminous outer region

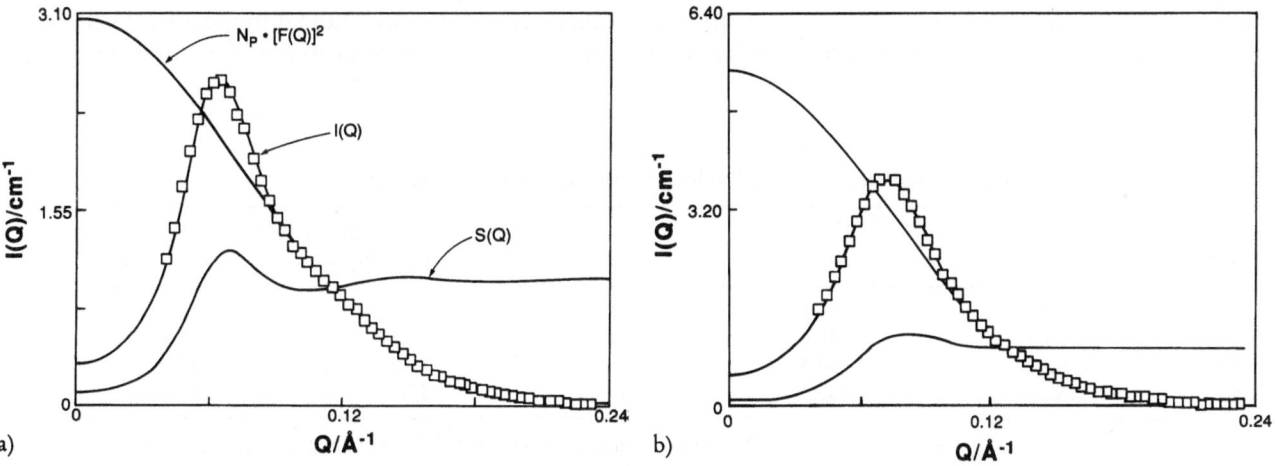

a)   b)

Fig. 3. Fits to the SANS curves for (a) 0.10 M SDS and (b) 0.10 M SDS : S 222

Table 3. Micellar parameters for SDS solution at 25 °C

[SDS], M	$\bar{n}$	Z	axial ratio	R_{tot}, Å	ϱ_{sh}, 10^{10} cm^{-2}
0.025	69.8 ± 0.7	12.6 ± 1.0	—	22.5	4.8
0.050	74.1 ± 0.5	18.7 ± 0.8	—	22.8	4.5
0.10	80.4 ± 0.5	23.0 ± 0.7	—	23.3	4.3
0.25	94.5 ± 0.4	26.7 ± 0.6	1.3 ± 0.1	24.6	5.2
0.50	106.3 ± 0.4	26.7 ± 0.5	1.2 ± 0.1	25.7	5.2

Table 4. Micellar parameters for SDS : C 222 solutions at 25 °C

[SDS], M	$\bar{n}$	Z	R_{tot}, Å	$\varrho_{sh}, 10^{10}$ cm^{-2}
0.05	49.5	13.3 ± 0.5	23.4 ± 0.2	3.18 ± 0.08
0.10	53.0	13.0 ± 0.2	25.3 ± 0.1	4.19 ± 0.03
0.25	51.5	12.9 ± 0.2	25.6 ± 0.2	4.09 ± 0.04
0.50	55.5	6.7 ± 0.7	24.8 ± 0.1	4.52 ± 0.03

	$I(0)/S(0)$	$[b_{mon,t} - V_{mon,t} \cdot \varrho_{D_2O}]^2, 10^{21}$ cm^2 (from $I(0)/S(0)$)	ϱ_{sh}, calc^a)
0.05	2.96	2.04	3.4
0.10	5.46	1.74	4.0
0.25	15.3	1.99	4.2
0.50	22.4	(1.34)	
		(avg: 1.92)	
		(1.88 for 60% of total C 222 visible)	

^{a}) Material visible: head groups, 60% of total C 222, remaining volume D$_2$O.

(consistent with the C 222 : Na$^+$ complex sitting out from the head groups). The values of ϱ_{sh} obtained agree with the V_{sh} values, provided 60% of the C 222 in the solution (81% of that associated with the micelles) is "seen" as part of the scattering entity. The

fitted values of $I(0)/S(0)$ give, using Equation (6), the same result: for the dry micelles, 60% of the total C 222 is contributing to $b_{mon,t}$ and $V_{mon,t}$.

One can now ask whether the changes in absolute intensities observed when C 222 is added to an SDS micellar solution are consistent with the micellar parameters reported above. We proceed by calculating the $I(0)/S(0)$ values expected with C 222 present from those obtained in its absence; Table 5 presents the results. The agreement between the expected and observed values is quite good.

Aggregation numbers for C_7SS and C_7SS : $C222$ in water

The data analysis for these solutions proceeded in a fashion similar to that described in detail for the SDS systems, so we will summarize only the differences here. The C$_7$SS micelles are ellipsoids [3–5], and they have a less sharphly-defined water-hydrocarbon interface than do the SDS micelles. As a result, only ca. 70% of the tails (this includes the ester moieties; [5]) reside in the dry core of the micelle, and ϱ_{sh} is substantially different from ϱ_{D_2O}, even without C 222 present.

Table 6 summarizes the micellar parameters obtained from the weighted nonlinear least squares fitting. At low concentrations of C$_7$SS, addition of C 222

Table 5. Prediction of absolute intensities for SDS : C 222 solutions using data for SDS solutions

[SDS], M	no C 222 $I(0)/S(0)$	$\bar{n}$	$(c - cmc), 10^{19}$ cm^3	with C 222 : $I(0)/S(0)$ from Eq. (6)	Observed
0.05	1.30	74.1	2.53	2.86	2.95
0.10	3.05	80.4	5.54	6.19	5.46
0.25	8.80	94.5	14.6	15.2	15.3
0.50	18.0	106.3	29.6	27.6	22.4

with C 222: $[b_{mon,t} - V_{mon,t} \cdot \varrho_{D_2O}]^2 = 1.91 \times 10^{-21}$ cm^2; without C 222: 6.03×10^{-22} cm^2; $\bar{n} = 51.5$ when C 222 is present

Table 6. Micellar parameters for C$_7$SS solutions at 45 °C

[C$_7$SS], M	$\bar{n}$	Z	axial ratio	R_{tot}, Å	$\varrho_{sh}, 10^{10}$ cm^{-2}
Without C 222:					
0.05	35.8 ± 0.6	13.7 ± 0.4	1.6 ± 0.1	19.3	3.2
0.10	50.4 ± 0.3	15.6 ± 0.2	2.1 ± 0.1	21.7	3.8
0.25	101.7 ± 0.8	15.3 ± 0.4	2.7 ± 0.1	27.7	4.0
With C 222:					
0.05	40	7.5 ± 0.2	2.2	23.8 ± 0.2	3.27 ± 0.06
0.10	56	7.4 ± 0.1	2.7	27.0 ± 0.2	3.37 ± 0.05
0.25	75	4.4 ± 0.5	3.3	28.7 ± 0.2	3.59 ± 0.04

increases the micellar aggregation numbers slightly, but at 0.25 M, a decrease is observed. The evolution of the absolute intensities with C_7SS concentration and with addition of C 222 is again completely consistent with the observed $\bar{n}$'s. Figure 4 contrasts the effect of C 222 on aggregation numbers in the two systems. As was the case for the SDS : C 222 system, the C_7SS : C 222 micelles have voluminous outer shells with about 60% of the total C 222 visible. The apparent micellar charge in the presence of C 222 is very low, but this is consistent with the results obtained with conductimetry.

The effect of added NaCl on micellar aggregation numbers

Figure 5 presents the evolution of $\bar{n}$'s with increasing (NaCl) for the two systems. SDS micellar solutions

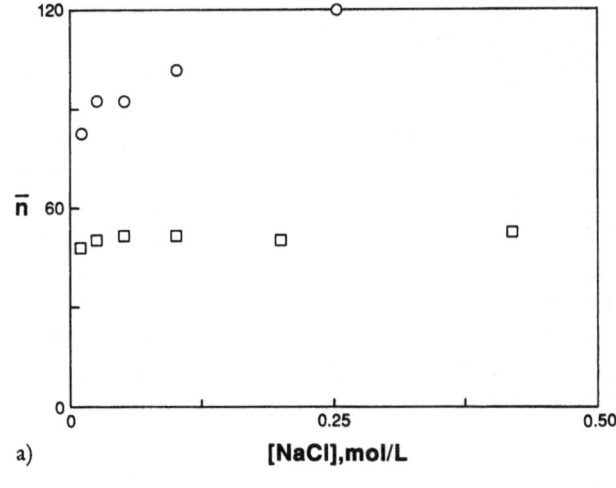

a)

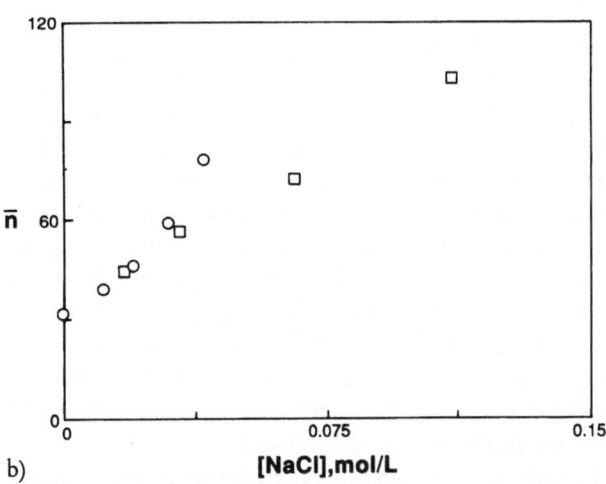

b)

Fig. 5. Effect of NaCl on the mean aggregation numbers for the micelles in (a) 0.025 M SDS and (b) 0.03 M C_7SS. Without C 222: (O); with (□)

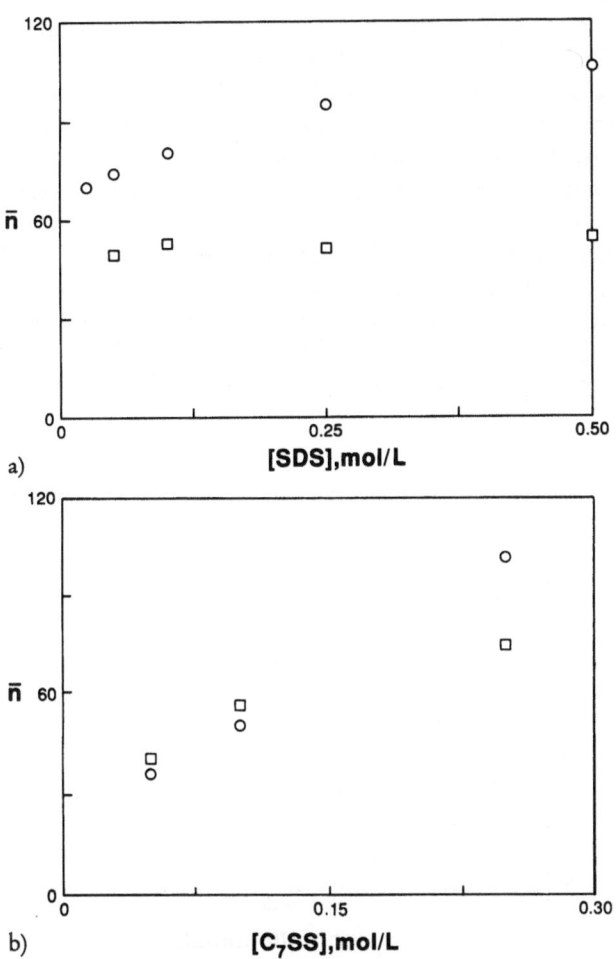

a)

b)

Fig. 4. Mean aggregation numbers for the micelles in (a) SDS and (b) C_7SS solutions. Without C 222: (O); with (□)

containing added NaCl have been investigated by many other experimental techniques [18, 19], and the $\bar{n}$'s obtained here agree well. For SDS : C 222, the addition of NaCl has no effect on $\bar{n}$ (see also Ref. [23] in which the same result was obtained using a fluorescence quenching technique), while for C_7SS : C 222, substantial increases (although smaller than in C 222's absence) are seen. We note also that the presence of C 222 greatly increases the solubility of C_7SS in aqueous NaCl.

The effect of C 222 on the self-assembly of $C_{10}SS$ in water

Evans and coworkers [24] have found in aqueous dispersions of double-chain cationic amphiphiles with highly hydrated counterions, a vesicle-to-micelle transition occurring at low concentrations (for example, in the range of 10^{-3} to 10^{-1} M for didodecyldimethylammonium acetate). We find that this behavior occurs for the sulfosuccinates as well. Addition of C222 to opaque dispersions of $C_{10}SS$ in water at 25 °C produces clear solutions. We observed, using video-enhanced differential interference contrast microscopy (VEDICM), that these solutions contain vesicles from concentrations of 10^{-4} M up to 10^{-2} M, where the vesicle-to-micelle transition is essentially complete. We are presently using a fluorescence quenching technique to obtain the surfactant inventory in the two kinds of aggregates.

Discussion

The introduction of the macrocyclic multidentate ligand C222 into SDS micellar solutions (1:1 molar ratio, SDS:C222) decreases the mean micellar aggregation numbers and makes them independent of SDS concentration (at least to 0.5 M), as well as of added NaCl concentration. The micellar shell containing the sulfate head groups, the C222:Na$^+$ inclusion complexes and the waters of hydration is several Å thick. This is consistent with the view [1] that formation of the large C222:Na$^+$ complex draws the Na$^+$ ions away from the sulfate head groups, thus producing greater curvature at the micellar surface and a resulting decrease in mean aggregation number.

In contrast, for C_7SS micelles, the decrease in mean $\bar{n}$ occurs only in the more concentrated micellar solutions. Also, with C222 present in a 1:1 molar ratio, the effect of added NaCl on micellar growth is muted somewhat but not eliminated. Why do the two systems behave differently?

First, recall that without C222 present, the C_7SS micellar growth is much more rapid with increasing amphiphile concentration than is the case for SDS. Because C_7SS is double-chained, the area per head group (and hence the extent of hydrocarbon-water contact) is large for the small micelles formed in dilute solutions. Minimization of the more extensive hydrocarbon-water contact for C_7SS as compared to SDS translates into more rapid micellar growth for the former amphiphile.

Now, let us simplify the actual micellar surface structure and consider a smooth spherical surface studded with head groups (having a diameter of 5.5 Å if they are sulfates, 4.9 Å if they are sulfonates). From the mean aggregation numbers, we compute the radii of the spheres containing all of the monomers moieties

except the head groups and counterions. Then we add the appropriate head group diameter and compute the area per head group available on the surface of this sphere. Figure 6 displays the results.

For SDS:C222 solutions just below the cmc, the area per molecule at the air-water interface (evaluated by surface tension, [1]) is 136 Å^2. The surface area available on the SDS micelles is insufficient to accommodate one C222 molecule per bound Na$^+$ (even at a somewhat lower area per molecule than 136 Å^2), unless changes in β and/or $\bar{n}$ occur. At the micellar surface, complexation of the sodium ions will increase the head group repulsion, even without a change in β, by increasing the head group-counterion separation and setting the stage for an increase in area per head group. Therefore, for SDS the C222 causes a decrease in the mean $\bar{n}$'s (to around 51.5) and an increase in head group area to 115 Å^2 at our idealized smooth surface. This translates into 156 Å^2 per C222:Na$^+$ complex if β is 0.74.

For the C_7SS micelles, the area per head group at 0.05 M and 0.10 M is sufficient for the binding of one C222 molecule per bound Na$^+$, without a change in $\bar{n}$. In fact, the mean $\bar{n}$'s increase slightly, and β increases substantially, compensating for the increased head group-sodium ion separation which arises from complexation. However, at 0.25 M C_7SS, the area per head group in the absence of C222 is too low to accomodate the C222 binding, and a decrease in aggregation number, with its concomitant increase in area per head group is observed. It should also be noted that at the lower C_7SS concentrations, because of the extensive

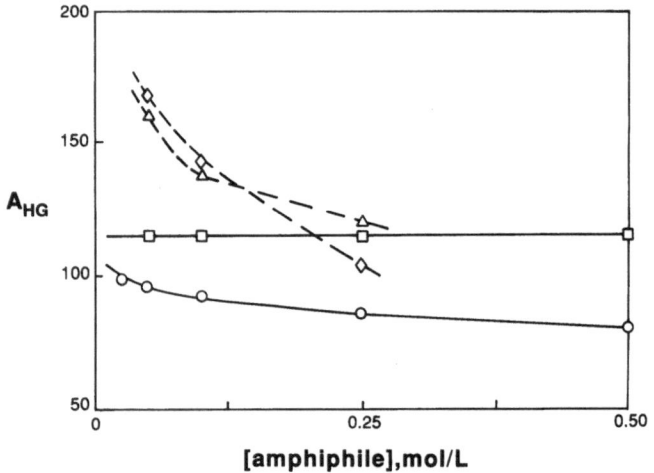

Fig. 6. Areas per head group for micelles in (O) SDS; (□) SDS:C222; (◇) C_7SS; (△) C_7SS:C222

hydrocarbon-water contact, interaction of the hydrocarbon patches of the C 222 : Na$^+$ inclusion complex with the micellar surface may mute the increasing head group repulsion. It is even possible that the ester carbonyls participate in local dipole-dipole interactions with the complex.

Acknowledgement

K.A.P. is a Sun Company graduate fellow. This work was funded by the National Science Foundation (grants to L.J.M.: CHE-8308362 and VPW-8600285). K.A.P. thanks D. D. Miller of the Department of Chemical Engineering and Materials Science, University of Minnesota, for his assistance with the VEDICM work.

References

1. Evans DF, Sen R, Warr GG (1986) J Phys Chem 90:5500
2. Brady JE, Evans DF, Grieser F, Warr GG, Ninham BW (1986) J Phys Chem 90:1853
3. Magid LJ, Daus KA, Butler PD, Quincy RB (1983) J Phys Chem 87:5472
4. Magid LJ, Daus KA, Butler PD, Triolo R, Caponetti E (1987) In: Magid LJ (ed) Static SANS Data for Micelles of Double-Tailed Surfactants, Magnetic Resonance and Scattering form Surfactant Systems, Plenum, New York, in press
5. Magid LJ, Payne KA (1986) ACS Colloid Symposium, June, Atlanta, Ga, USA
6. Evans HC (1956) J Chem Soc 579
7. Hayter JB, Penfold J (1983) Coll & Polym Sci 261:1022
8. Kotlarchyk M, Chen S-H (1983) J Chem Phys 79:2461
9. Magid LJ (1986) Coll Surf 19:129
10. Hayter JB, Penfold J (1981) Mol Phys 42:109
11. Hansen JP, Hayter JB (1982) Mol Phys 46:651
12. Nägele G, Klein R, Medina-Noyola M (1985) J Chem Phys 83:2560
13. Chao Y-S, Sheu EY, Chen S-H (1985) J Phys Chem 89:4862
14. Sheu EY, Wu C-F, Chen S-H, Blum L (1985) Phys Rev A 32:3807
15. Cabane B, Duplessix R, Zemb T (1985) J Phys 46:2161
16. Triolo R, Caponetti E, Graziano V (1985) J Phys Chem 89:5743
17. Berr SS, Coleman MJ, Jones RRM, Johnson JS (1986) J Phys Chem 90:6492
18. Huisman HF (1964) Proc K Ned Akad Wet Ser B Phys Sci 67:367, 376, 388, 407
19. Warr GG, Grieser F, Evans DF (1986) J Chem Soc Faraday Trans 1 82:1829
20. Kostorz G, Lovesey SW (1979) Treatise Mater Sci Technol 15:5
21. Immirzi A, Perini B (1977) Acta Crystallogr Sect A 33:216
22. Sauvage JP, Lehn JM (1975) J Am Chem Soc 97:6700
23. Evans DF and coworkers (1987) manuscript in preparation
24. Miller DD, Bellare JR, Evans DF, Talmon Y, Ninham BW (1987) J Phys Chem, in press

Received January 21, 1987;
accepted January 29, 1987

Authors' address:

Dr. Linda J. Magid
Department of Chemistry
University of Tennessee
Knoxville, Tennessee 37996-1600, U.S.A.

Progress in Colloid & Polymer Science   Progr Colloid & Polymer Sci 73:18–29 (1987)

Control of size and shape of micelles, of flow properties, and of pH-values in aqueous CTAB-solutions via photoreactions of solubilizates

T. Wolff, T. A. Suck, C.-S. Emming, and G. von Bünau

Physikalische Chemie, Universität Siegen, Siegen, F.R.G.

Abstract: The influence of in situ photochemical transformations of aromatic solubilizates on the flow properties of aqueous micellar solutions of cetyltrimethylammonium bromide was investigated, using Ostwald-viscometers in cases of Newtonian flow and a rotating viscometer in cases of non Newtonian flow behaviour, i. e. rheopexy, thixotropy, and viscoelasticity. Newtonian flow was observed in systems containing trans and/or cis-stilbene derivatives or nonpolar 9-substituted anthracenes as solubilizates which are photochemically cis-trans isomerized or photodimerized, respectively, whereby the viscosity changed. Non Newtonian flow was observed in systems containing 9-anthracene carboxylic acid which could be photodimerized while a change of flow behaviour occurs simultaneously. In these systems pH changes were observed upon irradiation. Static light scattering experiments at low angles indicated that spherical micelles differing in radius exist in the systems exhibiting Newtonian flow, while in the non Newtonian systems rod-like aggregates differing in length are present.

Key words: Micelles, photochemistry, rheology, static light scattering.

1. Introduction

The solubilization of certain aromatic compounds in aqueous micelles of cetyltrimethylammonium bromide (CTAB) causes a change of size and shape of the micelles, accompanied by an increase of the viscosity of the solution. We call this class of aromatic compounds "rheologically active". Other aromatic solubilizates do not show this effect and are consequently called "rheologically inactive". A common feature of the rheologically inactive compounds is a low solubility in micellar CTAB solutions. In some cases photochemical transformations of rheologically active compounds into inactive ones (and vice versa) can be performed in situ whereby the viscosity of the system is varied. This has been previously found [1, 2, 3] in:

1. 9-substituted anthracenes and acenaphthylene which form rheologically inactive photodimers on irradiation,

2. N-methyldiphenylamine which forms rheologically inactive N-methylcarbazole on irradiation, and

3. 4-hydroxystilbene which exists in photoisome-rizable cis and trans forms differing in rheological activity.

In this paper more detailed rheological analyses of the "photorheological effects" are presented which were performed using a rotating viscometer and the following compounds as solubilizates: nonpolar 9-substituted anthracenes (methyl, ethyl, n-propyl, n-butyl, n-pentylanthracene), the nonpolar 9,10-di-n-butylanthracene, the polar 9-anthracene carboxylic acid, and three stilbene derivatives (4-hydroxystilbene, 3- and 4-stilbene carboxylic acid). In addition some information on microscopic morphological changes of the micelles are reported which were obtained from static low angle light scattering experiments. Moreover, in situ changes of pH values observed in the systems containing 9-anthracene carboxylic acid and CTAB in connection with photorheological effects are described in the paper. All experiments were carried out in CTAB solutions at concentrations between 0.001 and 0.25 M, i.e. at concentrations at which small spherical micelles are present in the absence of additives and solubilizates [4].

2. Experimental

Materials

Cetyltrimethylammonium bromide (Merck, p.a.) was recrystallized from acetone. 9-methylanthracene (Riedel de Haen, 98 %) and 9-anthracene carboxylic acid (Janssen, 98 %) were used as supplied. 9-ethyl, 9-n-propyl, 9-n-butyl, and 9-n-pentylanthracene were prepared as described in the literature [14]. 9,10-di-n-butylanthracene was prepared in analogy to a literature procedure [15]; mp, 105 °C, ^{1}H NMR (CDCl$_3$) 0.8–2.1 ppm (m, CH$_3$ and CH$_2$), 3.4–3.7 ppm (t, CH$_2$), 7.3–7.6 and 8.1–8.4 ppm (2m, aromatic).

Solutions

Micellar solutions were prepared using triply distilled water. Solubilization procedures were accelerated by ultrasonic irradiation. Light scattering samples were filtered through pores of 50–100 nm radius.

Irradiations

Irradiations were performed using a high pressure mercury lamp and interference filters. Degrees of photochemical conversion were determined spectrophotometrically using a Beckman Acta MVII spectrometer. The absorption spectrum of cis-4-hydroxystilbene (needed for the data in Fig. 4) was determined according to the method of E. Fischer [16].

Rheological experiments

Dynamic viscosities of Newtonian liquids were measured using Ostwald-viscometers or a rotating viscometer. Non-Newtonian solutions were examined using a rotating viscometer (Haake CV 100) which was processed by a 68000 CPU microcomputer (SAM 68K, kws Computersysteme, D-7505 Ettlingen) which also served for recording, evaluating, and editing data. Samples were sheared using Mooney-Eward geometry throughout.

pH-measurements

pH values were determined using a Philips PW-9414-pH-meter.

Static low angle light scattering experiments

Light scattering intensities are usually measured as Rayleigh factors $R(\theta, r) = I(\theta) r^2 c(\theta)/I_0 V$ where $I(\theta)$ is the intensity measured at an angle θ (between the directions of incident and scattered light) and at a distance r between the sample and the light detector. I_0 is

the intensity of the incident light and V the scattering volume. $c(\theta)$ is called Cabannes factor which accounts for electrical anisotropies of the scattering particles that enhance the scattering intensity. According to the theory of light scattering [17] the Rayleigh factor is related to the molecular mass of scattering particles by

$$\frac{K\Delta C}{\Delta R(\theta)} = \frac{1}{M \cdot P(\theta)} + \frac{2A_2 \Delta C}{P(\theta)} + \cdots \tag{1}$$

where $\Delta R(\theta)$ is the difference of Rayleigh factors for solution and solvent, M the molecular mass, A_2 a virial coefficient, ΔC the difference of surfactant concentration and cmc, $P(\theta)$ the so called scattering function, and the constant K is given by

$$K = 2\pi^2 n^2 \left(\frac{dn}{d\Delta C}\right)^2 (1 + \cos^2\theta)/\lambda^4 N \tag{2}$$

with n: refractive index, λ: wavelength of the incident light, and N: Avogadro's number. $dn/d\Delta C$ has to be determined separately. Since at very low angles $P(\theta)$ approaches unity micellar masses may be obtained from Equation (1) by extrapolating a graph of $K\Delta C/\Delta R(\theta)$ vs. ΔC to zero ΔC.

For $\theta \rightarrow 0°$ the scattering function $P(\theta)$ may be expanded in a power series considering only first order terms and we obtain [17]:

$$P(\theta)_{\theta \rightarrow 0} = 1 - \frac{16\pi^2 R_g^2 \sin^2(\theta/2)}{3\lambda^2} \tag{3}$$

where R_g is radius of gyration of the micelle. Thus in the limit of zero concentration and at low angles Equation (1) can be rewritten as

$$\frac{K \cdot \Delta C}{\Delta R(\theta)} = \frac{1}{M - (16\pi^2/3\lambda^2) M R_g^2 \sin^2(\theta/2)} \tag{4}$$

and used for the determination of M and R_g (Zimm-plot: $K\Delta C/\Delta R(\theta)$ vs. $\sin^2(\theta/2)$ and extrapolation of $\Delta C \rightarrow 0$ and $\theta \rightarrow 0$). Because of the Cabannes factor $c(\theta)$ and the scattering function $P(\theta)$ Zimm plots deviate from linearity at larger angles. It follows that in our systems it is advantageous to measure at low angles so that the $\theta \rightarrow 0$ — extrapolation can be omitted with an error of less than 2‰ at $\theta < 5°$.

We therefore built a light scattering photometer which is operated at low angles (2–5°). The photometer was designed following the publication of Kaye and McDaniel [18] and is schematically represented in Figure 1. An improvement[1]) of the apparatus is pos-

---

[1]) developed by T. Suck Wisent, D-5900 Siegen.

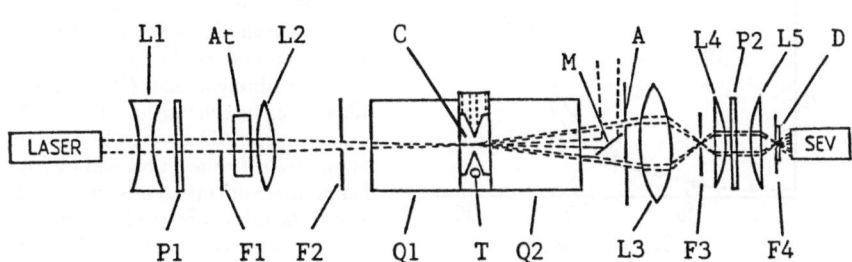

Fig. 1. Apparatus for static low angle light scattering in schematic representation. (A) annulus; (At) attenuator; (C) cuvette; (D) diffusor; (F1 – F4) field stops; (L1 – L5) lenses; (P1, P2) polarizers; (Q1, Q2) quarz windows; (T) temperature control.

sible by taking out the annulus A which is not needed as by the lens L3 being a thick lens only scattered light of a certain angle is focussed at the field stop F3 so that low angles can be chosen steplessly by moving F3 and F4. Thereby interferences arising from the annulus are eliminated. A 5 mW He-Ne-laser (Polytec PL 750) served as light source. The radiant power spectrum of the sample is measured by a photomultiplier (Hamamatsu R712) connected to a 12 bit ADC and to a 68000 CPU microcomputer (SAM 68K, kws Computersysteme GmbH, D-7505 Ettlingen) used as a multichannel analyser, recording the distribution of the scattered light intensity on line as shown in Figure 2. The maximum is used to calcualte the results. The contribution of dust particles to the intensity of the scattered light does not strongly affect the results since it is collected in the right side tail of the distribution diagram (Fig. 2).

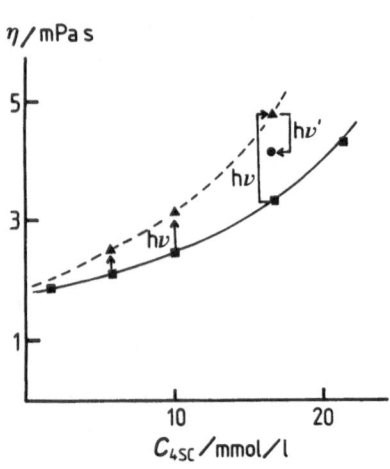

Fig. 2. Intensity distribution of scattered light at 3.1° of an aqueous solution containing 60 mM cetyltrimethylammonium bromide and 100 mM NaBr

3. Results

3.1 Rheological experiments

Stilbene derivatives

Dynamic viscosities of 0.25 M aqueous CTAB-solutions containing 3- and 4-stilbene carboxylic acid (3SC, 4SC) are displayed in Figure 3 as a function of solubilizate concentration. When these systems were tested by a rotating viscometer up to shear rates $\dot{y} = 300\ \mathrm{s}^{-1}$ Newtonian flow was obtained by a linear flow curve (shear stress vs. shear rate). In these experiments the same viscosity values were determined as in capillary (Ostwald) viscometers differing in capillary radius. This demonstrates that the linear flow curves can be extrapolated to the high shear rates of capillary viscometers. Therefore, viscosity data were mostly obtained from capillary viscometry. Within the accuracy of the method the kinematic viscosities measured by capillary viscometers agree with dynamic (i. e. absolute) viscosities, since the density of aqueous micellar solutions does not deviate significantly from that of water. Inspection of the figure shows that increasing concentrations of the trans-isomers cause an increase of viscosity (solid lines in Fig. 3) which can be increased further when the solutions are irradiated at a wavelength of 313 nm. Thereby the trans-isomers are transformed into cis-isomers according to Equation (1) until photostationary concentrations of cis-

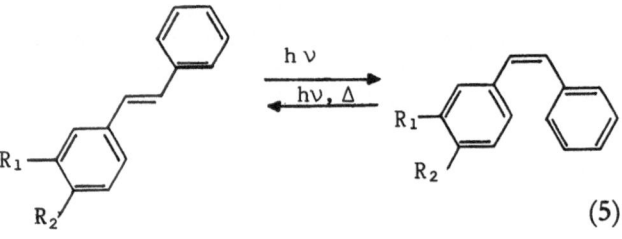

(5)

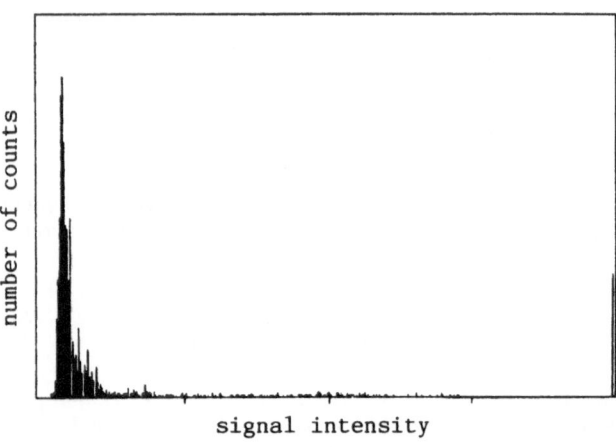

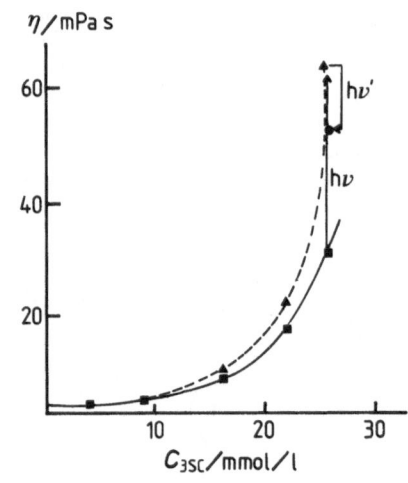

Fig. 3. Dynamic viscosity η at 25 °C of 0.25 M aqueous cetyltrimethylammonium bromide as a function of the concentration C of the solubilizates 3-stilbene carboxylic acid (3SC) and 4-stilbene carboxylic acid (4SC). (■) trans-isomers; (▲) cis-rich mixture of cis-and trans-isomers obtained by irradiating at 313 nm up to constancy of viscosity; (●) values obtained after re-irradiation of cis-rich mixtures at 254 nm

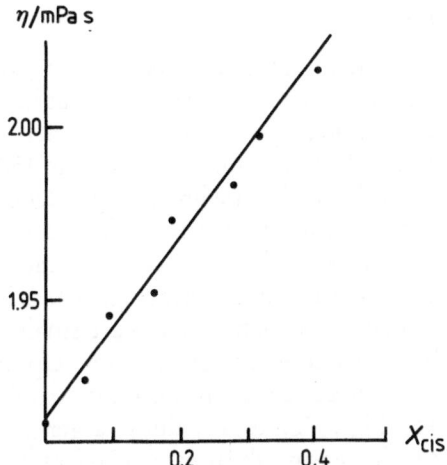

Fig. 4. Dynamic viscosity η at 30 °C as a function of the fraction X of the cis-isomer in 0.25 M aqueous cetyltrimethylammonium bromide solutions containing photochemically prepared mixtures of cis- and trans-4-hydroxystilbene

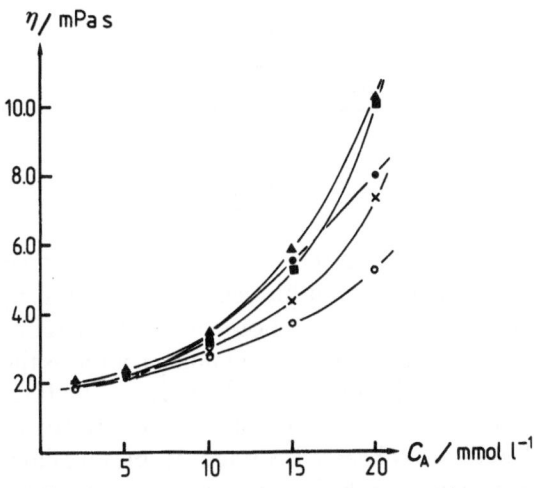

Fig. 5. Dynamic viscosity η at 25 °C of aqueous 0.25 M cetyltrimethylammonium bromide solutions as a function of the concentration C_A of the solubilizates 9-methylanthracene (O), 9-ethylanthracene (×), 9-n-propylanthracene (■), 9-n-butylanthracene (▲), and 9-n-pentylanthracene (●)

and trans-isomers are built up. The broken lines in Figure 3 correspond to these photostationary cis-trans-mixtures which contain more than 90 % of the cis-isomer at 313 nm (cf. [3]). Thus the cis-isomers are more rheologically active than the trans-isomers. Reirradiation of such cis-rich solutions at 254 nm leads to a reisomerization of cis- into trans-isomers until the photostationary cis-trans-ratio at this wavelength is reached, accompanied by a corresponding decrease of viscosity. This photochemical back reaction is accompanied by side reactions to a larger extent in 3SC than in 4SC. In both stilbene derivatives a thermal reisomerization of cis- into trans-isomers takes place when the solution is stored in the dark for several days after which a value of the viscosity characteristic of a solution containing only trans stilbene is restored.

Diagrams qualitatively agreeing with those of Figure 3, but pertaining to 4-hydroxystilbene are published in Reference [3]. The dependence of the viscosity on the extent of photochemical conversion, i.e. on the fraction of the cis isomer, is shown in Figure 4 for 4-hydroxystilbene. The dependence appears to be linear.

Nonpolar anthracene derivatives

A homologous series of anthracenes substituted by n-alkane chains in 9-position induces a viscosity increase when solubilized in 0.25 M aqueous CTAB solutions (Fig. 5). Thereby these compounds differ from unsubstituted anthracene [2], 9,10-dimethylanthracene [2], and 9,10-di-n-butylanthracene which have no or very little effect on the viscosity and exhibit low solubility limits (about or less than 20 mM in 0.25 M CTAB) like all rheologically inactive compounds tested so far. Newtonian flow is observed in all solutions containing these nonpolar anthracenes.

Irradiation of oxygen free solutions of the monosubstituted anthracenes at 366 nm leads to formation of photodimers according to Equation (6). In the presence of oxygen endoperoxides are also formed [Eq. (7)]. Both the products, dimers and endoperoxides,

$$\text{structure} \underset{\Delta,\,h\nu'}{\overset{h\nu}{\rightleftharpoons}} 2 \; \text{anthracene–R} \underset{\Delta,\,h\nu'}{\overset{h\nu}{\rightleftharpoons}} \text{structure} \qquad (6)$$

lished in Reference [3]. The dependence of the viscosity on the extent of photochemical conversion, i.e. on the fraction of the cis isomer, is shown in Figure 4 for 4-hydroxystilbene. The dependence appears to be linear.

belong to the rheologically inactive compounds, so that the viscosity of sufficiently irradiated solutions is decreased while the Newtonian flow behaviour is retained in all experiments. The decrease can lead to

$$\text{(anthracene)}-R + O_2 \xrightarrow{h\nu} \text{(dimer)}-R \qquad (7)$$

viscosities very close to the viscosity of pure CTAB solutions (see curve for 9-n-butylanthracene in Fig. 6) depending on the amount of photoconversion. In contrast to the 4-hydroxystilbene containing solutions the viscosity changes do not depend linearly on the fraction of photoconverted solubilizates and in some cases an initial increase is observed. Some examples are illustrated in Figure 6, which may be compared with Figure 4. It should be noted that solutions containing high fractions of photodimers can be thermodynamically labile with respect to precipitation of photodimers since, for these compounds being rheologically inactive, the solubility limits may be exceeded. Therefore the irradiation experiments displayed in Figure 6 were not carried on when precipitation was observed. The shapes of the curves of Figure 6 also depend on the CTAB concentration and on the solubilizate/surfactant ratio, i.e. the maxima need not necessarily occur in differently composed solutions.

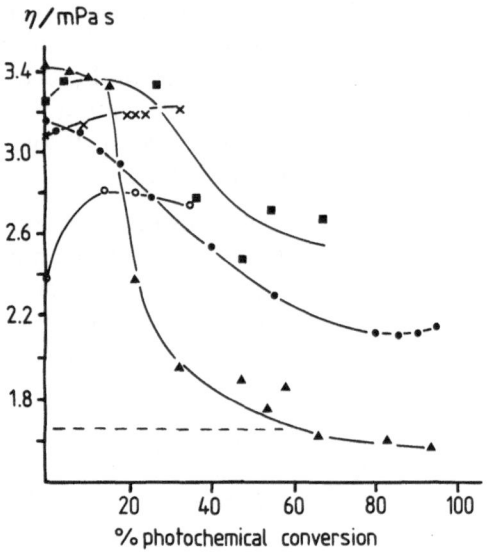

Fig. 6. Dynamic viscosity η at 25 °C of 0.25 M aqueous cetyltrimethylammonium bromide solutions in the presence of 10 mM of nonpolar 9-substituted anthracenes as a function of the degree of photochemical conversion of anthracene monomers into dimers. Symbols coincide with those of Figure 5. Dashed line: viscosity of pure 0.25 M aqueous CTAB

9-anthracene carboxylic acid

In the systems containing 9-anthracene carboxylic acid (9AC) Newtonian flow is observed only at low concentrations of solubilizates (for instance when a concentration of 25 mM 9AC is not exceeded in 150 mM CTAB). This is shown by the linear dependence of the shear stress τ on the shear rate $\dot{\gamma}$ in the flow curve of Figure 7a. On increasing the 9AC concentration, the flow behaviour becomes rheopectic (Fig. 7b), as is revealed by flow curves in which η-values are smaller for increasing than for decreasing shear rates. Upon further increasing the 9AC-concentration thixotropic flow is observed (Figs. 7c, d; η-values larger for increasing than for decreasing shear rate). On irradiation of the rheopectic solution of Figure 7b the Newtonian solution characterized by Figure 7c is obtained after photochemical conversion of 5% of the 9AC-monomers. The flow curves of Figure 7f and 7g correspond to photoconversion of 5% of the 9AC-monomers of Figures 7c and 7d, respectively.

The non-Newtonian liquids formed by aqueous CTAB solutions in the presence of 9AC exhibit viscoelastic behaviour. A characterisitic feature of viscoelastic surfactant solutions can be obtained when the viscosity η at a constant shear rate $\dot{\gamma}$ is measured as a function of time. After an initial increase such curves pass through a maximum η_{max} and then approach a constant value η_∞ [5, 6]. An example is given in Figure 8a. The ratio η_{max}/η_∞ decreases with decreasing shear rate and the maximum vanishes at low shear rates. In Figure 8b such ratios are plotted at $\dot{\gamma} = 65 \text{ s}^{-1}$ for three concentrations of CTAB as a function of the 9AC-concentration which is plotted as a percentage of the CTAB concentration. As η_∞ does not vary much in the concentration range of strongly increasing η_{max}/η_∞ (Fig. 8b) this ratio (at a fixed shear rate) can be taken as a rough measure of the elastic part of viscoelasticity since the maximum η_{max} is mainly due to elastic deformation [6]. The samples at 150 mM are the same as those used for measuring the flow curves of Figure 7. It can be seen that η_{max}/η_∞ drops after irradiation, i.e. when 9AC-monomers are removed and dimers are formed.

The non-Newtonian flow properties and the light scattering experiments described below indicate that rodlike aggregates are present in the solutions. Formation of the aggregates is induced by 9AC since in pure CTAB solutions only spherical micelles exist at the concentrations under investigation [4]. According to the theory of Doi and Edwards [7] the lengths of the

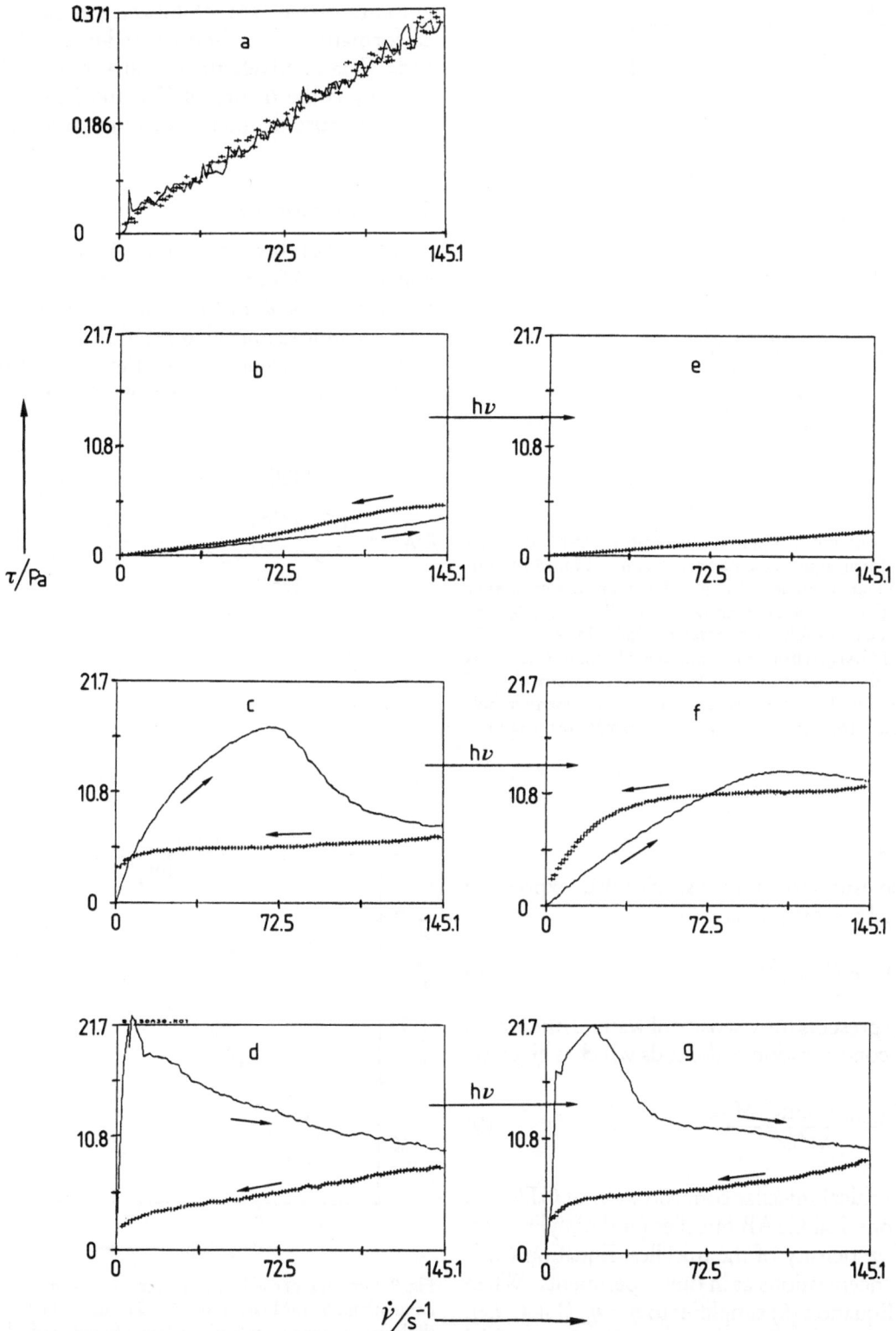

Fig. 7. Flow curves (shear stress τ vs. shear rate $\dot{\gamma}$) at 25 °C for aqueous 0.15 M solutions of cetyltrimethylammonium bromide containing (a) 22.5 mM, (b) 30 mM, (c) 37.5 mM, and (d) 45 mM of 9-anthracene carboxylic acid. The flow curves (e), (f), and (g) were obtained after photoconversion of 5% of the concentrations of 9-anthracene carboxylic acid present in (b), (c), and (d), respectively. The shear rate $\dot{\gamma}$ was increased linearly from 0 to 145 s^{-1} within 50 s, kept at 145 s^{-1} for 20 s, and then decreased linearly to 0 s^{-1} within 50 s

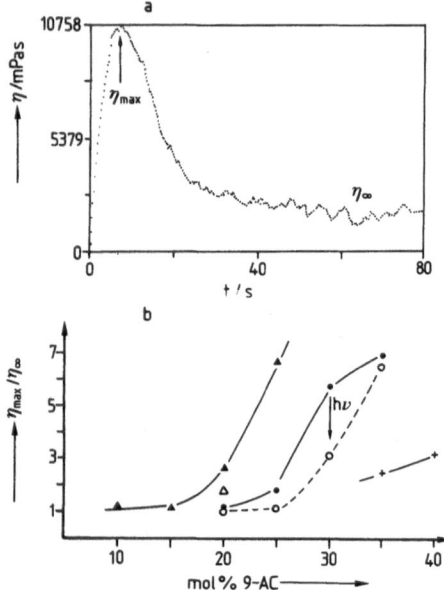

Fig. 8. (a) Viscosity η at 25 °C and at a shear rate $\dot{y} = 65\ \mathrm{s}^{-1}$ as a function of time t for an aqueous solution containing 150 mM cetyltrimethylammonium bromide and 37.5 mM 9-anthracene carboxylic acid (25 % of the surfactant concentration). (b) Ratios η_{max}/η_{∞} [cf. Fig. 8 (a)] for aqueous solutions containing 250 mM (▲), 150 mM (●), and 75 mM (×) cetyltrimethylammonium bromide (CTAB) as a function of the concentration of 9-anthracene carboxylic acid (9AC) plotted as mol % of the CTAB concentration. Open symbols refer to irradiated solutions. The samples at 150 mM are the same as in Figure 7

amounts of 9AC are solubilized and that photochemical formation of 9AC-dimers reduces the lengths of the rods. Quantitatively the L values may need some correction, as the theory of Doi and Edwards does not take electrostatic interactions into account.

3.2 pH measurements

When 9-anthracene carboxylic acid is solubilized in aqueous CTAB the pH value of the solution drops, because a part of the 9AC molecules dissociates. This is mainly due to ion pair formation of CTA-cations and 9AC-anions according to Equation (10), i. e. the anions replace some of the bromide counterions of the

$$\text{(10)}$$

rods can be estimated from viscosity data expressing the observed viscosity η as

$$\eta = \eta_w\left[1 + (C_r L^3)^3\right] \tag{8}$$

where η_w represents the viscosity of water, L the length and C_r the concentration of the rods which is given by

$$C_r = \frac{(C_{CTAB} - cmc)\, M_{CTAB}}{\pi r^2 L \varrho} \tag{9}$$

with cmc: critical micellar concentration of CTAB, r: radius of spherical CTAB micelles (and of cylindrical rods), and ϱ: density of the micelles. Equation (8) is valid for concentrations as in our experiments. When $C_r \cdot L \ll 1$, Equation (8) simplifies to $\eta = \eta_w\,(1 + C_r L^3)$. Using η-values at very low shear rates ($\dot{y} \to 0$, i.e. initial slope of flow curves) the rod lengths L given in Figure 9 can be obtained from Equation (8). The figure qualitatively shows that rod lengths grow when increasing

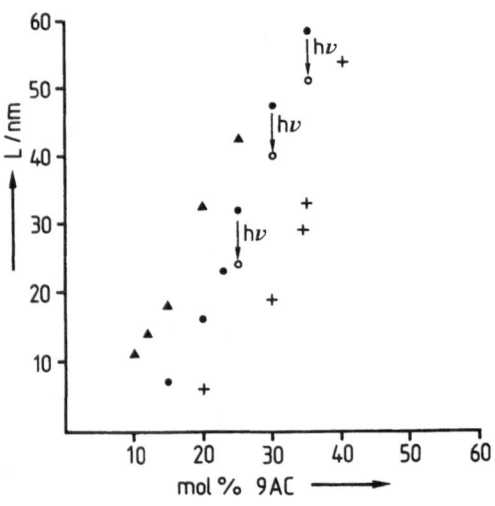

Fig. 9. Length L of rod like aggregates at 25 °C in aqueous solutions containing 250 mM (▲), 150 mM (●), and 75 mM (×) cetyltrimethylammonium bromide (CTAB) as a function of added 9-anthracene carboxylic acid in mol % of the CTAB concentration. Open symbols refer to irradiated solutions. The samples at 150 mM CTAB are identical with those of Figures 7 and 8. L-values were calculated using Equation (8) taking $r = 1.8$ nm and $\varrho = 0.75$ g/cm³

$$2\,HBr + \quad\begin{array}{c}\text{COO-CTA}\\[2pt]\text{COO}\\|\\\text{CTA}\end{array} \longrightarrow \quad\begin{array}{c}\text{COOH}\\[2pt]\text{COOH}\end{array} \quad + 2\,CTAB \qquad (11)$$

micelles. A contribution of solvolytic dissociation of the acid in water to the pH value can be neglected. From the pH values concentrations of hydrobromic acid formed in Equation (10) can be calculated which give also the degree of dissociation. Four examples of such solutions (agreeing with the solutions of Fig. 7) are given in Table 1. The table shows that between 20% and 30% of the 9AC molecules are dissociated, so that two states of solubilized 9AC must exist: AC-anions acting as large counterions at the micellar surface and the neutral acid solubilized somewhere within the micelles. Photodimerization of 5% of the 9AC-monomers in these solutions leads to a substantial pH increase. Obviously the anions of the photodimers do not act as counterions so that HBr is consumed according to Equation (11). However, stoichiometrically, more HBr vanishes than photodimers are formed, indicating that the ratio 9AC-anions/neutral 9AC is strongly decreased by the presence of photodimers. It should be mentioned that the pH changes are connected with variations of the overall electrolyte concentration which influences size and shape of the micelles and affects light scattering intensities (see below).

3.3 Static low angle light scattering experiments

When light scattering samples are placed between two perpendicularly crossed filters for linearly pola-

rized light, one observes scattered light only from solutions that contain anisometric (rod-like) particles. In this way it is possible to find out that in solutions containing aqueous CTAB and sufficient 9-anthracene carboxylic acid, rod-like micelles are present, while in all other cases globular micelles must exist. This is the only qualitative information to be obtained from solutions as described in the previous sections. For more detailed light scattering experiments such solutions are not suitable because of strong electrostatic interactions of the highly concentrated charged micelles which inhibit diffusion. Generally two possiblilities exist to overcome this problem: (i) addition of sufficient salt to ensure an underground of electrolyte that suppresses electrostatic interactions and (ii) diluting the solutions until the charged particles are sufficiently separated so that electrostatic interactions can be neglected. This is the case at CTAB concentrations below 5 mM since Imae et al. [8] determined correct micellar weights of CTAB from light scattering data within this concentration range. Although it is possible to have the same surfactant/solubilizate-concentration ratios a direct comparison of light scattering results in dilute solutions and viscosity experiments at higher concentrations is problematic. The addition of salt also leads to results not comparable with the viscosity measurements described above as the salt influences size and shape of the micelles.

The results of some experiments in dilute solutions ($C_{CTAB} < 80$ mM) are illustrated in Figure 10. The Rayleigh differences $\Delta R(\theta)$ (see Experimental section) of pure CTAB solutions increase as a function of concentration only at CTAB concentrations below 0.5 mM. At higher concentrations, a decrease of $\Delta R(\theta)$ is observed, as expected from electrostatic interactions and in agreement with the results of other authors [8]. In the presence of 9AC the range of increasing $\Delta R(\theta)$ is extended, probably due to the formation of H^+ and Br^- in the aqueous phase which reduce electrostatic interactions (as does the presence of 0.1 M NaBr). Within the range of increasing $\Delta R(\theta)$ a sudden change of the slope of the curves take place at concentrations at which the viscosity increases strongly and the flow

Table 1. pH variations of 150 mM aqueous solutions of cetyltrimethylammonium bromide in the presence of various amounts of 9-anthracene carboxylic acid (9AC) and after photochemical conversion of 5% of the 9AC-monomers to dimers

before irradiation				irradiated	
C_{9AC} mM	pH	C_{HBr} mM		pH	C_{HBr} mM
—	6.85	—			
22.5	2.28	5.25			
30.0	2.03	9.33		2.17	6.76
37.5	1.83	14.79		2.12	7.59
45.0	1.88	13.18		2.04	9.12

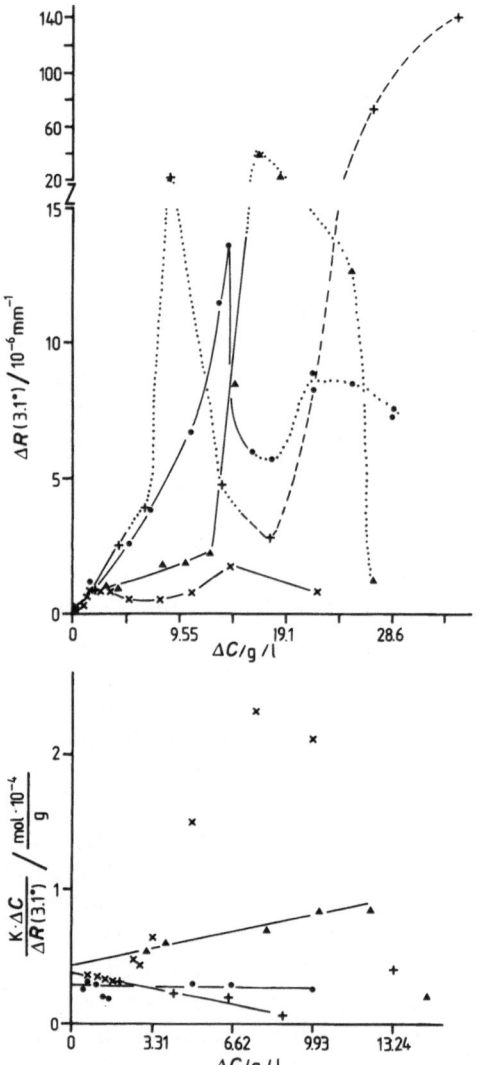

Fig. 10. (a) Differences $\Delta R(\theta)$ of Rayleigh factors of solution and solvent in aqueous cetyltrimethylammonium bromide (CTAB), measured at $\theta = 3.1°$ and at 25 °C as a function of concentration differences $\Delta C = C_{CTAB} + C_{9AC}$ — cmc. A part of the solutions contains 9-anthracene carboxylic acid (9AC). ($\times$) pure CTAB; ($\bullet$) CTAB + 0.1 M NaBr; ($\blacktriangle$) CTAB + 9AC (25 mol % of CTAB); (+) CTAB + 9AC (40 mol % of CTAB); (———) Newtonian, (······) rheopectic, (————) thixotropic flow behaviour. (b) Ratios $K\Delta C/\Delta R(\theta)$ at $\theta = 3.1°$ and at 25 °C as a function of differences $\Delta C = C_{CTAB} + C_{9AC}$ — cmc. For K see Experimental section. Symbols are the same as in Figure 10a

properties of the solution change from Newtonian to rheopectic flow. This is shown in Figure 10a for solutions containing 25 % and 40 % 9AC (with respect to the CTAB concentration). The latter system becomes thixotropic at $C_{CTAB} > 40$ mM; at 40 mM CTAB the Rayleigh differences also start increasing again.

Thereby a connection between the microscopic solution structure and the macroscopic viscosity is indicated, and the flow behaviour of more concentrated solutions (see Fig. 7) is paralleled.

However, because of the electrostatic interactions, the determination of microscopic micellar dimensions from the data is limited to very low concentrations. We used values at < 5 mM for pure CTAB, < 30 mM for CTAB + 0.1 M NaBr, for CTAB + 25 % 9AC, and for CTAB + 40 % 9AC, i.e. values well in the concentration range of increasing $\Delta R(\theta)$. The $\Delta R(\theta)$ values were plotted as $K\Delta C/\Delta R(\theta)$ vs. ΔC in Figure 10b (cf. Experimental section) to obtain molecular weights of the micelles from extrapolation to zero concentration. The results given in Table 2a were then calculated assuming

i) a 20 % micelle — bromide counterion dissociation,

ii) that rods are formed in the presence of 9AC and NaBr which cannot be concluded from light scattering data alone in this concentration range,

iii) a rod radius of 1.8 nm and a micelle density of 0.7 g/cm^3, and

iv) that rod lengths do not vary within the concentration ranges.

Alternatively to (ii) one might assume that at concentrations below the sudden increase of $\Delta R(\theta)$ (see Fig. 10a) in the presence of 9AC globular micelles exist. The globes then had radii of 2.2 and 2.3 nm for the 25 % 9AC-probe and the 40 % 9AC-probe, respectively. These values agree with the maximal length of a surfactant monomer. An attempt was made to evaluate data from the $\Delta R(\theta)$-values of the 40 % 9AC-probe in the concentration range 75–85 mM CTAB, i.e. in a concentration range in which the $\Delta R(\theta)$ values increase for the second time (see Fig. 10a). The rod length obtained ($L = 21$ nm, last line in Table 2a) agrees with the corresponding value derived from viscosity data ($L = 50$ nm, cf. Fig. 9) by an order of magnitude which is satisfying when the assumptions made in both evaluations are considered.

Table 2b gives results evaluated for spherical micelles for a system containing a non polar anthracene (9-n-butylanthracene). It can be seen that the micelle radius by far exceeds the maximal length of a surfactant monomer (ca. 2.3 nm).

4. Discussion

The results are discussed on the basis of the following scheme in which five different states of micelles are

Table 2. (a) Data from low angle light scattering experiments in the system aqueous CTAB – 9-anthracene carboxylic acid (9AC). $M_{CTAB + 9AC}$: micellar masses; M_{CTAB}: part of $M_{CTAB + 9AC}$ pertaining to CTAB; n: aggregation number of CTAB-molecules (for 20 % $CTA^+ - Br^-$ dissociation); L: length of rodlike micelles; A_2: second virial coefficient (cf. Eq. (1))

C_{CTAB} mM	C_{9AC} mol % of CTAB	C_{HBr}[a]) mM	$M_{CTAB + 9AC}$ g/mol	M_{CTAB} g/mol	n	L nm	A_2 10^{-6} mol dm³/g²
1– 5	–	–	–	25250	74	–	3.0
1–30[b])	–	–	–	36200	106	8.4	0.019
10–30	25	0.6–3.5	22250	19100	56	5.2	1.65
7–27	40	2.5–5.6	24570	19600	58	5.7	6.5
75–85	40	10	90000	72000	210	21.0	0.22

[a]) formed via Equation (10), calculated from pH values; [b]) in the presence of 0.1 M NaBr

Table 2. (b) as a) but for the system CTAB – 9-n-butylanthracene (NBA). r: radius of spherical micelles

C_{CTAB} mM	C_{NBA} mol % of CTAB	$M_{CTAB + NBA}$ g/mol	M_{CTAB} g/mol	n	r nm	A_2 10^{-7} mol dm³/g²
1–2.5	3.8	$5.0 \cdot 10^5$	$4.9 \cdot 10^5$	1440	6.1	-7
1–2.5	7.4	$6.7 \cdot 10^6$	$6.3 \cdot 10^6$	$1.9 \cdot 10^4$	14.3	6

considered, some of which contain solubilizates (not drawn proportional to the micelle size).

substituted anthracene: since no rheological effect can be observed [2] the size and shape of the micelle is not

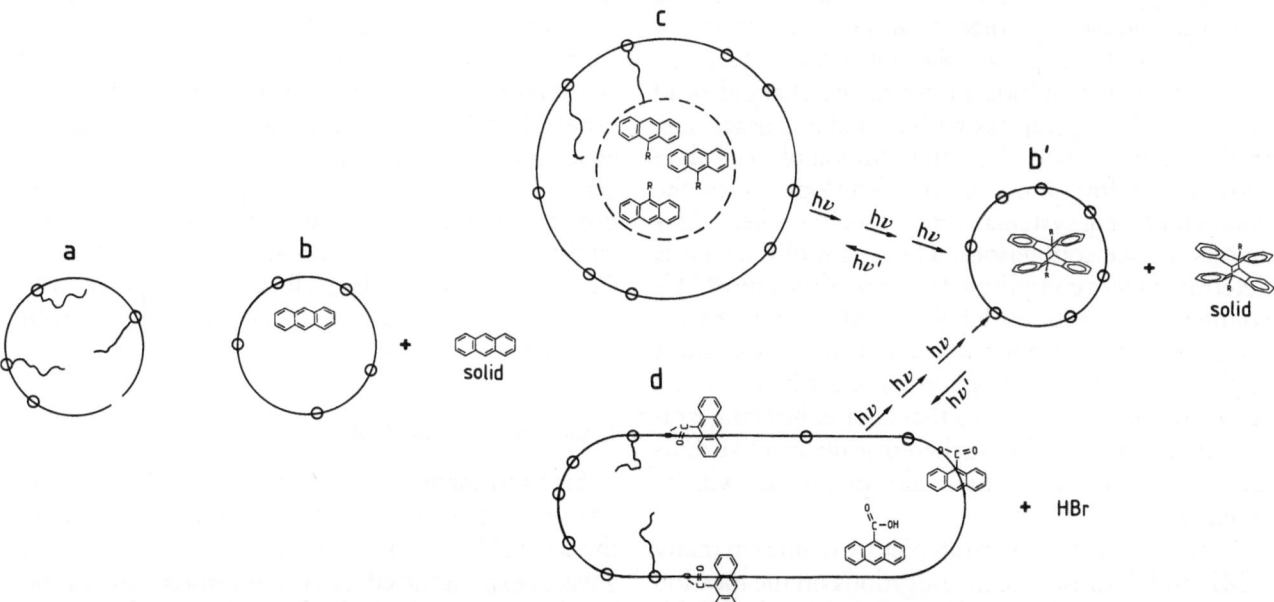

Picture a represents an ordinary CTAB micelle with a radius not exceeding twice the length of a monomer; some of the head groups and surfactant monomers are indicated. Picture b illustrates the situation in the presence of rheologically inactive solubilizates like non-changed and amounts of anthracene exceeding the solubility limit will not be solubilized. Picture c is a hypothetical representation of a micelle containing large amounts of rheologically active nonpolar anthracenes. The picture is in keeping with the facts that

(i) the micelles are large because of a high viscosity of the solution, (ii) the micelles are globular having a radius exceeding the length of a surfactant monomer, and (iii) Newtonian flow is observed under all conditions. If the globular shape were a consequence of appropriately folding flexible rods, one would expect deviations from Newtonian flow at least at high shear rates. The situation in picture c can be interpreted as a microcrystalline or precrystalline assembly of 9-substituted anthracenes surrounded by a layer of surfactant molecules. Features occurring in the UV-spectra of such solutions [2] support an aggregation of anthracenes. For the systems containing stilbene derivatives, similar considerations must be made since only Newtonian flow is observed. Upon photochemically transforming the 9-substituted anthracenes into rheologically inactive photodimers the micelles shrink gradually approaching a situation like picture b' which represents a solution containing micelles of ordinary size out of which the photodimers precipitate when the solubility limit is exceeded. Picture d represents a rod-like micelle which bears anions of 9-anthracene carboxylic acid as counterions and contains neutral 9AC-molecules in a solubilized state. Hydrobromic acid formed via Equation (10) is present in the aqueous phase. That large and strongly binding counterions support rod formation in micellar systems was found previously by Angel et al. [9] using salicylate anions. The effect appears plausible since large counterions reduce repulsion forces of the charged head groups so that aggregates with a smaller surface area (rods) can be formed. Also, the addition of electrolyte favours rod formation. Both effects operate in the 9AC-containing systems as HBr is set free when 9AC-anions replace counterions. The length of the rods is determined by a complicated balance of the ratio 9AC-counterions / neutral solubilized 9AC and the corresponding HBr-concentration. This balance is sensitive to the presence of 9AC-photodimers which cause a reduction of rod lengths so that after exhausting irradiation, i.e. when only the rheologically inactive dimers are present, a situation like picture b' will be reached.

Discussion may arise from picture c, since in many NMR and fluorescence investigations on the location of aromatic solubilizates it is concluded that the solubilizates are located in the vicinity of the head groups (cf. [10, 11]). However, these conclusions do not pertain to solubilizate/micelle ratios as high as in some of our experiments. More spectroscopic experiments are needed (NMR, etc.) to obtain information on the states be-

tween c and b'. Geometric considerations [12, 13] suggest that the radius of micelles like those in picture c should be given by

$$R = 3/2 \, r \, (1 + x) \tag{12}$$

where x is the ratio of the volumes of solubilizate and surfactant. Quantitative light scattering experiments are in progress to prove whether this relation holds for CTAB micelles containing nonpolar anthracenes. Hoffmann et al. [13] recently found that Equation (12) describes experimental results when n-decane was solubilized in tetradecyltrimethylammonium salicylate micelles which were transformed from rods to globes by the solubilizates. These authors also found that globe formation was not induced in the same micelles by the aromatic compound toluene. This might contrast with our results in nonpolar anthracenes. However, the experiments are not strictly comparable, as the anthracenes are solids at room temperature.

The question remains of why closely related compounds like anthracene, 9-methylanthracene, and 9,10-dimethylanthracene differ so much in rheological activity. In previous papers [1, 2] we have noticed that rheologically active nonpolar anthracenes generally have much lower melting points than inactive anthracenes (anthracene: mp 216 °C; 9.10-dimemthylanthracene: 194 °C). The melting points of the rheologically active compounds of Figure 3 are 81 °C (methylanthracene), 59 °C (ehylanthracene), 49 °C (butylanthracene), 70 °C (propylanthracene), 85 °C (pentylanthracene). However, the low melting point is not the only prerequisite for rheological acitvity as 9,10-di-n-butylanthracene (mp, 105 °C) is not rheologically active. Common to the rheologically active nonpolar anthracenes is a lower symmetry compared to inactive anthracenes.

Conclusion and outlook

Size and shape of CTAB micelles can be changed in situ via photoreactions of solubilizates. Especially in the cases of Newtonian flow behaviour, work is in progress to exploit the effect for actinometry and measuring the kinetics of micelle formation (via flash experiments). Progress towards the characterization of the geometry of the particles involved in the photorheological effects will arise from light scattering data examining nonionic micelles in which no electrostatic interactions occur.

Acknowledgements

Financial support of the Deutsche Forschungsgemeinschaft and the Fonds der Chemischen Industrie is gratefully acknowledged. We thank Mrs. H. Christian, Miss S. Weber, and Mr. F. Schmidt for technical assistance.

References

1. Müller N, Wolff T, von Bünau G (1984) J Photochem 24:37
2. Wolff T, von Bünau G (1984) Ber Bunsenges Phys Chem 88:1098
3. Wolff T, von Bünau G (1986) J Photochem 35:239
4. Reiss-Husson F, Luzzati V (1964) J Phys Chem 68:3504
5. Gravsholt S (1979) Naturwissenschaften 66:263; Hoffmann H, Platz G, Rehage H, Schorr W, Ulbricht W (1981) Ber Bunsenges Phys Chem 85:255
6. Schurz J (ed) (1974) Einführung in die Strukturrheologie, Verlag W. Kohlhammer GmbH, Stuttgart
7. Doi M, Edwards SF (1978) J Chem Soc Faraday Trans II 74:560, 918
8. Imae T, Kamiya R, Ikeda S (1985) J Coll Interf Sci 108:215
9. Angel M, Hoffmann H, Löbl M, Reizlein K, Thurn H, Wunderlich I (1984) Progr Coll & Polym Sci 69:12
10. Ulmius J, Lindman B, Lindblom G, Drakenberg T (1978) J Coll Interf Sci 65:88
11. Zachariasse KA, Kozanciewicz B, Kühnle W (1984) In: Mittal KL, Lindman B (eds) Surfactants in Solutions, Plenum Publishing Corporation, New York, Vol 1, p 565
12. Israelachvili IN, Mitchell DJ, Ninham BW (1976) J Chem Soc Faraday Trans II 72:1525
13. Hoffmann H, Platz G, Ulbricht W (1986) Ber Bunsenges Phys Chem 90:877
14. Krollpfeiffer F, Branscheid F (1923) Chem Ber 56B:1617; Sieglitz A, Marx R (1923) Chem Ber 56B:1619; Martin RH, van Hove L (1952) Bull Soc Chim Belg 61:504
15. Harvey RG, Arzadon L, Grant J, Urberg K (1969) J Am Chem Soc 91:4535; Harvey RG, Davis CC (1969) J Org Chem 34:3607
16. Fischer E (1967) J Phys Chem 71:3704
17. Kerker M (ed) (1969) The Scattering of Light, Academic Press, New York
18. Kaye W, McDaniel JB (1974) Appl Opt 13:1934

Received January 21, 1987;
accepted February 11, 1987

Authors' address:

Prof. Dr. T. Wolff
Universität Gesamthochschule Siegen
Fachbereiche Chemie-Biologie
Postfach 10 12 40
D-5900 Siegen, F.R.G.

Progress in Colloid & Polymer Science

Progr Colloid & Polymer Sci 73:30–32 (1987)

Interaction of 1:1 electrolytes with nonionic surfactants in methanol: effect of changing the alkyl chain from a hydrocarbon to a fluorocarbon nature

C. Burger-Guerrisi and C. Tondre

Laboratoire d'Etude des Solutions Organiques et Colloïdales (L.E.S.O.C.), Unité Associée au CNRS n° 406, Université de Nancy, Vandoeuvre-les-Nancy, France

Abstract: The binding constant of 1:1 electrolytes with a series of hydrogenated as well as fluorinated nonionic surfactants having short polyoxyethylated chains have been obtained from electric conductivity measurements. The results for the surfactants with hydrogenated alkyl chains are systematically lower than for glymes having the same number of ethyleneoxide units.

When changing the alkyl chain from a hydrocarbon to a fluorocarbon nature, the electrophilic character of the fluorine atoms is responsible for a neat decrease of the complexing ability of the surfactant molecules.

Key words: Nonionic surfactants, alkali metal ions, fluorinated nonionic surfactants, complexing ability.

Introduction

Alkali metal ions are known to bind more or less strongly with the coordinating oxygen atoms of polyethers [1–3]. A great deal of work has been done not only on the complexing ability of macrocyclic polyethers [4] but also on that of open chain polyethyleneoxide (glymes [5] as well as long chain polymers [1–3]). Much less is known regarding the interaction of nonionic surfactants having short polyoxyethylated hydrophilic chains. The aim of this work was to compare the complexing ability of hydrogenated nonionic surfactants with that of fluorinated surfactants, which have potential biomedical applications [7].

For this purpose conductivity measurements have been performed in methanolic solutions of different 1:1 electrolytes in the presence of different surfactants whose hydrophilic chains contained from 4 to 8 ethylene oxide units.

Experimental

The hydrogenated nonionic surfactants of general formula $C_{12}EO_n$ (with $n = 4$, 6 and 8) were obtained from Nikko Chemicals (Japan). The fluorinated surfactants of general formula C_m-$F_{2m+1}CH_2EO_n$ (with $m = 6$ or 7 and $n = 4$, 5 and 6) were synthesized in our laboratory according to a previously published procedure [8]. The electrolytes of analytical grade were dried under vacuum and methanol was freshly distilled after refluxing over magnesium.

An autobalance precision bridge (Wayne-Kerr) was used for the conductivity measurements carried out at 25 °C. The data were interpreted assuming a 1:1 complex formation between the alkali metal ion and the surfactant molecule. Details of the theoretical treatment and of the fitting procedure used to obtain the apparent binding constants K_b have been given elsewhere [9].

Results and discussion

The values of the apparent binding constants for the different surfactant-salt complexes investigated are collected in Tables 1 and 2. Some values taken from the literature for comparable situations are also given in Table 1 for the sake of comparison.

An increase of the binding constant with the number of ethylene oxide units is clearly observed both for the surfactants $C_{12}EO_n$ and $C_mF_{2m+1}CH_2EO_n$, in agreement with previous observations by Chaput et al. on a series of glymes [5]. In addition, the values obtained for complexation with K^+ ions are systematically larger than those for Na^+ ions. A more detailed study of the effect of the anion on complexation by

Table 1. Hydrogenated compounds. Results in methanol at 25 °C (Apparent binding constant K_b in $1 \cdot mole^{-1}$)

number of EO units	surfactants (this work)	electrolyte	$\log K_b$	comparison with literature data	electrolyte	$\log K_b$
4	$C_{12}EO_4$	NaCl	0.75	$C_{12}EO_4$ (B 30)[a]	NaNO$_3$	not detectable[a]
		KSCN	1.30		KNO$_3$	3.76[a]
		KI	1.23	$CH_3(EO)_4CH_3$	NaCl	1.28[d]
				(tetraglyme)	KI	1.72[d]
5				CH_3EO_5 (PEGM)[a]	NaNO$_3$	not detectable[a]
					KNO$_3$	1.98[a]
				$CH_3(EO)_5CH_3$ (PG)[a]	NaNO$_3$	1.54[a]
				(pentaglyme)	KNO$_3$	2.07[a]
					NaSCN	1.00[a]
					KSCN	2.10[b]
					NaCl	1.52[c]
					KCl	2.20[c]
					NaCl	1.47[d]
					KI	2.20[d]
6	$C_{12}EO_6$	NaCl	1.10	$CH_3(EO)_6CH_3$	NaCl	1.60[d]
		KSCN	2.41	(hexaglyme)	KI	2.55[d]
		KI	2.13			
7				$CH_3(EO)_7CH_3$	NaCl	1.67[d]
				(heptaglyme)	KI	2.87[d]
8	$C_{12}EO_8$	NaCl	1.26			
		KSCN	2.74			
		KI	2.57			

[a]) Buschmann [10]; [b]) Früh and Simon [12]; [c]) Frensdorf [13]; [d]) Chaput et al. [5]. Methods used: calorimetry for [a]) and [b]), potentiometry with selective electrodes for [c]) and [d]).

$C_{12}EO_8$ had shown that although the binding is in the order in $SCN^- > I^- \gtrsim Cl^-$, the differences for the same cation are always within a factor of two whereas factors larger than 15 exist between Na^+ and K^+ [9].

Table 2. Fluorinated surfactants. Results in methanol at 25 °C (Apparent binding constant K_b in $1 \cdot mole^{-1}$)

number of EO units	surfactant	electrolyte	$\log K_b$
4	$C_6F_{13}CH_2(EO)_4$	NaCl	0.44
		KSCN	0.97
		KI	0.54
5	$C_6F_{13}CH_2(EO)_5$	NaCl	0.66
		KI	1.38
6	$C_7F_{15}CH_2(EO)_6$	NaCl	0.65
		KSCN	1.78
		KI	1.73

To our knowledge, the only study of complexing properties of a nonionic surfactant with a short hydrophilic chain is that of Buschmann [10], whose results for $C_{12}EO_4$ (B 30 in his paper) are indicated in Table 1. Considering that the NO_3^- anion does not significantly change the values obtained for pentaglyme comparatively to Cl^- (see Table 1), the value of 3.76 obtained by this author for the binding of K^+ to $C_{12}EO_4$ does not appear to be reliable. It would indicate that the long nonpolar chain of the surfactant facilitates complex formation, whereas our results show that complexation by such surfactants is systematically weaker than for glymes having the same number of ethylenoxide units. The large value, obtained by Buschmann from calorimetric measurements, is also very unlikely when compared with the values measured by Ono et al. for long chain polyethyleneoxides [2]: 2.03 for NaI and 2.67 for KI, i. e. K_b values lower by more than one order of magnitude.

When substituting the hydrogenated alkyl chain of a nonionic surfactant by a fluorinated one, one may

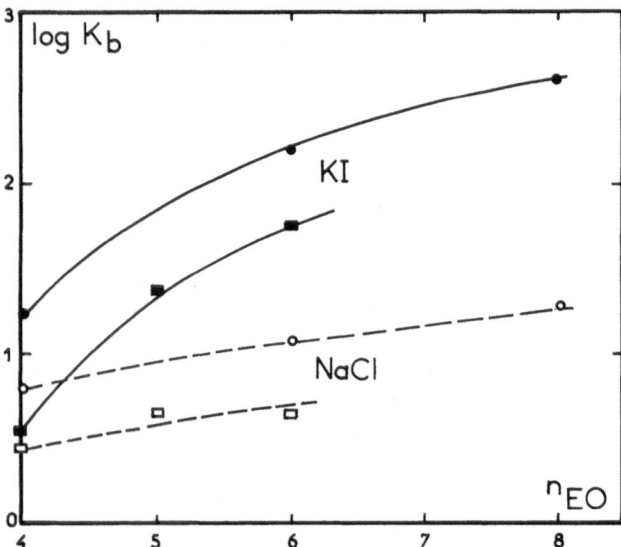

Fig. 1. Plots of the log K_b versus the number of ethyleneoxide units n_{EO}: (●, ○) hydrogenated surfactants; (■, □) fluorinated surfactants (closed symbols = KI; open symbols = NaCl). Reproduced form Reference [9] with permission of the Journal of Colloïd and Interface Science

wonder if the complexing ability of the hydrophilic head remains unaffected or not. A weakening of the basicity of the polyether moiety is expected when the ether groups themselves are fluorinated [11], but in the present case the fluorine atoms are far away from the electron donor oxygen atoms. In addition there is a CH_2 group separating the fluorinated alkyl chain from the hydrophilic chain. Despite these geometrical and chemical factors it is clear from a comparison of the log K_b values given in Tables 1 and 2 that the electrophilic character of the fluorine atoms brings about a neat decrease of the complexing properties for the same number of ethylene oxide units.

A graphical representation of this observation is given in Figure 1 for both a K^+ salt (KI) and a Na^+ salt (NaCl). From these plots, empirical expressions can be deduced, relating the values of K_b^H and K_b^F (where the H or F index refers to "hydrogenated" and "fluorinated" respectively):

$$\log K_b^H \simeq \log K_b^F + \alpha$$

with $\alpha = 0.56 \ (\pm \ 0.12)$ for KI

and $0.36 \ (\pm \ 0.08)$ for NaCl.

For the kind of products investigated here, the value of K_b^F can thus be approximated from the knowledge of K_b^H characterizing a surfactant molecule with the same number of ethyleneoxide units.

The conclusion from these data is that, in spite of a distinct decrease of their complexing ability compared with H-alkylated surfactants, the fluorinated polyethoxylated surfactants show still significant complexing properties towards K^+ ions when their hydrophilic moiety includes more than four ethyleneoxide units.

References

1. Liu KJ (1968) Macromolecules 1:213
2. Ono K, Konami H, Murakami K (1979) J Phys Chem 83:20
3. Buschmann H-J (1986) Makromol Chem 187:423
4. Izatt RM, Bradshaw JS, Nielsen SA, Lamb JD, Christensen JJ, Sen D (1985) Chem Rev 85:271
5. Chaput G, Jeminet G, Juillard J (1975) Can J Chem 53:2240
6. Dietrich B (1985) J Chem Ed 62:954
7. Mathis G, Leempoel P, Ravey JC, Selve C, Delpuech J-J (1984) J Am Chem Soc 106:6162
8. Selve C, Castro B, Leempoel P, Mathis G, Gartiser T, Delpuech J-J (1983) Tetrahedron 39:1313
9. Burger-Guerrisi C, Tondre C (1987) J Coll Int Sci, March
10. Buschmann H-J (1985) Polyhedron 4:2039
11. Lin W-H, Bailey WI, Lagow RJ (1985) J Chem Soc Chem Commun 1350
12. Früh PU, Simon W (1973) In: Peeters H (ed) Protides of the Biological Fluids, 20th Colloquium, Pergamon Press, Oxford New York, p 505
13. Frensdorf HK (1971) J Am Chem Soc 93:600

Received December 24, 1986;
accepted January 23, 1987

Authors' address:

C. Tondre
L. E. S. O. C.
Faculté des Sciences
Université de Nancy I
Boite Postale 239
F-54506 Vandoeuvre les Nancy Cedex, France

Progress in Colloid & Polymer Science   Progr Colloid & Polymer Sci 73:33–36 (1987)

Copolymerization of acrylamide and sodium acrylate in microemulsions

F. Candau[1]), Z. Zekhnini and J. Durand[2])

[1]) Institut Charles Sadron (CRM-EAHP), CNRS-ULP, France
[2]) Institut Français du Pétrole, Division de Recherche 92, Rueil-Malmaison, France

Abstract: We have investigated the thermal copolymerization of acrylamide and sodium acrylate in nonionic microemulsions. The formation of microemulsions is strongly related to the acrylate content in the comonomer feed due to a salting out effect of the ethoxylated surfactant. After polymerization, clear and stable inverse latices are formed which contain high solid contents (up to 23 %) dispersed in the oil-continuous medium. The dimensions of the particles are rather low (450 Å $< d <$ 700 Å) with a narrow distribution. They decrease with increasing emulsifier or sodium acrylate concentrations.

Key words: Poly(acrylamide-co-acrylates), microemulsion polymerization, quasi-elastic light scattering.

Introduction

Although microemulsions have been known for a long time, polymerization of various monomers in these media has only developed around 1980 [1–4]. The aim of the investigations was mainly to prepare stable latexes of small size which cannot be obtained easily with classical emulsion polymerization processes. Recently, one of us [5, 6] reported the polymerization of an inverse microemulsion of acrylamide which produced small stable latex particles with high molecular weights.

For most applications, water soluble polymers must have a polyelectrolyte character and this can be achieved by incorporation of an ionic monomer in the comonomer feed. It was of interest to us, then, to extend the first polyacrylamide microemulsion polymerization study to the synthesis of copolymers of acrylamide and sodium acrylate. Also, in keeping with the fact that industrial applications of these systems require a high polymer content and a low surfactant concentration, we have sought to optimize these parameters in our polymerizations.

Experimental

Microemulsions were prepared by adding the aqueous solution of acrylamide (Am) and sodium acrylate (Aa) (neutralized at pH ~ 9) to the mixture of emulsifiers, AIBN (azobisisobutyronitrile) and the oil, Isopar M.

The polymerization experiments were carried out in water jacketed reaction vessels, after bubbling purified nitrogen through the microemulsion to eliminate oxygen. The comonomer feed was initiated thermally at 45 °C with AIBN as the hydrophobic initiator. Total conversion to copolymer was achieved within less than 30 min.

The transmittance of the microemulsions and of the final latices were measured at 25 °C spectrophotometrically using a helium-neon laser as a monochromatic source of light.

The determination of the size of the latex particles was obtained by quasi-elastic light scattering as described elsewhere [5]. The latices were diluted with pure Isopar M down to a volume fraction of the dispersed phase of around 0.5 to 2 %. The value of the translational diffusion coefficient D ($D = kT/6M \eta_o R_H$, where R_H is the hydrodynamic radius of the particle and η_o the viscosity of the continuous medium) was extrapolated to zero volume fraction. The latter is defined as $\phi = (V_E + V_{Am} + V_{Aa} + V_{H_2O})/V_{total}$ where V_E, V_{Am}, V_{Aa} and V_{H_2O} represent the volumes of emulsifier, acrylamide, sodium acrylate and water dispersed in the system, respectively.

Table 1. Solubility parameters of the oil and emulsifiers

Compounds	Solubility parameter (Hildebrand)			
	δ_d	δ_p	δ_h	δ
hydrophilic part of G 1086	8.70	0.85	4.89	10.02
hydrophilic part of Arlacel 83	9.00	3.80	14.60	17.57
lipophilic part of G 1086	7.87	0	0	7.87
lipophilic part of Arlacel 83	7.87	0	0	7.87
Isopar M	7.79	0	9	7.79

Results and discussion

1 Formation of microemulsions

The selection of a system suitable to the formation of microemulsion was based on the same approach as that developed by Beerbower and Hill [7] for the stability of emulsions. Their approach was to assume that there is a perfect chemical matching between the lipophilic moiety of the emulsifier (L) and oil (O) on one side and the hydrophilic moiety (H) and water (W) on the other side. The corresponding cohesive energy ratio R_o which determines the nature and the stability of the emulsion is then directly related to the HLB of the emulsifier by:

$$R_o = E_{LO}/E_{HW}$$
$$= (d_H/d_L)\left(\frac{20}{\text{HLB}} - 1\right)\left(\frac{\delta_L^2}{\delta_H^2}\right) \qquad (1)$$

where E_{LO} and E_{HW} are the cohesive energies of the lipophile for oil and of the hydrophile for water respectively. The quantities d_H and d_L are the densities of the hydrophilic and lipophilic moieties respectively; the corresponding solubility parameters δ_H^2 and δ_L^2 are expressed as:

$$\delta_H^2 = \delta_d^2 + 0.25\,\delta_p^2 + 0.25\,\delta_h^2$$
$$\delta_L^2 = \delta_d^2 + 0.25\,\delta_p^2 + 0.25\,\delta_h^2$$

The subscripts d, p and h refer to the London dispersion forces, polar forces and hydrogen bonding forces respectively.

Because of the equivalence between the lipophile and the oil, Equation (1) allows one to predict the required HLB for a given oil whose solubility parameters are known.

These criteria led us to select the following system:

The emulsifier consists of a blend of two nonionic emulsifiers : a sorbitan sesquioleate (Arlacel 83, HLB

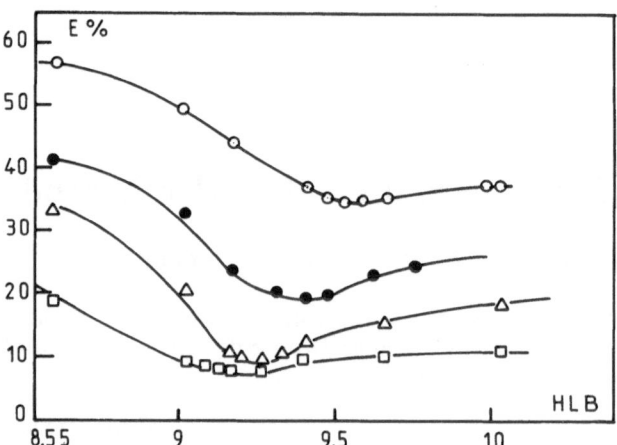

Fig. 1. Percentage of emulsifiers required for the formation of microemulsions versus HLB at different comonomer compositions (Am + Aa)/water and oil to aqueous phase weight ratios ~ 1 (T = 24.6 °C); □ 22.35 % Aa; △ 35.8 % Aa; ● 46.1 % Aa; ○ 56.42 % Aa

= 3.7) and a polyoxyethylene sorbitol hexaoleate (G1086, HLB = 10.2). The oil is a narrow cut isoparaffinic mixture Isopar M. The aqueous phase consists of water and a mixture of acrylamide and sodium acrylate in variable proportions (water/comonomer feed : 1/1).

Table 1 gives the solubility parameters of the different components. They were calculated from laws based on the additivity of the cohesive energies of the different chemical groups which constitute the molecules [8]. The solubility parameters of the lipophilic moieties of the emulsifiers are equal as expected since they both consist of an oleate chain. These values are very close to that obtained for Isopar M indicating a good compatibility between the species. On the other hand, the solubility parameters of the hydrophilic moieties of the emulsifier blend are much larger than the corresponding values of the lipophilic parts or that of the oil; this will favor the segregation of the emulsifier molecules at the water-oil interface.

Figure 1 represents the minimum percentage of emulsifiers required for the formation of a microemulsion versus their HLB value at different comonomer compositions. Microemulsions can be formed in an HLB range going from 8.5 to 10.2. The curves exhibit a minimum for an optimum HLB_{opt} value which depends on the ionic monomer concentration. This result can be attributed to a salting out effect of the hydrophilic moiety of the emulsifier by this electrolyte,

Table 2. Compositions and characteristics of the systems

| System | Composition (% w/w) | | | | | copolymer (% w/w) | % transmittance (25 °C) | |
	Isopar	Am	Aa	H₂O	Emuls.		Microemulsion	Microlatex
Th-6	37.92	13.67	10.13	24.86	13.42	23.80	1.4	82.0
Th-21	37.76	13.57	9.63	23.57	15.47	23.21	80	81.8
Th-4	36.70	12.90	9.33	23.03	18.04	22.23	90	84.7
Th-1	35.54	12.45	9.02	21.91	21.08	21.47	93	90.9
A-2	38.62	15.39	5.06	19.25	21.68	20.45	90	75.0
A-8	38.35	13.45	7.50	19.14	21.56	20.95	95	94.5
A-14	37.38	12.24	8.57	20.80	21.01	20.81	93	75.8
A-9	36.76	11.02	9.63	20.37	22.22	20.63	97	87.6
A-17	36.66	10.11	10.76	21.74	20.73	20.87	1.8	74.0

the blend being made more lipophilic on addition of sodium acrylate in the system [9].

The range of HLB which allows the formation of microemulsions (8.5 to 10.2) suggests that the initial systems – which contain equivalent amounts of oil and aqueous phase – correspond to a phase inversion [7]. In this case, it is known that the microemulsions exhibit a bicontinuous structure, where oily and aqueous domains are interconnected [10] or as an alternative a transient lamellar structure [11]. Such a possibility was confirmed by looking at the systems located in the polyphasic domain and containing lower emulsifier contents, all other parameters being kept constant. After equilibrium, the systems consisted of three phases, a middle phase with a bicontinuous character coexisting with oil and aqueous phases.

2. Polymerization

We have prepared two series of copolymer latices (see Table 2). In the first series, (Th), the emulsifier concentration (HLB = 9.3) was varied from 13 % to 21 %. The final copolymer content was about 23 % with respect to the total weight and the percentage of sodium acrylate in the comonomer feed was 42 %. The second series (A) corresponds to variable amounts of ionic monomers in the feed, the other parameters being kept constant. In this series the emulsifier and copolymer concentrations were both approximatively equal to 21 %. The polymerization reactions are characterized by a very high rate and most of the yields of conversion are above 90 %. The values of the transmittance of the systems before and after polymerization are reported in Table 2. The final latices are clear and very stable with optical transmittances of between 75 % and 95 %. The starting systems are clear micro-

emulsions except for systems Th-6 and A-17. In these cases, the turbidity is not due to an insufficient amount of emulsifier but is related to the high electrolyte content which produces a salting out of the ethoxylated surfactant as shown above. During the polymerization, the salt saturated aqueous phase is consumed and a clear latex is subsequently obtained.

In the course of polymerization, the bicontinuous structure of the initial microemulsions is eventually destroyed and turns into a concentrated dispersion of spherical particles. An investigation by quasi-elastic light scattering of the dilute latexes leads to the determination of the hydrodynamic radii of the particles and of their variance. The latter, which provides an estimate of the particle distribution is low for all the systems investigated and ranges between 0.02 to 0.08.

Figures 2 and 3 show the variation of the hydrodynamic radius R_H of the particle vs. the emulsifier con-

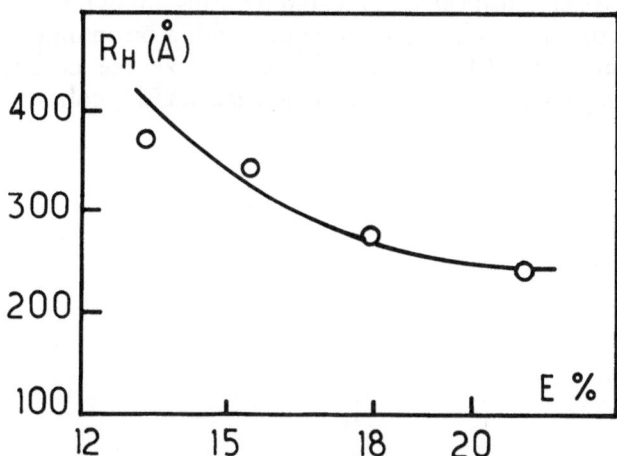

Fig. 2. Variation of the radius of the latex particle with the emulsifier concentration (series Th)

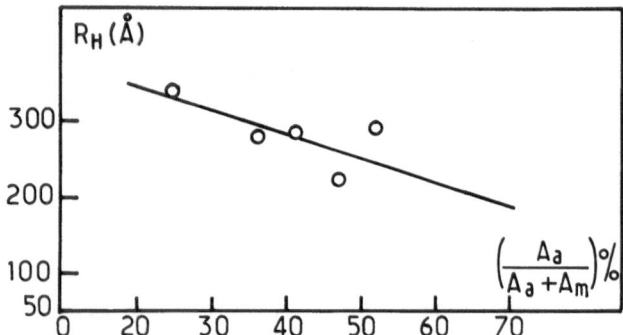

Fig. 3. Variation of the radius of the latex particle with the sodium acrylate content in the comonomer feed (series A)

centration and the copolymer composition, respectively. The decrease of R_H with increasing emulsifier concentration is usually observed in colloidal dispersions because of an increase of the particle number [5]. The linear decrease of R_H with increasing the acrylate content can be related to the dehydration of the hydrophilic moieties of the emulsifier by the salt together with a lowering of the water activity in the particles. The combination of these two effects accounts for the observed diminution of the radius of curvature.

These results corroborate other data obtained for systems prepared photochemically under UV irradiation [9].

Conclusion

The study presented here shows that stable inverse latices containing high amounts of poly(acrylamide-co-acrylates) can be prepared by using a polymerization process in microemulsions with a bicontinuous structure. Our results also emphasize the critical importance of the choice of the system. Using cohesive energy ratio and HLB concepts, we have been able to establish simple rules of selection leading to a good matching between oils and emulsifiers.

Acknowledgements

We thank F. Dawans (Institut Français du Pétrole) and C. Holtzscherer for helpful discussions. We are indebted to S. Candau (Université Louis Pasteur, Strasbourg) for his assistance in quasi-elastic light scattering experiments. This work was partially supported by Institut Français du Pétrole.

References

1. Schauber C (1979) Thèse Docteur-Ingénieur, Université de Mulhouse France
2. Stoffer JO, Bone T (1980) J Disp Sci Technol 1:37
3. Leong YS, Riess G, Candau F (1981) J Chim Phys 78:279
4. Atik SS, Thomas KJ (1982) J Am Chem Soc 104:58
5. Candau F, Leong YS, Pouyet G, Candau J (1984) J Coll Int Sci 101:167; Candau F, Leong YS, Kohler N, Dawans F (1984) French Patent (to IFP-CNRS) 2, 524:895
6. Candau F, Leong YS, Fitch RM (1985) J Pol Sci Chem Ed 23:193
7. Beerbower A, Hill MW (eds) (1971) McGutcheon's Detergents and Emulsifier Annual, Allured Publ Co, Ridgewood, NJ, p 223
8. Van Krevelen DW, Hoftyzer PJ (1976) Properties of Polymers: their estimation and correlation with chemical structure, Elsevier, New York, Ch 7
9. Candau F, Zekhnini Z, Durand JP (1986) J Coll Int Sci 114:398; Durand JP, Nicolas N, Kohler N, Dawans F, Candau F (June 1984) French Patent Application (to IFP) 84/08906 and 84/08907
10. Talmon Y, Prager S (1978) J Chem Phys 69:2984
11. Shah DO, Hamlin Jr RM (1971) Science 171:483

Received February 9, 1987;
accepted February 9, 1987

Authors' address:

Françoise Candau
Institut Charles Sadron
(CRM-EAHP)
6, rue Boussingault
67083 Strasbourg Cedex, France

Progress in Colloid & Polymer Science   Progr Colloid & Polymer Sci 73:37–47 (1987)

Interfacial and colloidal properties of cosmetic emulsions containing fatty alcohol and fatty alcohol polyglycol ethers

F. Schambil, F. Jost, and M. J. Schwuger

Laboratories of Henkel KGaA, Düsseldorf, F.R.G.

Abstract: The phase behavior of multicomponent systems, especially of emulsions containing cetostearyl alcohol and fatty alcohol polyglycol ethers has been investigated. At the phase inversion from o/w to w/o emulsions, one-phase regions containing lamellar liquid crystals have been observed. In appropriate ratios of the amphiphiles, extremely low interfacial tensions ($\approx 10^{-3}$ mN/m) are obtained, which lead to very easy emulsification and narrow particle size distribution. Thermoanalysis (DSC) and microscopy of binary, ternary and quaternary mixtures demonstrate the existence of hydrated crystals or gels at room temperature. Anisotropic phases could be identified around oil droplets of more coarse emulsions. Emulsions prepared at low temperatures possess structural viscosity. Increasing the temperature leads to newtonian behavior. At medium emulsifier concentrations fluid and stable emulsions are obtained. There is a strong dependence of the properties of the emulsions on the polarity of the oil phase. It can be concluded that the ability to form fine droplets, the viscosity and the long term stability of cosmetic emulsions do not depend on the properties of the fatty alcohol alone but on the mixed phases formed by the emulsifier and the alcohol.

Key words: Emulsion, microemulsion, cetostearyl alcohol, phase behavoir.

Introduction

Cosmetic oil-in-water emulsions often contain cetostearyl alcohol (1/1 mixture of cetyl and stearyl alcohol) to increase viscosity and stability. In the literature, the stabilization mechanism of the alcohol mixture is controversial and has been discussed. It is known that a very disadvantageous behavior is observed when cetyl alcohol or stearyl alcohol alone is used instead of their mixture. The difference in stability at constant temperature is supposed to be connected with the polymorphism of the alcohol [1, 2]. Also, the formation of a viscoelastic gel network with hydrated alcohol has been postulated [3–5]. Further, it is assumed that only the crystals of cetostearyl alcohol are able to contribute to the emulsion stability by accumulation at the interface [6]. Rheological measurements from several long chain alcohols have shown that the highest yield values and the highest apparent viscosities of oil-in-water emulsions have been obtained with cetostearyl alcohol [7].

In this paper, results are reported of investigations regarding phase behavior, interfacial tension, temperature and rheological behavior of oil-in-water emulsions containing cetostearyl alcohol and several fatty alcohol polyglycol ethers in combination with different oils.

Experimental

Substances

The following fatty alcohol polyglycol ethers of Henkel KGaA were used as emulsifiers:
  Cetostearyl alcohol-12 EO (Eumulgin B 1)
  Cetostearyl alcohol-20 EO (Eumulgin B 2)
  Cetostearyl alcohol-30 EO (Eumulgin B 3)
  Tallow alcohol-5 EO (Dehydol TA 5)
as well as a pure C_{14} ethoxylate with an EO-distribution (C_{14} – 4 EO). Cetostearyl alcohol (Lanette O) of Henkel KGaA was used as a co-emulsifier.

As organic phase, heavy liquid paraffin oil DAB 8 (Wasserfuhr Co., Bonn, West Germany) as well as 1,3-di-isooctyl cyclohexane

(Cetiol S), decyl oleate (Cetiol V) 2-octyl dodecanol (Eutanol G) and almond oil (all from Henkel KGaA) were used.

Methods of investigation

Electrical conductivity

The electrical conductivity of emulsions versus temperature was determined with a conductivity measuring bridge (Radiometer Co., Copenhagen, Denmark). The heating and cooling rate was 0.5°/min and was controlled by means of a temperature control device connected to a cooling thermostat (Haake Company, Karlsruhe, West Germany). As a rule, the emulsions were prepared by blending the components in a stirring device (LM 34, Pendraulik Co., Bad Münder, West Germany) at 85°C for 10 min. Subsequently, they were cooled down to room temperature by stirring for 20 min.

Polarization microscopy

The phase behavior of the mixtures was determined by means of a polarization microscope (Zeiss Co., Oberkochen, West Germany) combined with two different heating plates (Reichert Co., Vienna, Austria, and Mettler Co., Greifensee, Switzerland). To avoid effects of evaporation of water at higher temperatures, the upper slide was stuck to the specimen slide by means of a special adhesive for glass (Profix, Henkel KGaA) at room temperature for 1 h. The evidence of the liquid crystalline phases or gels was furnished by comparison with micrographs from the literature [8].

Thermoanalysis

The heating and cooling thermograms were determined by means of a differential scanning calorimeter (DSC-2, Perkin Elmer Co., Überlingen, West Germany). The heating and cooling rates were between 0.3 and 10°/min.

Interfacial tension

The interfacial tension between aqueous and oily phases was determined by means of a spinning-drop tensiometer (Krüss Co., Hamburg, West Germany). To determine the interfacial tension at equilibrium, the water soluble emulsifier was dissolved in water, and the co-emulsifier in the oil. The oily phase was subsequently poured on the aqueous phase. The system was allowed to adjust to equilibrium at the measuring temperature in a drying cabinet for 24 h.

Viscosity measurements

The viscosity measurements were conducted by means of the rheomate RM 30 (Contraves Company, Zürich, Switzerland) at shearing rates of from 5 to 400 s^{-1} at room temperature.

Results

Phase behavior

Emulsions of the oil-in-water type that contain fatty alcohol polyglycol ether can invert to water-in-oil emulsions when the temperature is raised. The inversion range is frequently designated as phase inversion temperature (PIT) [9]. This designation is not quite correct as the inversion always takes place in a more or less broad temperature zone. As the electrical conductivities of oil-in-water emulsions with small quantities of electrolytes, and water-in-oil emulsions, differ by powers of ten, the temperature dependence of the electrical conductivity was used for determining the PIT ranges.

Figure 1 shows a typical conductivity versus temperature curve of a ternary mixture. In the range up to 25°C an oil-in-water emulsion exists due to the high electrical conductivity. The increase of the conductivity form 18° to 25°C is connected with the melting

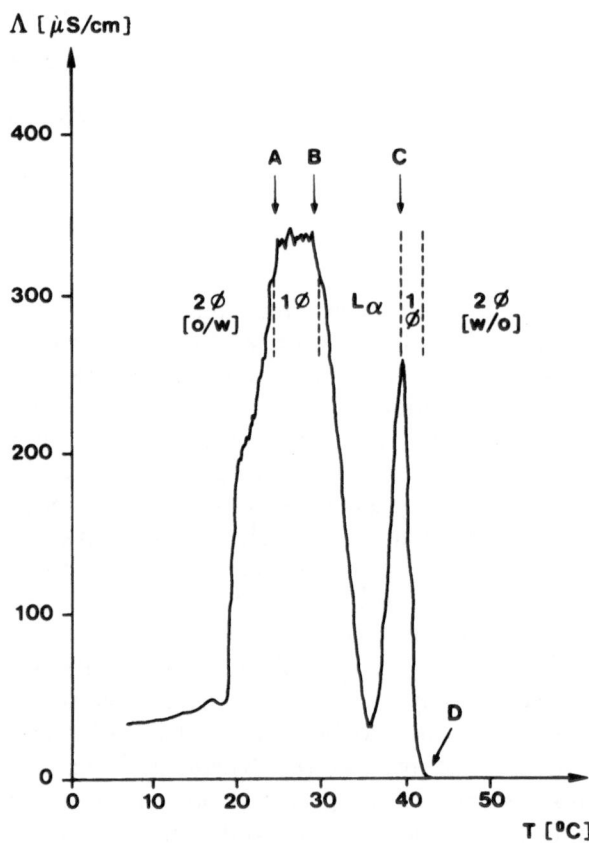

Fig. 1. Conductivity vs temperature of a ternary mixture (water/paraffin oil: 1/1; 15% C$_{14}$ — 4EO; heating rate 0.5°/min)

range of the nonionic surfactant. Above 42 °C the conductivity of the system is nearly zero since oil is now the outer phase (water-in-oil emuslion). A peak is recognized in the inversion zone. This peak appears independent of the rate of heating (0.1° to 2°/min) and the rate of stirring. The conductivity versus temperature curve in the inversion range has been explained both visually and by means of microscopic observation of emulsions. For this, the emulsions were placed on an object slide in a thin layer and covered by a second slide. Both slides were then stuck together. Gravimetric measurements showed that even after repeated slow heating to 95 °C no measurable loss of water occurred by evaporation of these specimens.

In the polarization microscope, several zones can be recognized in the phase inversion range between A and D (Fig. 1). In the temperature zone between A and B a

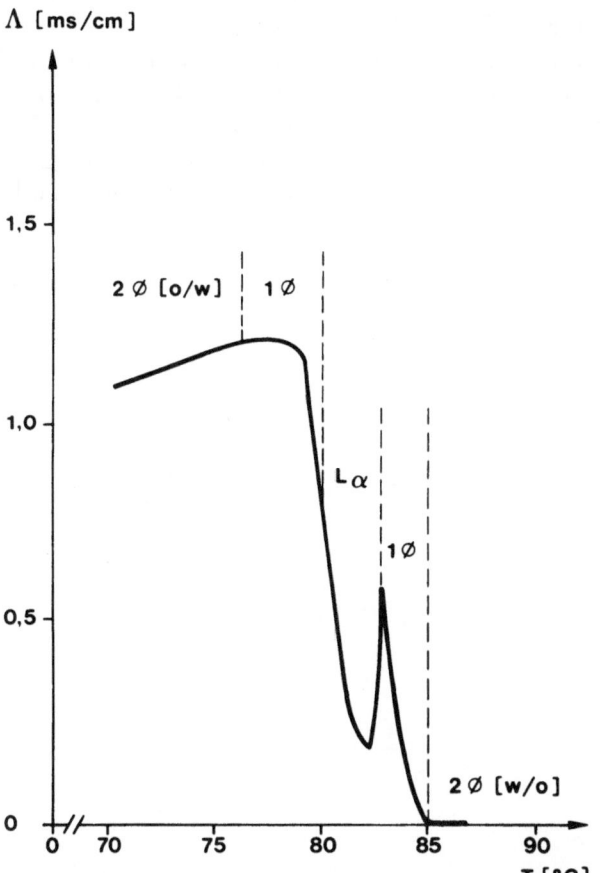

Fig. 2. Conductivity vs temperature of a quaternary mixture (water/paraffin oil/$C_{16/18}$ − 12 EO/$C_{16/18}$ − OH; heating rate: 0.5°/min)

transluscent isotropic phase is formed. At the beginning of the decrease of conductivity (point B) the formation of lamellar liquid crystals (L_α) takes place. At the first minimum and at the increase of the conductivity up to the peak (point C) liquid crystals still exist. The course of the conductivity between B and C is a consequence of the formation and degradation of liquid crystals. The decrease of the conductivity in a zone of several degrees from the peak down to the value $\Lambda = 0$ (C–D) is interlinked with the formation of a further phase. In the two temperature ranges between A–B and C–D no changes of the conductivity were observed in 24 h tests. Also, the optically transluscent phases did not show any indication of separation nor anisotropy in the polarization microscope at all. Hence, these two temperature ranges are to be assigned to a one-phase zone. It lies between the oil-in-water and water-in-oil zones and encompasses the liquid crystalline phase.

The conductivity versus temperature curves of mixtures that contain both cetostearyl alcohol, water, oil and emulsifier show the same qualitative course as the ternary mixture (Fig. 2). In the polarization microscope a liquid crystalline phase is observed in the range of the peak. It is delimited by two isotropic phases. Figure 3 depicts the results of the investigations of the phase behavior of quaternary mixtures versus the concentration of the emulsifier. The water/oil/fatty alcohol ratio is equal for all mixtures. Between the oil-in-water and water-in-oil emulsion ranges an isotropic area with a liquid crystalline range (L_α) exists. At low emulsifier concentrations a three-phase area is found. The three phases show rapid separation.

Figure 4 shows that the existence range of the phase inversion area is shifted to lower temperatures with increasing concentrations of amphiphilic substances at a constant ratio of water to oil and emulsifier to co-emulsifier. Here, too, a lamellar liquid crystal (L_α) is observed in the isotropic area.

The emulsifier in Figure 4 is completely water soluble, even at room temperature, in the concentration range investigated. Contrary to this, the co-emulsifier as a 1/1-mixture of cetyl and stearyl alcohol is practically water insoluble. The behavior of the mixture of both amphiphilic substances was investigated through DSC-measurements. The DSC-curves show that cetostearyl alcohol in the non-aqueous phase (Fig. 5, curve A) has two endothermal peaks. In a 10% aqueous dispersion (curve B) the lower peak that corresponds to the polymorph phase transition of the β-into the α-modification [10], is clearly shifted to lower

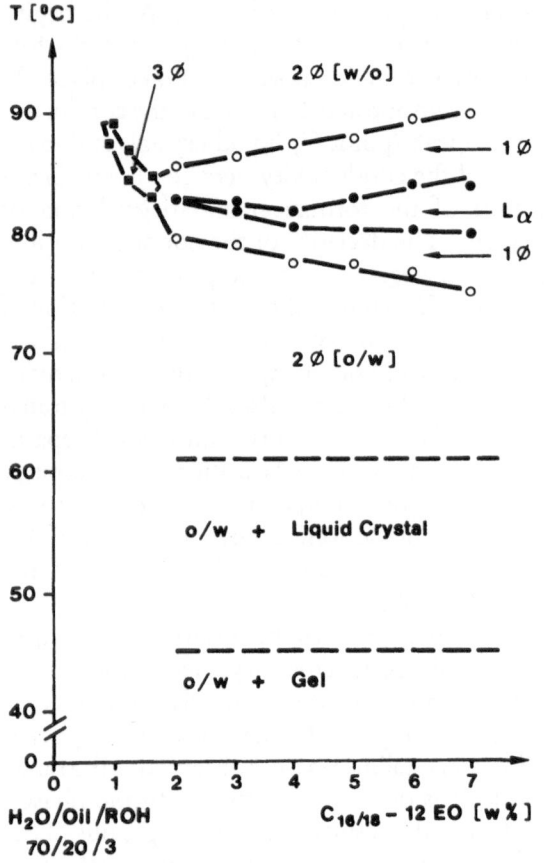

Fig. 3. Phase diagram of a quaternary system

temperatures. The position of the upper peak, which represents the melting peak, is practically unchanged. The DSC-curves of the aqueous dispersion of pure cetyl or stearyl alcohol (Fig. 5, curves C, D) do not show the shift of the lower peak to lower temperatures.

The ternary mixtures water/cetostearyl alcohol/fatty alcohol ethoxylate were studied in various mixing ratios.

Figure 6 shows that in these oil-free systems two melting peaks appear at higher temperatures. One recognizes at both upper curves A and B that the area under the lower peak depends strongly on the concentration of the fatty alcohol. The temperature corresponds to the melting peak of the fatty alcohol in water (Fig. 5). The upper peak at approximately 60 °C was identified in a polarization microscope as the melting area of a liquid crystal. Hence, liquid cetostearyl alcohol and a liquid crystal phase exist between both peaks. Above 60 °C, two liquid phases exist. The position of the melting peak of the liquid crystal is only a little dependent on the ratio of emulsifier/co-emulsifier. A further indication, with respect to the formation of liquid crystals, is given by the comparison of the heating and cooling curves (B and C in Fig. 6). The upper peak appears in both curves at the same temperature. The main crystallization peak of cetostearyl alcohol, however, lies 10 °C lower than the main melting peak,

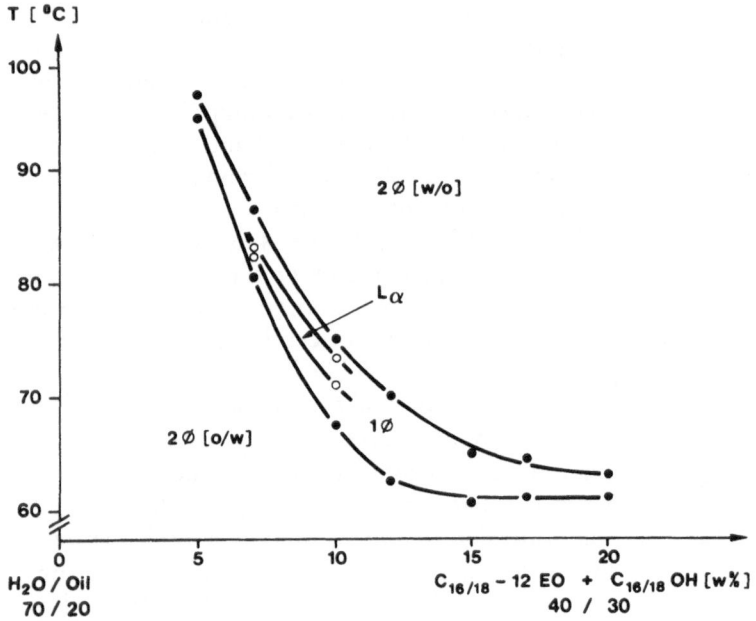

Fig. 4. Phase diagram of a quaternary system

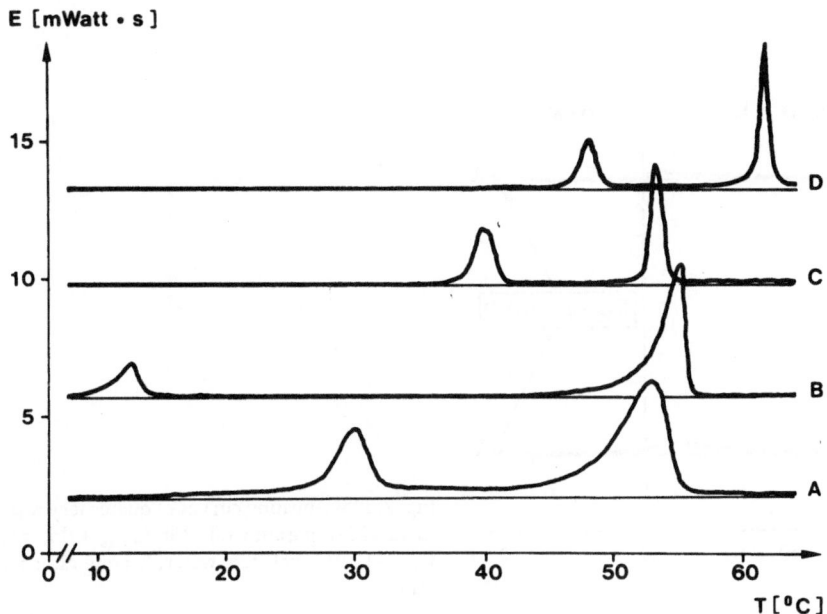

Fig. 5. DSC-heating curves (A): $C_{16/18}$–OH; (B): 10% $C_{16/18}$–OH in water; (C): 10% C_{16}–OH in water; (D): 10% C_{18}–OH in water; heating rate: 10°/min

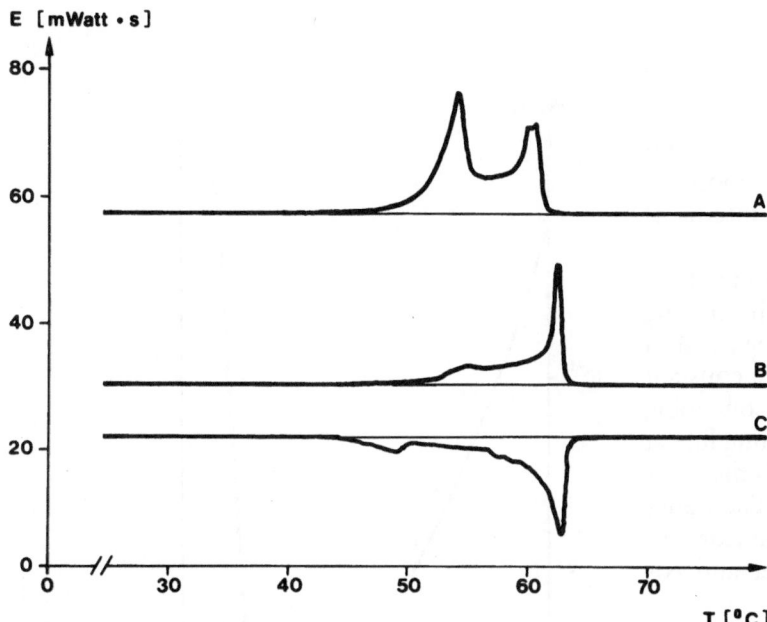

Fig. 6. DSC-curves of ternary mixtures (A): 17% $C_{16/18}$–OH + 3% $C_{16/18}$ 12–EO in water, heating curve; (B): 10% $C_{16/18}$–OH + 10% $C_{16/18}$ 12–EO in water, heating curve; (C): as B, but cooling curve (heating and cooling rate: 10°/min)

due to undercooling. As is known, solutions from which liquid crystals form do not show the undercooling phenomenon, in contrast to solutions from which solid crystals form.

Also, quaternary systems were thermoanalytically investigated in the range from 25° to 80°C. Figure 7 shows that two melting ranges occur with emulsions,

too. The lower area corresponds to the melting peak of cetostearyl alcohol in oil, whereas the upper peak is interlinked with a transition of the liquid crystal into an isotropic phase. The formation of the liquid crystalline phase in the isotropic zone, which takes place just above 80°C (Fig. 3) fo the systems investigated here, is not visible in the DSC-curves.

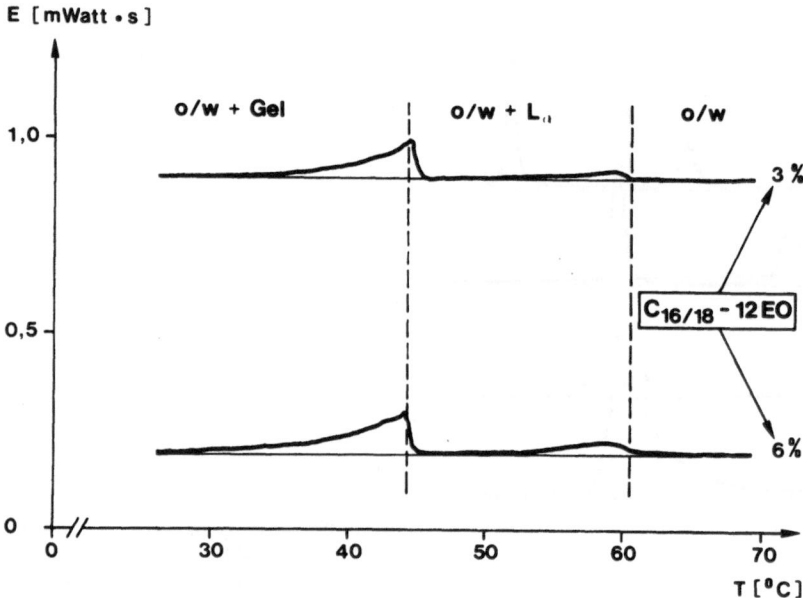

Fig. 7. DSC-heating curves of quaternary mixtures (20% paraffin oil, 3% $C_{16/18}$–OH, x% $C_{16/18}$–12 EO, balance: water; heating rate: 5°/min)

Interfacial tension

Since the co-emulsifier cetostearyl alcohol does not melt under 50°–60°C (depending on the composition of the emulsion) all measurements of the interfacial tensions were carried out at 75°C.

Figure 8 shows the results of measurements of the interfacial tension at equilibrium versus the mixing ratio of emulsifier and co-emulsifier, keeping the sum of the concentrations of both substances at a constant level. The interfacial tension of the mixture containing 11% fatty alcohol, without emulsifier (limiting line at right) was so high that a measurement with the spinning-drop tensiometer was not possible. The figure shows that extremely low interfacial tensions exist in a broad range of mixing ratios of emulsifier and co-emulsifier.

Figure 9 shows the results of interfacial tension measurements versus the concentration of the emulsifier at a constant concentration of the fatty alcohol. Here, too, the ratio of water to oil is constant. Again, extremely low values are observed beyond a certain concentration of the emulsifier (2 wt.%) in a broad range. Partly, the values are even under 10^{-3} mN/m.

The values of the interfacial tension are strongly dependent on the degree of ethoxylation $\bar{m}$ of the emulsifier. Figure 10 shows that a dramatic decrease of the interfacial activity takes place with increasing hydro-

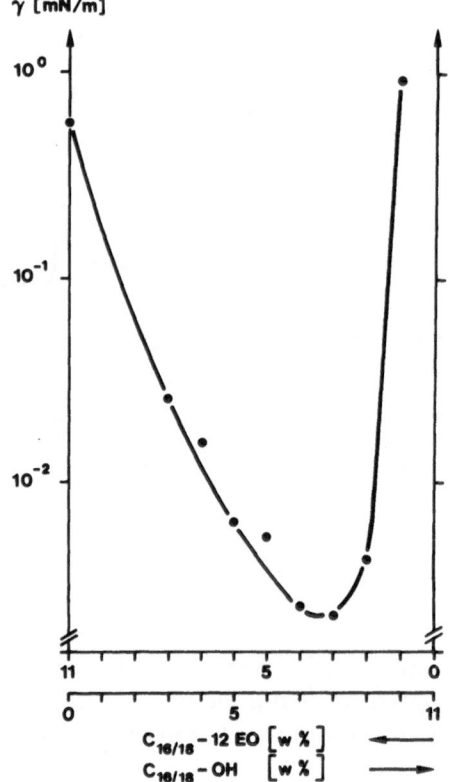

Fig. 8. Interfacial tension γ between aqueous phase and oily phase vs mixing ratio of emulsifier/co-emulsifier (20% paraffin oil, $T = 75$°C)

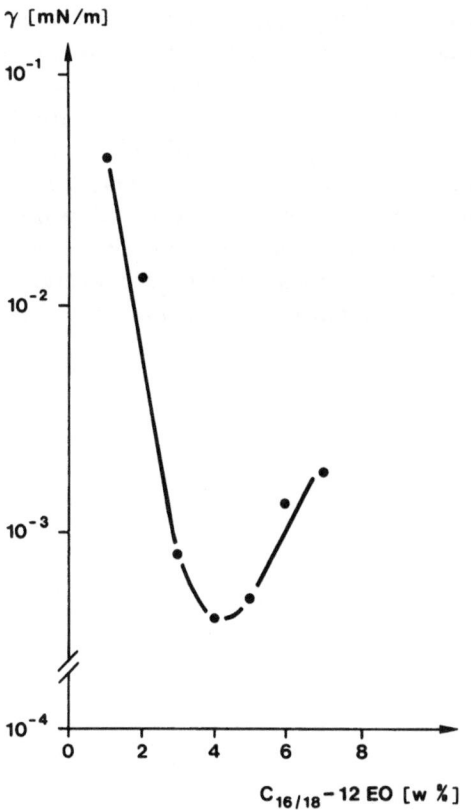

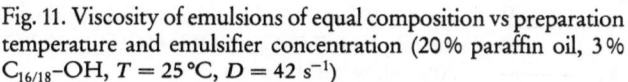

Fig. 9. Interfacial tension γ between aqueous phase and oily phase vs concentration of the emulsifier (20 % paraffin oil, 3 % $C_{16/18}$–OH, T = 75 °C)

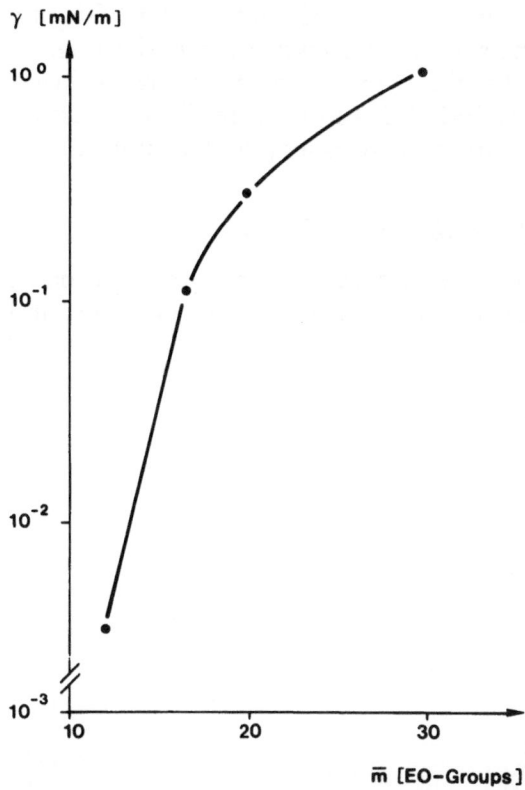

Fig. 10. Interfacial tension γ between aqueous phase and oily phase vs degree of ethoxylation $\bar{m}$ of the emulsifier (equilibrium phases: water/$C_{16/18}$-12 EO: 12/1; paraffin oil/$C_{16/18}$–OH: 6/1; T = 75 °C)

philicity of the emulsifier at constant concentration of the co-emulsifier. The interfacial tension with $\bar{m} = 30$ lies about 300 times higher than the value with $\bar{m} = 12$. A 1/1-mixture of both low ethoxylated emulsifiers was also used. It is recognized that the value of the interfacial tenison of this emulsifier blend lies on the curve that connects the values of the three individual emulsifiers.

Flow behavior

Figure 11 shows the results of viscosity measurements versus the concentration of the emulsifier and the temperature.

At preparation temperatures equal or higher than 80 °C emulsions were found that at ambient tempera-

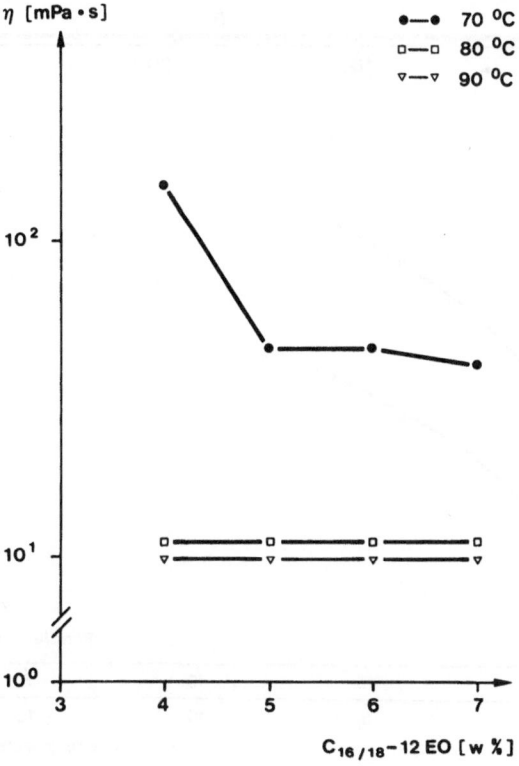

Fig. 11. Viscosity of emulsions of equal composition vs preparation temperature and emulsifier concentration (20 % paraffin oil, 3 % $C_{16/18}$–OH, T = 25 °C, $D = 42\ s^{-1}$)

ture show long term stability and very low viscosity. These emulsions show Newtonian behavior (Fig. 12, curve B). Emulsions that were made below the inversion range ($T = 70\,°C$) were substantially more viscous and show structural viscosity (Fig. 12, curve A).

Oil specifity of the phase behavior

To determine the effect of the polarity of the oil on the phase behavior, mixtures of water, oil, cetostearyl alcohol and fatty alcohol polyglycol ether were investigated at constant ratios (70/ 20/ 3/ 4). Only the emulsions with both non-polar oils (paraffin oil and 1,3-diisooctyl cyclohexane) invert under the given conditions. The systems with polar oils (decyl oleate, 2-octyldodecanol and almond oil), however, do not show inversion. In mixtures of a polar with a non-polar oil the inversion temperature increases with increasing proportion of polar oil. In Figures 13 and 14, results from two oil blends are presented. For these

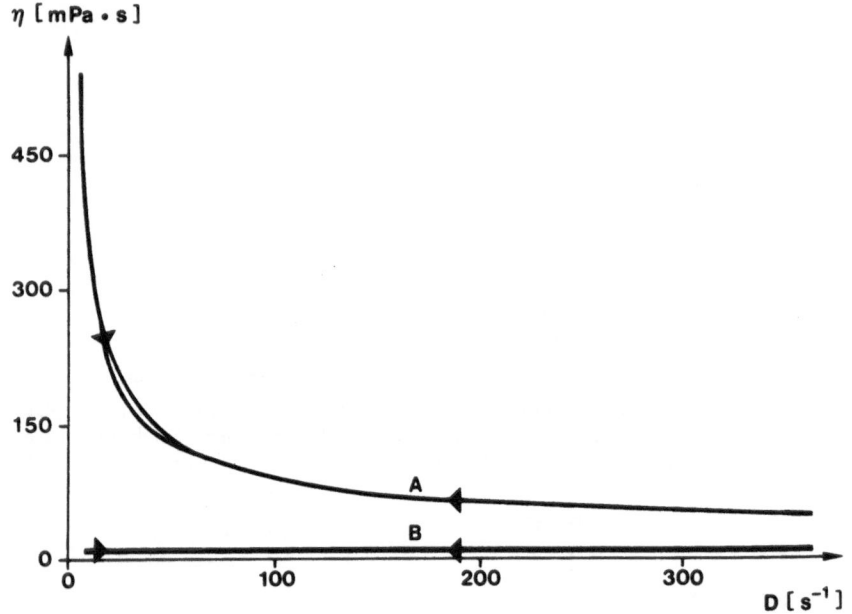

Fig. 12. Viscosity curves of two emulsions of equal composition and different preparation temperature at 25 °C (20% paraffin oil, 3% $C_{16/18}$–OH, 4% $C_{16/18}$-12 EO, water: balance; (A) 70 °C, (B) 90 °C)

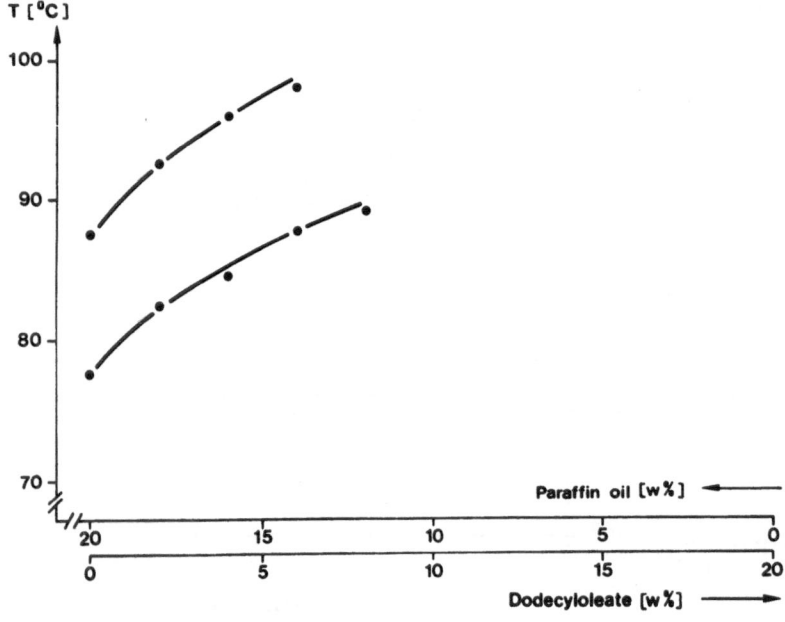

Fig. 13. Phase inversion zone vs composition of the oil phase (3% $C_{16/18}$–OH, 4% $C_{16/18}$-12 EO, 20% oil, water: balance)

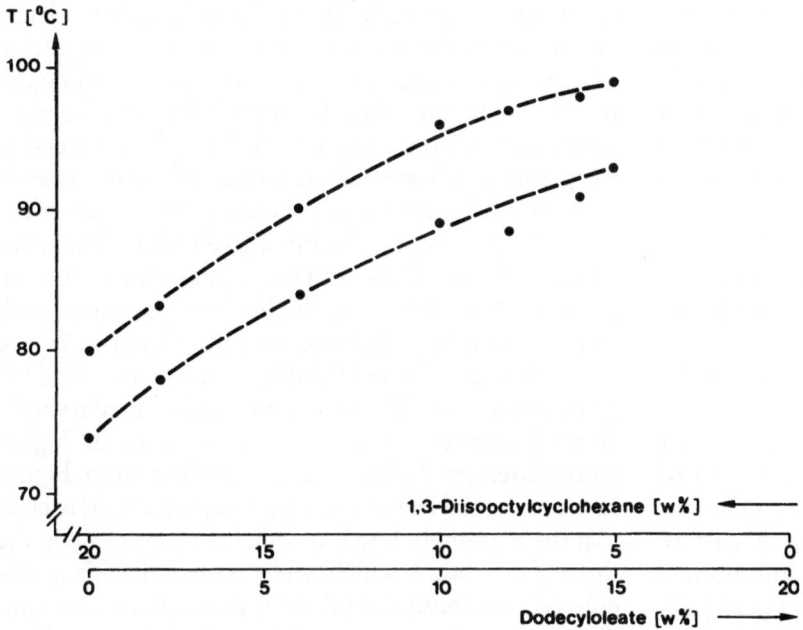

Fig. 14. As Figure 13, but with 1,3-diiosoctylcyclohexane

mixtures the conductivity versus temperature curves show the same picture as in Figure 1. Hence, it follows that also in this case isotropic and liquid crystalline phases exist in the inversion zones. The hydrophilicity of the emulsifier affects the temperature range of the phase inversion in a way that is similar to the polarity of the oil. Figure 15 shows the results of measurements from mixtures with two emulsifiers of equal chain length distribution, but with different degrees of ethoxylation (5 and 12 EO). Also, in this case, a liquid crystalline area (L_α) exists within the inversion range.

Discussion

The phase behavior of multicomponent systems of the water-oil-amphiphil type has been determined by

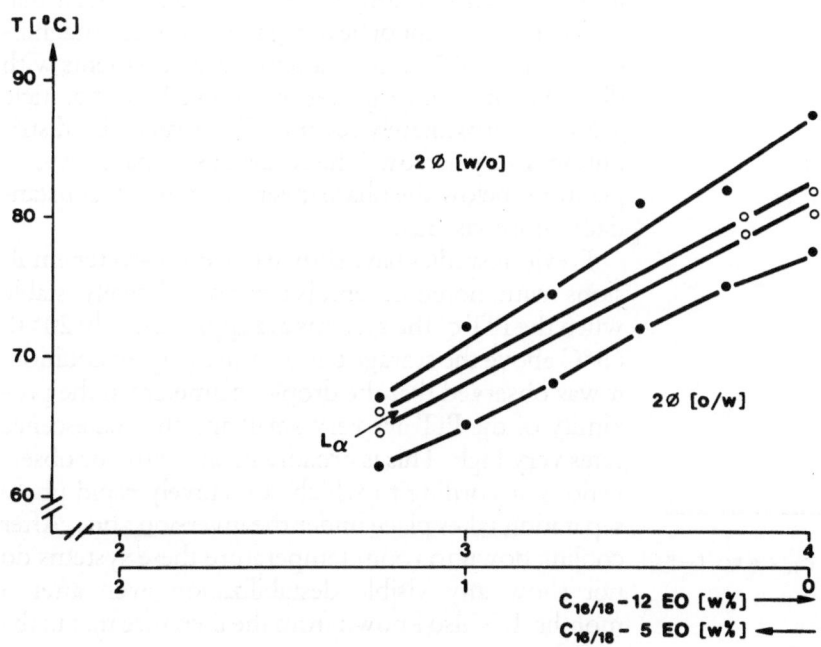

Fig. 15. Phase inversion zone vs mixing ratio of two emulsifiers (20% paraffin oil, 3% $C_{16/18}$-OH, 4% emulsifier, water: balance)

numerous working groups in recent years. Investigations with mixtures of pure substances showed that the phase behavior and the interfacial tension between two partly miscible phases (e. g., water and oil) are very closely interlinked [11, 12]. Attention has been particularly drawn to the fact that mixtures with liquid three-phase bodies show extremely low interfacial tensions in the existency range of these three phases. The interfacial tension is a measure of the emulsifying efficiency and of the particle size distribution of the inner phase. Therefore it is to be expected that the consistency and stability of emulsions also strongly depend on the phase behavior.

Most of the quaternary mixtures investigated here show an inversion from oil-in-water to water-in-oil when the temperature is increased. The inversion passes through an isotropic area (Fig. 3). A narrow three-phase area is found with small concentrations of the emulsifier. Similar phase diagrams have been published for ternary mixtures of pure components (water/ alkane/ nonionic surfactant). In those cases three-phase areas exist in addition to the one-phase area at low surfactant concentrations (Fig. 16). From a comparison of Figure 3 with Figure 9 it is seen that the condition for extremely low interfacial tensions is apparently the proximity of a three-phase body.

From the values of the interfacial tensions depicted in Figure 8 it can be deduced that the phase behavior of the quaternary mixture depends only slightly on the mixing ratio of the fatty alcohols and the emulsifier in a broad range. From 5 to 9 wt. % fatty alcohol the interfacial tension are under 10^{-2} mN/m. The dependence of the interfacial tension on the average degree of ethoxylation of the emulsifier as determined in Figure 10 can also be linked to the phase behavior of ternary mixtures from pure substances. The phase inversion temperature of water/oil/surfactant mixtures depends strongly on the number of ethylene oxide groups of the nonionic surfactants [13]. For a given oil, the existence zones of the one-phase and three-phase bodies in such mixtures are shifted to higher temperatures with increasing hydrophilicity of the emulsifier [12]. Figure 15 shows that this is also valid for mixtures of technical-grade substances. The inversion area is shifted to higher temperatures with increasing portion of higher ethoxylated emulsifier. As the measurements in Figure 10 were carried out at constant temperature, it is clear that the extremely low interfacial tensions close to the three-phase body which were observed with a low ethoxylated emulsifier ($\bar{m} = 12$) do neither occur with the mixture from two emulsifiers ($\bar{m} = 16$) nor with the highly ethoxylated emulsifier ($\bar{m} = 20$ respectively 30).

From Figures 3 and 11 it follows that the phase behavior decisively affects the consistency of the emulsions. When the emulsions are made within or above the phase inversion zone, stable emulsions with viscosities as low as that of water are formed. The probable cause for this is that these systems show extremely low interfacial tension. In addition, the oil in these systems is molecularly dissolved in the inversion zone. Therefore, optimal dispersion conditions result. Microscopic observations at ambient temperature showed that the oil droplets cannot be recognized despite a magnification of × 1000. Light scattering measurements with diluted emulsions (e. g., 2 % oil) showed mean particle sizes of approximately 100 nm. The particle size distribution is very narrow. The emulsions prepared at temperatures below the phase inversion zone are substantially more viscous.

Previous studies have shown that oil-in-water emulsions with nonionic emulsifier are relatively stable when the PIT of the systems are approximately 20° to 65 °C above the storage temperature [13]. In addition, it was observed that the droplet diameters in the proximity of the PIT are very small and the coalescence rates very high. This is broadly in line with our observations, according to which, a relatively rapid phase separation takes place under the inversion zone. After cooling down to room temperature these systems do not show any visible destabilization even after 6 months. It is also known from the literature that in the

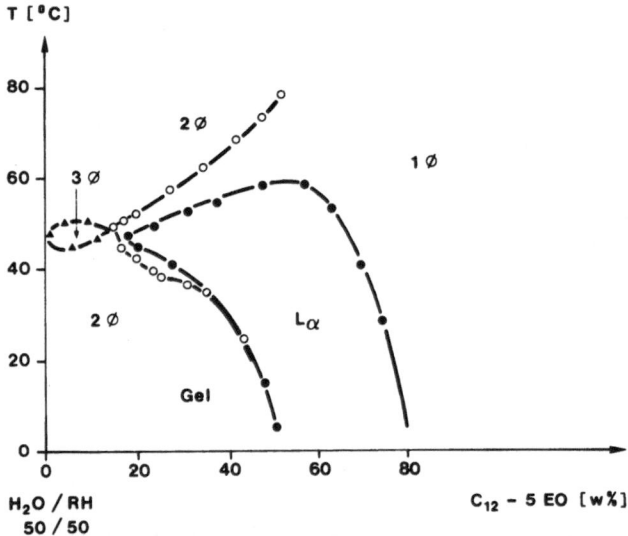

Fig. 16. Phase diagram of a ternary system [12]

existence range of three liquid phases easily emulsifiable but rather instable emulsions are formed. However, the stability of the emulsion is significantly increased when a liquid crystal exists as a third phase [14]. In the three-phase zone of the mixture (Fig. 3) no liquid crystals but three liquid phases exist. According to our own findings a high coalescence rate has been observed for our system in the three-phase zone which leads to a rapid phase separation.

Observations in the polarization microscope at ambient temperature show that the oil droplets of emulsions prepared below the inversion temperature are surrounded by anisotropic phases. Probably a gel phase is involved here, which consists of emulsifier and co-emulsifier. Gels are metastable phases that consist of bimolecular layers of surfactants with arranged hydrocarbon chains which are separated from each other by layers of water [15]. Also, in emulsions with very low viscosities and very fine oil droplets, the existence of gels surrounding the droplets is possible, even when these are not visible under the microscope. Independent of the consistency and the concentration of the emulsifier, the DSC-curves (Fig. 6) show a broad lower melting peak to be assigned to melting of the gel phase. The independence of the flow behavior on the upper limit of the preparation temperature, as shown in Figure 11, indicates that the existence of the anisotropic phase in the inversion zone is of no importance.

Previous x-ray investigations have shown the existence of liquid crystals in ternary mixtures from cetostearyl alcohol, nonionic emulsifier and water [2]. Investigations with systems that contained saturated monoglycerides as emulsifier showed that liquid crystals exist during the preparation of the emulsion [16]. It is known that liquid crystals and gels act as a rheologic barrier against the coalescence of oil droplets [17]. In addition, liquid crystals may increase the viscosity of the outer phase by forming network structures [18]. Based on their amphiphilic nature the liquid crystals cover the droplets of the inner phase. On cooling down, the transition of the liquid crystal into a gel may take place. The long term stability of the investigated systems is hence very probably due to the formation of gel layers around the oil droplets.

The phase inversion temperature strongly depends on the nature of the oil used. Previous investigations have shown that the PIT values of oil mixtures can be calculated from the PIT values of the individual components [19]. In this paper, the inversion zone itself has not been investigated further. Figures 13 and 14 show that the broad inversion zone is shifted to higher temperatures with increasing polarity of the oil. In both systems, liquid crystalline areas exist in the inversion zone. They have not been plotted. For these mixtures, too, it is true that stable emulsions with very low viscosity are formed when they are prepared in or above the inversion zone. As a consequence, there is a possibility of affecting the consistency of the emulsions not only by the preparation temperature and the selection of the emulsifiers but also by using appropriate oil mixtures.

References

1. Fukushima S, Takahashi M, Yamaguchi M (1976) J Coll Interf Sci 57:201
2. Fukushima S, Yamaguchi M, Harusawa F (1977) J Coll Interf Sci 59:159
3. Barry BW (1968) J Coll Interf Sci 28:82
4. Barry BW (1971) Rheol Acta 10:96
5. Barry BW, Saunders GM (1972) J Coll Interf Sci 38:616
6. Mapstone GE (1974) Cosmetics and Perfumery 89:31
7. Talman FAJ, Rowan EM (1970) J Pharm Pharmacol 22:338
8. Rosevear FB (1968) J Soc Cosmetic Chem 19:581
9. Shinoda K, Saito H (1968) J Coll Interf Sci 26:70
10. Junginger H, Führer C, Ziegenmeyer J, Friberg S (1979) J Soc Cosmetic Chem 30:9
11. Saito H, Shinoda K (1970) J Coll Interf Sci 32:647
12. Kahlweit M, Strey R (1985) Angew Chem, Int Ed Engl 24:654
13. Shinoda K, Saito H (1969) J Coll Interf Sci 30:258
14. Madani K, Friberg S (1978) Progr Coll & Polym Sci 65:164
15. Luzzati V et al (1960) Acta Cryst 13:660
16. Krog N, Lauridsen JB (1976) In: Friberg S (ed) Food Emulsions, Dekker, New York, p 67
17. Friberg S (1976) In: Friberg S (ed) Food Emulsions, Dekker, New York, p 1
18. Friberg S (1979) J Soc Cosmetic Chem 30:309
19. Akai H, Shinoda K (1967) J Coll Interf Sci 25:396

Received January 21, 1987;
accepted January 23, 1987

Authors' address:

Dr. F. Schambil
c/o Henkel KGaA
Postfach 11 00
D-4000 Düsseldorf, F.R.G.

Progress in Colloid & Polymer Science   Progr Colloid & Polymer Sci 73:48–56 (1987)

Lipid swelling and liposome formation on solid surfaces in external electric fields

D. S. Dimitrov and M. I. Angelova

Central Laboratory of Biophysics, Bulgarian Academy of Sciences, Sofia, Bulgaria

Abstract: External electric fields can induce or prevent lipid swelling and liposome formation on solid surfaces. These effects depend on the type of the lipid and the surface, the medium parameters (temperature, osmolarity, ionic strength), the dried lipid layer thickness, the type and parameters of the electric field (direct current (DC) or alternating current (AC), amplitude, frequency, current), the place of the lipid (on the electrode surface itself or on another surface) and the time of exposure. This paper reviews experimental results and theoretical estimates for these effects and presents new data and estimates for the effects of lipid charge and frequency of the field. The basic conclusion is that the electric fields can affect lipid swelling and liposome formation by at least four mechnisms: (1) direct electrostatic interaction, (2) redistribution of the double layer counter-ions, (3) membrane surface and line tension changes and (4) electroosmotic effects.

Key words: Lipid electroswelling, liposome electroformation, intermembrane interactions.

Introduction

After the pionering work of Bangham et al. [1], a number of methods for preparation of liposomes have been proposed; the liposomes have been thoroughly investigated and in several cases applied in technology and medicine (see e.g. [2–4]). Cell-size liposomes have attracted much attention for several basic reasons: (1) historically they were the first obtained liposomes, (2) the procedure for their preparation is rather simple – they form spontaneously by swelling of lipids on solid surfaces, (3) they serve as a simple model of cells and membranes. One of the difficulties with the classic method of their preparation [1] is that they are commonly multilayered. Another disadvantage is that the size distribution is wide. In addition, many of the lipids simply do not form such liposomes by swelling. Lipid mixtures are better than pure lipids in such cases. To avoid these problems, other methods have been used which are not based on the process of swelling of lipids on solid surfaces. These methods, however, have their own disadvantages, e.g., formation in non-water solutions, etc.

Certainly, understanding the mechanisms of liposome formation can significantly improve the methods of their preparation. In addition, it is very important when considering fundamental biophysical and biological problems, e.g. origin of life, etc. Surprisingly, in spite of the volumous literature on lipids and liposomes, mechanisms of liposome formation, in particular those for cell-size liposomes, have received relatively little attention. Israelachvili et al. [5–8] and Mitchell and Ninham [9] suggest that lipid aggregates can spontaneously form small unilamellar vesicles (SUV), which represent a stable equilibrium state. Helfrich [10, 11], Fromherz [12, 13] and Lasic [14] suppose that the distortion of a planar membrane to form a SUV requires energy and therefore the SUV are inherently unstable. Fromherz found experimental evidence for the existence of transient discs, after sonication in detergent solutions, and showed that edge energy drives the disc-to-vesicle transformation. The role of the detergent is to decrease the edge energy and to facilitate in this way the disc formation [12]. Cornell et al.

[15] proposed that the hydration energy tends to eliminate the exposed hydrocarbon edges of the bilayer by inducing curvature, while the configurational entropy and energy of the lipid opposes curvature. On the basis of their theoretical calculations, they conclude that the formation of SUV from zwitterionic lipids represents a metastable state induced by a disruption of the bilayer. Spontaneous formation of SUV is possible only for a limited class of lipid molecules, e.g. the dodecyldimethylammonium ion [16,17]. Cell-size liposomes can, however, form spontaneously by swelling of egg lecithin in water solutions [18,19]. Harbich and Helfrich [18] used an original procedure to make the swelling reproducible. They showed that there is no equilibrium spacing and that undulation forces can be dominant in lipid swelling. They observed tube formation and tube-to-vesicle transformation. The swelling and liposome formation was suppressed in NaCl solutions. Other investigators have also observed suppresion of liposome formation in ionic solutions [20].

Presently, the exact mechanism of cell-size liposome formation is not known. Evidently, liposome formation requires membrane separation and bending. Electric fields can affect both. They can change intermembrane forces and destabilize membranes. In addition, they can be controlled precisely. Therefore, electric fields can help in elucidating mechanisms of liposome formation. A knowledge of the mechannisms will help in the preparation of liposomes of predetermined properties.

This paper presents experimental results and theoretical estimates that DC and AC electric fields can strongly affect lipid swelling and liposome formation. The degree of the electric field effects depend on the dried lipid layer thickness and other parameters of the system [21–25].

Materials and methods

We used L-α phosphatidylcholine (EggL-) (Sigma P-5394) which contains 71% phosphatidylcholine, 21% phosphatidylethanolamine, and 8% phosphatidylserine, i.e., it is negatively charged; dodecyl amine (DA+) (Merck)/L-α phosphatidylcholine (PC) 99% (Sigma P5763), mixtures of mol ratios 1:10 and 1:100, which is positively charged; dimyristoilphosphatydilcholine (DMPC) 99% (Fluka 41803), synthetic, which is neutral; and the negatively charged phosphatidylserine (PS) 99% (Sigma P6641), from bovine brain. For the experiments on the glass surfaces we used PC; dilauroylphosphatydilcholine (DLPC) 99% (Fluka 38484), synthetic; and phosphatidylserine PS 99% (Serva 32516), from bovine brain. The lipids were dissolved in chloroform-methanol (9:1) solvent. Two drops of this solution (1µl each) were deposited on two

parallel platinum electrodes (diameter 0.48 mm, separation distance, 0.5 mm) or on two plane-parallel Ni-covered electrodes (the same separation distance), or on glass coverslips above the electrodes. The solvent was then evaporated under nitrogen. Electric fields were applied and distilled water or water solutions added. The observations were performed under phase contrast. In some cases Ficoll was added to improve visualising the thin-walled liposomes.

Experimental Results

Surface effects

Figure 1 shows cell-size liposomes formed by swelling of EggL- on glass beads [21]. For this picture Ficoll 400 is added for better visualisation of the thin-walled liposomes. The number of liposomes prepared in the presence of these beads, which increased more than 100 times the surface area of the lipid-glass contact, was more than five times larger than in the control experiments without glass beads for the same amount of lipid. The yield of liposomes in the presence of another type of glass beads was almost the same as in the control experiments. Liposome size distribution is shown in Figure 2. The average diameter is 3–5 µm. The shape of the distribution curve does not depend significantly on the surface type and area. It depends mainly on the temperature of lipid film swelling and on shaking rate of the liposome suspension [21, 22].

We also carried out experiments on flat (Fig. 3) and cylindrical glass surfaces (Fig. 4) for better observation of the lipid swelling and liposome formation [23]. The dried lipid film deposited on the substrate surface after

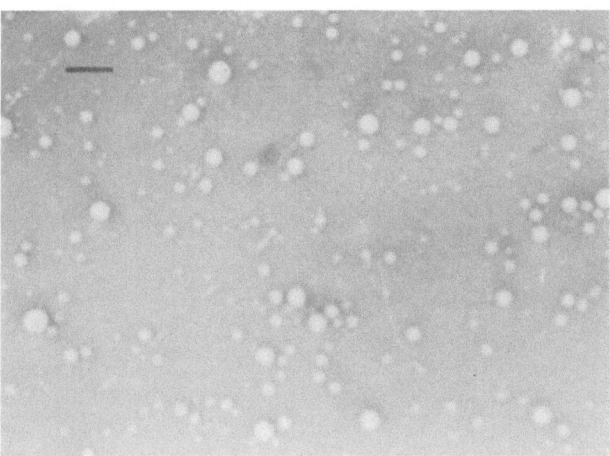

Fig. 1. Liposome suspension prepared by swelling of EggL- on glass beads; swelling in distilled water at 70 °C for 2 h; bar = 10 µm. Ficoll was added for better observation of the thin-walled liposomes

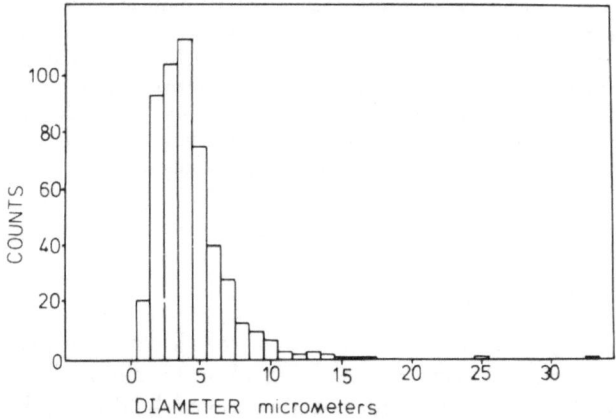

Fig. 2. Liposome size distribution for Figure 1. More than 50 % of the liposomes are of diameters 4 ± 1 μm

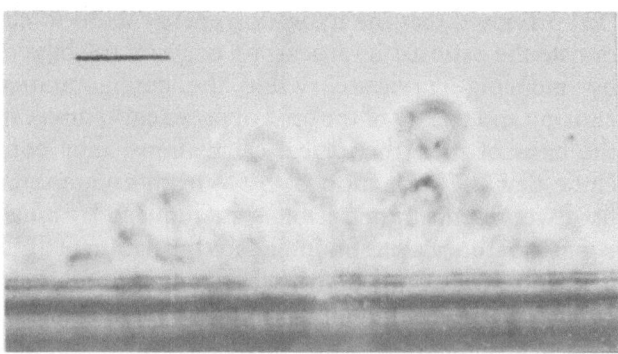

Fig. 4. Liposome formation on cylindrical glass surface; DMPC; $N = 20$; swelling in distilled water at 30 °C for 1 h; bar = 50 μm

evaporation of the solvent is not uniform and regions of different shape and size can be seen (Fig. 5). No liposomes formed when the lipid/surface area ratio is too small (corresponding to average number of bilayer $N = 2$–5 bilayers). When this ratio is too large, much of the lipid yields thick-walled lipid vesicles, aggregates and clumps. For certain value of this ratio the liposome production is most effective.

Lipid films deposited on teflon are quite unstable. The adhesion is bad and the film detaches easily from the surface after adding water. This is due to the hydrophobicity of the teflon surface. Liposome formation on metal surfaces is similar to that on glass surfaces.

DC electric field effects

All the experiments were performed at a DC voltage below 2.5 V because above this voltage we observed formation of gas bubbles. The temperature was kept at 30 °C.

The main results for EggL- are [25]: (1) The liposome yield on the negative electrode is much higher than on the positive one. (2) For N larger than 500 we did not observe any difference between both electrodes. (3) On the positive electrode and without field, liposome formation was not observed when N was below 90 (Fig. 6). (4) The voltage, applied to induce

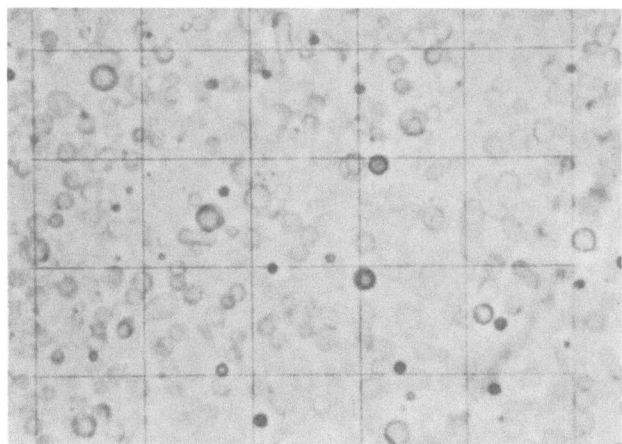

Fig. 3. Liposome formation on flat glass surface; DMPC, $N = 100$; swelling temperature in distilled water at 30 °C and swelling time 10 min; set division = 100 μm

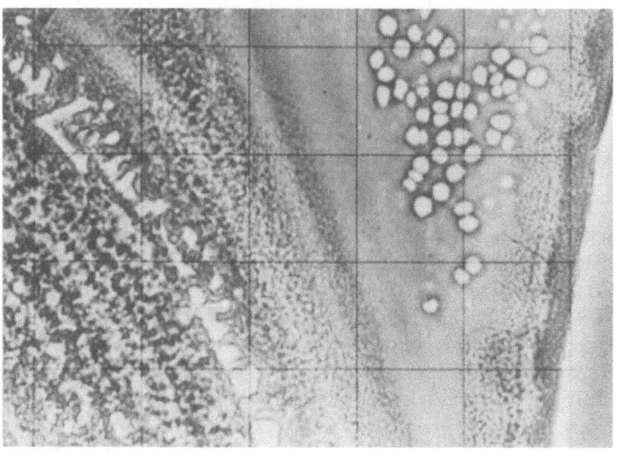

Fig. 5. Dry lipid film deposite on flat glass surface. The film is not uniform and regions of different shape and size can be seen. Set division = 100 μm

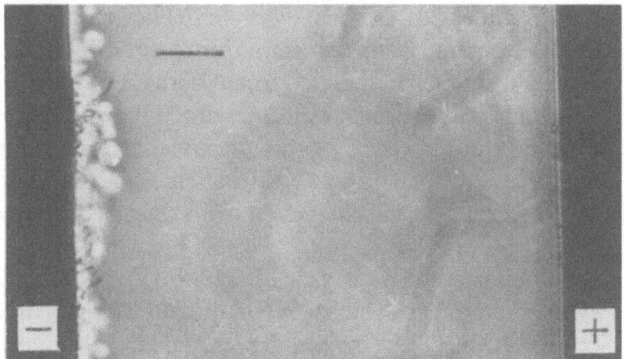

Fig. 6. Effect of DC fields on liposome formation from EggL- (negatively charged mixture); vesicles formed only at the negative electrode. $N = 40$; $T = 30\,^{\circ}\text{C}$; $U = 0.6$ V; swelling in distilled water for 20 min. Ficoll was added to improve the phase contrast. Bar = 100 μm

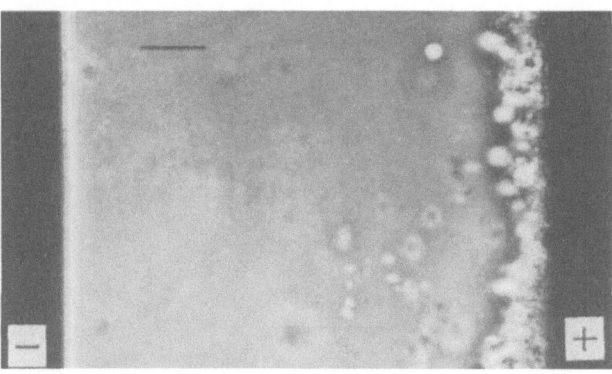

Fig. 7. Effects of DC fields on liposome formation from DA+/PC 1:10 (positively charged mixture); vesicles formed predominantly at the positive electrode. Fast electrophoretic motion of the liposomes from (+) to (−) electrode was observed. $N = 105$; $T = 22\,^{\circ}\text{C}$; $U = 1$ V; swelling in distilled water for 30 min. Ficoll was added to improve the phase contrast. Bar = 100 μm

liposome formation on the negative electrode when N is smaller than 90, increases linearly with decreasing ln N. (5) Decreasing N leads to decrease of the number of tubes and multilamellar liposomes. (6) Below $N = 10$, only very thin-walled liposomes were observed. (7) The attached liposomes grew in size by ca. 10 % for 30 min, while the separated ones did not. (8) The liposomes were a little elongated. (9) Liposomes did not form when N was smaller than 2–3, even in fields of rather high amplitude (up to 3V). (10) The rate of swelling on the positive electrode was smaller than on the negative one and without field for N up to 500. (11) Equilibrium thickness of swelling exist for which the rate of swelling is zero. (12) The equilibrium thickness in 1 M sucrose solutions (50 μm) is considerably smaller than that in distilled water (130 μm). (13) There is no lipid swelling and liposome formation in 2.5 M sucrose solution and PBS.

For the positively charged lipid mixture (DA+/PC 1:100) the phenomena are similar to those for EggL- but the positive electrode plays the role of the negative one. Elecrophoretic motion of the liposomes was observed. When the amount of the positively charged component (DA+) was increased (DA+/PC = 1:10) (Fig. 7) the rate of swelling on the positive electrode increased almost one order of magnitude.

The effects of DC fields on lipid swelling and liposome formation from DMPC (a noncharged lipid) are much smaller and rather different from those for EggL- (a charged lipids mixture). We observed formation of liposome-like structures, which appear differently on both electrodes. The number and size of DMPC liposomes was larger in DC fields, especially at the negative electrode for voltages above 1.2 V, than without field.

We also observed electroosmotic effects, which caused *the lipid* to move. The motion, however, was for very short period of time immediately after applying the DC field because it was due to charging of the double layer at the electrode surface. The mechanical stresses caused by this motion may affect liposome formation.

AC field effects

We studied the effects of AC fields for two cases [25]: firstly on formed liposomes and second, on liposome formation. We observed that liposomes and other structures, which formed during lipid swelling in water, vibrate with the same frequency as the applied AC field, following the electroosmotic motion of the medium.

With increasing the amplitude of the field, the amplitude of the vibrations increases. Unfortunately, we did not have appropriate equipment to record the amplitude as a function of the voltage and the frequency. Instead, we used the observation that at fixed voltage the amplitude decreases with increasing the frequency and above certain frequency one cannot see any change of the position of the liposome, i.e., above this frequency the vibration amplitude is very small, at least smaller than the optical resolution of our micro-

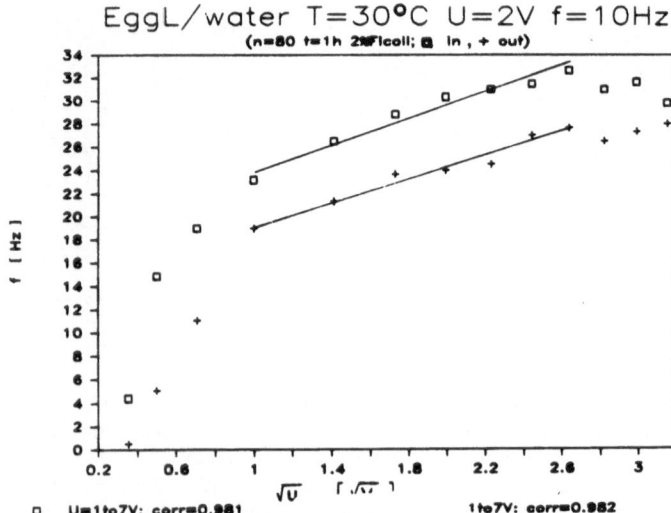

Fig. 8. Lipid vibrations in AC fields – a typical dependence of the characteristic frequency on the square root of the applied voltage on the inner (□) and outer (+) side of the electrode. Experimental points lie well on straight lines (correlation coefficients abore 0.98) for U between 1 and 7 V. Gas bubble formation was observed for U > 7 V. Preparation: EggL- layer swollen in distilled water, $N = 80$; $T = 30\,°C$; $U = 2\,V$; $f = 10$ Hz; $t = 1$ h; 2 % Ficoll

scope, in the worst case, let us say 1 μm. This characteristic frequency depends on the applied voltage. The increase of the voltage leads to increase of the frequency.

Figure 8 shows a typical dependence of the characteristic frequency on the square root of the applied voltage for EggL-. The lipid structures were formed at $T = 30\,°C$ at $U = 2$ V, frequency $f = 10$ Hz in distilled water for 1 h. The vibration amplitude between the electrodes is larger than outside them. The main conclusion is that this functional dependence does not depend on the lipid, the solution osmolarity, the time of swelling and the lipid layer thickness. This is probably a pure electroosmotic effect. When a rigid wall was placed between the electrodes, the vibration amplitude strongly decreased. The amplitude decrease was larger as the wall was closer to the respective electrode. Placing the wall at the outer sides of the electrodes did not change appreciably the vibration amplitude between the electrodes. Electroosmotic effects when the lipid was placed on plane-parallel electrodes for voltages up to 2 V were hardly visible.

One of the most important observations is that without field there was no liposome formation from EggL-. The application of AC field (frequency 10 Hz, voltage 2 V) led to formation of giant thin-walled lipo-somes and a lot of small vesicles. The neutral lipid DMPC formed just a few giant thin-walled liposomes. Application of AC field led to an increase of the number of these liposomes at least ten times. The negatively charged lipid PS did not form liposomes by swelling neither with nor without electric field. The lipid layer broke into flakes and detached from the electrode surface.

When the lipids (EggL-, DLPC, PS (Serva 32516)) swelled on the glass surfaces (on the internal side of the coverslip), i.e., not on the very electrodes, the vibration amplitude was much smaller than that near to the electrode surface. When Ficoll 400 was added, the vibration almost stoped. The main results are that, in this case, there are no expressed effects of the AC fields. The swelling rates and formation of liposomes were the same as without field.

Theoretical estimates

Lipid swelling

The lipid can swell by increasing the separation distance between plane-parallel bilayers and/or by forming different structures like spheres, tubes, etc. Plane-parallel model is considered in [24]. Swelling of spherical membranes is discussed in details in [25]. We assume that the swelling of the lipid is due to water penetration into the interbilayer space through the membranes. Therefore, the rate of swelling v can be calculated as

$$v = dl/dt = L\Delta p, \tag{1}$$

where l is the distance from the electrode surface to the front of the swelling lipid, t = time, L = hydraulic permeability and Δp is the driving pressure difference across the bilayer. The driving pressure can be represented as a sum of osmotic pressure (P_{osm}) electrostatic pressure (P_{el}) pressure, due to membrane tension effects (p_t), hydration component (p_h), and other pressures (p_o) due to other effects, in particular, to van der Waals and undulation forces

$$\Delta p = p_{osm} + p_{el} + p_t + p_h + p_o. \tag{2}$$

The osmotic pressure can be calculated as

$$p_{osm} = R_g T_k (c_i - c_o), \tag{3}$$

where R_g is the gas constant, T_k – absolute temperature, c_i and c_o are the solute concentrations inside and outside the separating membrane.

The internal concentration can be expressed as

$$c_i = c_{i0}V_0/V, \qquad (4)$$

where c_{i0} is the inside concentration at $t = 0$. V_0 and V – respective volumes of the interbilayer space.

The *simplest* expression for the pressure due to external electric fields is

$$p_{el} = U\sigma/l_s, \qquad (5)$$

where U is the applied voltage, σ = surface charge of the lipid bilayer, and l_s = distance between the electrode surfaces. It should be noted that for cylindrical electrodes the Equation (5) is an approximation.

The pressure, due to membrane tension, appears only when considering swelling of spherical membranes. For that case

$$p_t = -2T/R = -4K\,\Delta R/R_0^2, \qquad (6)$$

where T is membrane tension, R = radius of the spherical membrane in particular liposome, K = elastic constant, $\Delta R = R - R_0$, R_0 being the radius of a free-tension liposome, for which the liposome gets its spherical shape. Equation (6) can be used only when $\Delta R > 0$. For $\Delta R \leq 0$, the membrane tension effects can be neglected.

For egg lecithin the hydration pressure is [26]

$$p_h = P_0 \exp(-h/\lambda), \qquad (7)$$

where h is the separation distance between two bilayers, $P_0 = 10^{9.76}$ dyn/cm^2, and $\lambda = 0.26$ nm.

The hydration force is important for very small separations. The stage, where the hydration forces dominate, is very short. For example, it takes 10^{-4} s to increase the bilayer separation h from 0 to 1 nm.

The numerical estimates have shown that for the purposes of present consideration p_o can be neglected, compared to osmotic forces. In other cases the undulation forces can be significant [19].

For the stage of separating *plane* membranes, considering only p_{osm} and p_{el}, and substituting (2, 3, 5) into (1), we obtain

$$dl/dt = \alpha\,(1/l - 1/l_e), \qquad (8)$$

where

$$\alpha = LR_gT_kc_{i0}h_0n$$

$$l_e = \alpha/L(R_gT_kc_o - U\sigma/l_s)$$

i.e., α and l_e are constants depending on the parameters of the system (8). l_e is equilibrium distance of swelling; n, the number of bilayers; h_0, the initial separation between plane-parallel bilayers.

The solution of Equation (8) is

$$ln\,(1 - x) + x = -\alpha t/l_e^2, \qquad (9)$$

where $x = l/l_e$, and an initial condition $l = l_0 = 0$ at $t = 0$ was used.

These formulae, derived for a plane-parallel model of the lipid layer structure are in functional and semi-quantitative agreement with the experimental data for the rate of swelling of egg phosphatidylcholine (a negatively charged mixture as mentioned above) in DC electric fields.

Liposome growth

For the case of swelling of spherical membranes, taking into account p_t and assuming that the changes of the liposome radius are small, i. e., that $\Delta R = R - R_0 \lll R$, Equations (1)–(6) lead to

$$\Delta R = \Delta R\infty\,[1 - \exp(-t/\tau)], \qquad (10)$$

where

$$\Delta R\infty = [R_gT_k(c_{i0} - c_o) + U\sigma/l_s]/(3R_gT_kc_{i0}/R_0 + 4K/R_0^2)$$

$$\tau^{-1} = L(3R_gT_kc_{i0}/R_0 + 4K/R_0^2)$$

i.e., $\Delta R\infty$ and τ are constants depending on the parameters of the system and an initial condition $\Delta R = 0$ at $t = 0$ was used.

The last stages of liposome electroformation on surfaces involve increase in size and detachment. After detachment liposomes do not change their dimensions, but they become a little elongated. This demonstrates the effect of the negative surface charge of the liposome membrane. Another confirmation of this statement is the observed electrophoretic motion of the liposomes.

The maximum relative area change observed $\Delta A/A$ is 0.2 – 0.3 which corresponds to membrane tensions $T = 100$ dyn/cm for one bilayer. These are rather high values (the critical membrane tension of rupture is of the order of 3–5 dyn/cm [27]). This result may indicate that during liposome growth material can come from the lipid layer or that liposomes are formed from several bilayers, which may rupture and slip over each other. Numerical estimates based on Equation (10) [25], show very good agreement with the experimental data: we obtain $\Delta R\infty = 1.9$ μm and $\tau = 23$ min.

Vibration in AC fields

By estimating the separate terms in the Navier-Stokes equations

$$\varrho(\partial v_x/\partial t) = -\partial p/\partial x + \mu \, (\partial^2 v_x/\partial z^2) , \tag{11}$$

$$v\varrho\omega = cU + \mu v/H^2 , \tag{12}$$

where

$$v = A\omega, \quad \omega = 2\pi f ,$$

we derived the following formula for the vibration amplitude A

$$A = U/af(1 + bf) \tag{13}$$

where

$$a = 8\pi^2\mu l_s^2/\varepsilon\kappa\varphi H^2 , \qquad b = \varrho H^2/2\pi\mu .$$

Here μ is the liquid viscosity, $\varepsilon = $ permitivity, $\varphi = $ surface potential, $l_s = $ distance between the electrodes, $H = $ characteristic length normal to the glass surface (of the order of half of the chamber height, and $\varrho = $ liquid density. The vibration amplitude decreases with increasing the frequency. When A becomes less than a certain value, A_c, the vibration cannot be observed.

$$U_c = aA_c f_c(1 + bf_c) \tag{14}$$

which is relevant to our experimental observations. It describes well the experimental curves, e.g., these shown in Figure 8.

Time of liposome formation

We also estimated the rate of liposome formation assuming that the driving force is the line tension k.

The forces which resist the line tension are due to the membrane viscosity and curvature elasticity. Then an approximate balance of forces leads to the following expression for the time of liposome formation

$$\text{time} = \text{viscosity x radius} /$$
$$(\text{line tension} -$$
$$\text{curvature elasticity/radius}) , \tag{15}$$

If we suppose that due to hydration, electric and other forces, the membranes are separated and destabilized to have boundaries of higher energy, then the time of liposome formation can be estimated by using Equation (15). We get times of the order of 100 s.

Conclusions

Bilayer separation and bending are prerequisites for liposome formation from hydrating lipids. Therefore, a possible molecular mechanism is that membranes should be destabilized to bend and fuse to form liposomes. This requires the right proportion between structured regions (in the form of bilayers), and defects and/or non-bilayer structures.

Two basic steps are involved in preparation of cell-size liposomes by swelling: (i) Drying of the lipids, dissolved in a mixture of polar and nonpolar solvent. During evaporation of the solvent, the lipids should adsorb onto the substrate surface and form a multilayer structure. The type of this structure will depend on the surface/lipid and lipid/lipid interactions. It may be expected that near the surface the lipid layer will have different properties than the lipids far from the surface. (ii) Swelling of the lipid in water. When water or a solution is added it goes through the bilayers and/or through the defects, driven by the hydration forces and in less extent by undulation forces. The time and swelling distance of the hydration stage are very short: it takes $\sim 10^{-4}$ s to increase the interbilayer separation from 0 to 1 nm. For larger separations, electrostatic membrane-membrane, and van der Waals membrane-membrane and membranes-substrate surface interactions as well as undulation forces can be important up to membrane separations ~ 0.1 μm. Lipid swelling to macroscopic distances (of the order of μm-s per lipid bilayer) can be driven by osmotic forces or external influence; in particular, electric field.

The electric fields can affect these processes by at least four mechanisms:

(1) Electrostatic interaction between the field and the lipid charges,

(2) Redistribution of the counter-ions, which changes the intermembrane forces,

(3) Decreasing of membrane surface and line tensions,

(4) Electrokinetic effects.

The direct elelctrostatic interaction leads to electrophoretic motion of the liposomes. This effect is, however, relatively small and cannot explain the observed dependence of the potential to induce liposome formation on the lipid layer thickness. It is important for the last stages of lipid swelling for relatively thick lipid layers. It can be also important for the growth of the liposomes; the observation that the liposomes are little elongated can be explained on the basis of electrostatic interaction of the field with the lipid charge and the induced potential.

Redistribution of the double layer ions between the lipid bilayers due to DC field can be the reason for the observed dependence of the potential to induce liposome formation and the lipid layer thickness. It is interesting to point out that the double layer thickness of distilled water for pH between 5 and 7 ranges from 100 nm to 1000 nm. This is just the range of thicknesses (10 to 100 bilayers) for which liposomes do not form from EggL- without fields. The increased interbilayer repulsion should overcome van der Waals attraction between membranes and between the membranes and the semi-infinite electrode material. The membrane/electrode interaction decreases with increasing the distance from the electrode surface. Probably, at lipid layer thickness corresponding to 90 bilayers, this interaction becomes equal to the sum of the other forces. One possibility is that the applied electric field sucks out the counterions between the membranes, thus increasing the interbilayer repulsion (on the negative electrode for negatively charged lipid mixtures). Increasing the potential will increase the intermembrane repulsion and therefore decrease the lipid layer thickness for liposome formation. Unfortunately, the theoretical problem for ion distribution in multilayered systems in external electric fields is rather complicated and still unresolved. It is presently under consideration.

The electric field can decrease the membrane surface tension, thus destabilizing the membrane. This can lead on one hand to membrane rupture and more defects, and on the other hand to instability of bending. As pointed by Fromherz [13], electric field can also decrease the edge energy (respectively line tension). This will facilitate the initial stage of formation of lines of defects and holes and therefore the liposome formation.

Electroosmotic effects are dominant in AC electric field. Unlike the case of DC fields, where the electroosmotic effect lasts very short time, AC fields induce significant periodic motion for the period of action of the field. There is a certain similarity between the effects in sonication and in AC fields. In both cases mechanical stresses are induced and they cause rupture of bilayers and formation of pieces of different size, which can bend to form liposomes. The electroosmotic vibrations are, however, in some aspects more gentle and lead to formation of larger thin-walled liposomes.

It must be pointed out that most of the above concepts are still hypotheses which need further experimental and theoretical work.

References

1. Bangham AD, Standish MM, Watkins JC (1965) J Mol Biol 13:238
2. Bangham AD (ed) (1983) Liposome Letters, Academic Press, London
3. Gregoriadis G (ed) (1984) Liposome Technology, CRC Press, Inc, Florida
4. Schmidt KH (ed) (1986) Liposome as Drug Carriers, Georg Thieme Verlag Stuttgart, New York
5. Israelachvili JN, Mitchell DJ, Ninham BW (1976) J Chem Soc, Faraday Trans 2, 72:1525
6. Israelachvili JN, Mitchell DJ (1975) Biochim Biophys Acta 389:13
7. Israelachvili JN, Mitchell DJ, Ninham BW (1977) Biochim Biophys Acta 470:185
8. Israelachvili JN, Marcelja S, Horn J (1980) Q Rev Biophys 13(2):121
9. Mitchell DJ, Ninham BW (1981) J Chem Soc, Faraday Trans 2, 77:601
10. Helfrich W (1973) Z Naturforsch, Teil C 28:693
11. Helfrich W (1983) Phys Lett 50A:115
12. Fromherz P (1983) Chem Phys Lett 94:259
13. Fromherz P (1986) Faraday Disc Chem Soc 81:39; 81:347
14. Lasic DD (1982) Biochim Biophys Acta 692:501
15. Cornell BA, Middlehurst J, Separovic F (1986) Faraday Disc Chem Soc 81:163
16. Brady JE, Evans DF, Kachar B, Ninham BW (1984) J Am Chem soc 106:4279
17. Hashimoto S, Thomas JK, Evans DF, Mukerjee S, Ninham BW (1983) J Coll Int Sci 95:594
18. Harbich W, Helfrich W (1984) Chem Phys Lipids 36:39
19. Servuss AM, Harbich W, Helfrich W (1976) Biochim Biophys Acta 436:900
20. Mueller P, Chien TF, Rudy B (1983) Biophys J 44:375
21. Dimitrov DS, Li J, Angelova MI, Jain RK (1984) FEBS Lett 176:398
22. Angelova MI, Dimitrov DS (1985) Biophys J 47:163a

23. Dimitrov DS, Angelova MI (1985) Proc Biotech '85 Geneva 1:655
24. Dimitrov DS, Angelova MI (1986) Studia Biophysica 113:15
25. Angelova MI, Dimitrov DS, Faraday Disc Chem Soc 81:303
26. Lis LJ, McAlister M, Rand RP (1982) Biophys J 37:657
27. Kwok R, Evans E (1981) Biophys J 35:637

Received January 21, 1987;
accepted February 11, 1987

Authors' address:

Dr. Dimiter S. Dimitrov
Central Laboratory of Biophysics
Bulgarian Academy of Sciences
Acad. G. Bonchev Str., Bl. 21
Sofia 1113, Bulgaria

Progress in Colloid & Polymer Science

Progr Colloid & Polymer Sci 73:57–65 (1987)

Electrochemical properties of barium sulfate precipitation membranes

R. Forke, R. Ortmann, D. Woermann, and O. Mohamed[1])

Institut für Physikalische Chemie, Universität Köln, Köln, F.R.G.
[1]) Department of Chemistry, University of Khartoum, Khartoum, Sudan

Abstract: A structural model of precipitation membranes proposed by Hirsch-Ayalon (see [4]) is tested by the following experiments: (a) Measurements of the zeta potential of suspended $BaSO_4$ crystals are used to estimate the sign and the absolute value of the charge density of the adsorption charges of the crystals forming the membrane, (b) the concentration limits for maintaining the membrane properties of $BaSO_4$ precipitation zones, as well as for generating precipitation zones with membrane properties are established. Measurements of the non-ohmic current-voltage characteristics of a $BaSO_4$ precipitation membrane demonstrate that Ba^{++} ions can be partially replaced by La^{+++} ions to maintain the membrane properties of $BaSO_4$ precipitation zones.

Key words: Membrane, precipitation, transport properties, barium sulfate.

1. Introduction

Anorganic crystalline precipitation zones with membrane properties can be formed by interdiffusion within a gel or porous matrix (e. g. cellophane skin) of two aqueous electrolyte solutions containing ions which can form a sparsely soluble precipitate. This was observed first by Terazawa (1954) [1] and independently by Hirsch-Ayalon (1956) [2]. It was found that most precipitation zones formed by 2,-2 valent precipitates acquire membrane properties [3] (i. e. non-ohmic current/voltage characteristics, characteristic membrane potentials). Recently, two reviews have been published summarizing the physical chemical properties of this type of membranes [4,5].

The properties of precipitation membranes can be explained by a model proposed by Hirsch-Ayalon [2]. Taking the $BaSO_4$ membrane as an example, this model can be described in the following way: A $BaSO_4$ precipitation zone with membrane properties is composed of three distinct layers of precipitate. Layer 1, which is in contact with the bulk phase containing Ba^{++} ions, carries positive surface charges caused by a preferential adsorption of Ba^{++} ions. Layer 2, which is in contact with the bulk phase containing SO_4^{--} ions, carries negative adsorption charges caused by a preferential adsorption of SO_4^{--} ions. Both layers are separated by a layer with no net charge. The adsorbed charges cause the formation of an electric double layer around the crystallites. If the average distance between the crystallites is assumed to be smaller than the extension of the electrical double layer, the double layers of adjacent crystallites can overlap. Under this condition, mobile ions having the same sign of charge as that of the adsorbed charges are excluded from the space between the crystallites by electrical repulsion. Therefore, layer 1, in contact with the solution containing Ba^{++} ions, has the properties of an anion exchange membrane. Layer 2, in contact with the bulk phase containing SO_4^{--} ions, has the properties of a cation exchange membrane. Recently new experimental results have been reported [3, 6] which give support to this model. In this study electrochemical experiments are described which give further insights into the functioning of the $BaSO_4$ precipitation membrane.

2. a) Adsorption charges on $BaSO_4$ crystallites suspended in aqueous solutions of $BaCl_2$ and Na_2SO_4, respectively

Experiments are carried out to estimate the sign and magnitude of the adsroption charges of the $BaSO_4$ crystals forming the layer with anion exchange properties (layer 1) and the layer with cation exchange properties (layer 2), respectively, in the above described structural model of a BaSO$_4$ membrane. These quantities are determined by suspending BaSO$_4$ crystals in aqueous solutions of BaCl$_2$ and Na$_2$SO$_4$, respectively, and measuring their electrophoretic mobility using a commercial zetameter (Rank Brothers, Bottisham, Great Britain, Type Mark II). The BaSO$_4$ crystals are produced by rapidly mixing equal concentrated solutions of BaCl$_2$ and Na$_2$SO$_4$:

$$c(BaCl_2) = c(Na_2SO_4) = 0.2 \ c^+, \ c^+ = 1 \ mol \ dm^{-3} \ .$$

After 12 h the precipitate is separated from the supernatant solution by decantation. The precipitate is treated with a methanol/water mixture containing ammonia, a methanol/water mixture and finally with distilled water. Thereafter the precipitate is separated from the equilibrium solution by filtration. It is dried at 70 °C and kept in a desiccator over P$_2$O$_5$. The crystals have an average size of about (0.27 ± 0.10) μm (scanning elelctron microscope, Stereoscan 180). For the measurements of electrophoretic mobility, 5 mg BaSO$_4$ are suspended in 100 cm^3 of an aqueous solution of BaCl$_2$ and Na$_2$SO$_4$, respectively, of known composition. Part of the solution containing suspended BaSO$_4$ particles is transfered into the glass cell of the zetameter and is used for the measuremnts.

From the measured electrophoretic velocity (applied electric field E: 2 V cm^{-1} < E < 12 V cm^{-1}); electrophoretic mobility (u_e: 1.3 10^{-4} cm^2 s^{-1} V^{-1} < u_e < 3.7 10^{-4} cm^2 s^{-1} V^{-1}) and the knwon viscosity of the medium in which the crystals are suspended, the zeta potential is calculated using the Smoluchowsky equation

$$\zeta = 3\eta/(2\varepsilon \varepsilon_o) \cdot u_e \ ,$$

with ζ: zeta potential; ε_o: influence constant; ε: dielectric constant of water. The experimental results are shown in Figure 1. They allow a determination of the electrolyte concentration at which BaSO$_4$ crystals carry no net charge. BaSO$_4$ crystals carry a small positive net charge when they are suspended in water saturated wit BaSO$_4$.

On the basis of data shown in Figure 1, it can be predicted that there must be a characteristic lower concentration c^* of Ba^{++} ions and SO$_4^{--}$ ions in the bulk phases to maintain the membrane properties of a BaSO$_4$ precipitation zone. If the charge density of the adsorbed charges becomes too small the counterion clouds sur-

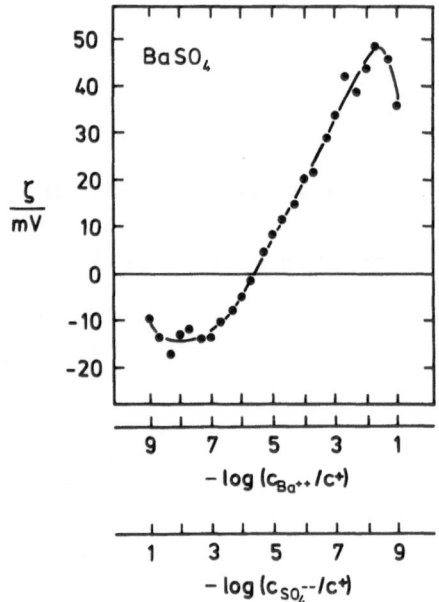

Fig. 1. Zeta-potential ζ of BaSO$_4$ crystallites suspended in aqueous solutions of BaCl$_2$ and Na$_2$SO$_4$ respectively. Qualitatively, the results are similar to those reported in the literature [7, 8]

rounding the crystallites can no longer overlap effectively enough to prevent the transport of mobile Ba^{++} and SO$_4^{--}$ ions across the precipitation zone.

2. b) Maintainance of membrane properties

What is the value of the concentrations of the Ba^{++} and SO$_4^{--}$ ions in the bulk phase of a BaSO$_4$ precipitation membrane to maintain its membrane properties? To answer this question the following experiments are carried out.

A BaSO$_4$ precipitation membrane is formed by interdiffusion of aqueous solutions of Ba(OH)$_2$ and H$_2$SO$_4$ (c'(Ba(OH)$_2$) = 10^{-2} c$^+$/cellophane skin/ c''(H$_2$SO$_4$) = 10^{-2} c$^+$), thickness of the cellophane skin in the swollen state skin: 50 μm; water content 61 %. The membrane potential ($\Delta\varphi = \varphi' - \varphi''$) is measured as a function of time using two identical calomel electrodes. It is assumed the precipitation zone has membrane properties if the measured value of the membrane potential is in agreement with the value predicted on the basis of the proposed layered structure (see Eq. (1)). The concentration ($c' = c'' = c$) of Ba(OH)$_2$ and H$_2$OH$_4$ in the bulk phases is lowered stepwise, keeping the cellophane skin with the BaSO$_4$ precipitation membrane in place. The membrane potential decreases with decreasing values of c. But the ratio $\Delta\varphi/\Delta\text{pH}$ has the theoretically expected time independent value of 58 mV as long as the concentration c is higher than $c > 5 \cdot 10^{-4}\ c^+$. At a concentration of about $c^* \simeq 3 \cdot 10^{-4}\ c^+$, the membrane potential no longer has the theoretically expected value and it decreases with time (see Fig. 2a) indicating that the precipitation zone is losing or has lost its membrane properties. Increasing again the electrolyte concentration of the bulk phases ($c > c^*$) regenerates the membrane properties of the precipitation zone. A similar value of c^* ($c^* \simeq 1 \cdot 10^{-4}\ c^+$) has been reported by Hirsch-Ayalon [9].

A concentration value of about $c^* \simeq 5 \cdot 10^{-4}\ c^+$ is found in similar experiments with the system

$$c'(\text{BaCl}_2)/\text{cellophane skin}/c''(\text{Na}_2\text{SO}_4)\ (c' = c'' = c)$$

(see Fig. 2b). In that case the membrane potential is also a function of c. It decreases with decreasing c but does not change with time as long as the $c > 5 \cdot 10^{-4}\ c^+$. The dependence of the membrane potential on the concentration c will be discussed in Section 3.

Combining these experimental results with the data shown in Figure 1, it can be concluded that at concentrations $c'(\text{Ba}^{++}) = c''(\text{SO}_4^{--}) = c^* \simeq 10^{-4}\ c^+$, the charge density of the adsorption charges has reached such low values that the negatively charged counter ion clouds, surrounding the crystallites forming layer 1 and the positively charged counter ion clouds surrounding the crystallites forming layer 2, can no longer overlap effectively enough to prevent the diffusion of SO_4^{--} ions and Ba^{++} ions across the precipitation zone. It is observed that under these conditions ($c < c^*$) the thickness of the precipitation zone increases with time. Finally the precipitation zone reaches the surface of the cellophane skin and large BaSO_4 crystals begin to form there. This happens usually at the phase boundary cellophane skin/$\text{Na}_2\text{SO}_4(\text{aq})$.

3. Characteristic time to acquire membrane properties

Measurements of the electrical potential difference between the bulk phases of a cellophane skin as a function of time after the skin has been brought into contact with the aqueous solutions containing Ba^{++} and SO_4^{--} ions, show that there exists a characteristic time interval Δt_{del} (delay time) between the start of an experiment and the onset of the build-up of a membrane potential, leading finally to an equilibrium value given

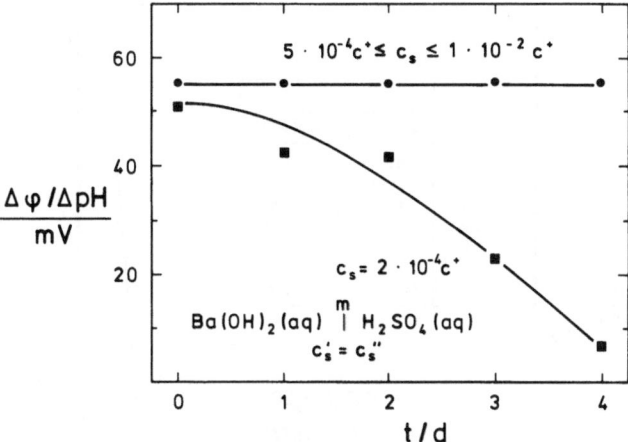

Fig. 2. Determination of the a characteristic concentration $c^* = c''(\text{Ba}^{++}) = c''(\text{SO}_4^{--})$ to maintain the membrane properties of a BaSO_4 precipitation zone. $\Delta\varphi$: membrane potential ($\Delta\varphi = \varphi' - \varphi''$); t: time. (a) Ratio $\Delta\varphi/\Delta\text{pH}$ as a function of time. System: $\text{Ba(OH)}_2(\text{aq})/m(\text{BaSO}_4)/\text{H}_2\text{SO}_4(\text{aq})$; parameter: $c = c''(\text{Ba(OH)}_2) = c''(\text{H}_2\text{SO}_4)$

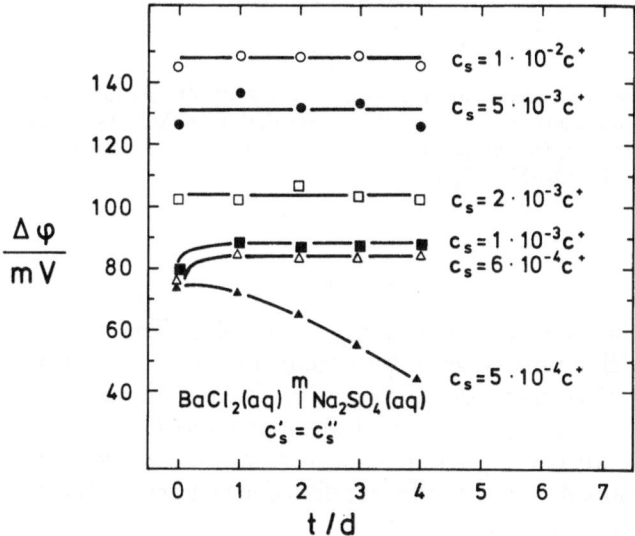

Fig. 2. (b) Membrane potential $\Delta\varphi$ of the system $\text{BaCl}_2(\text{aq})/m(\text{BaSO}_4)/\text{Na}_2\text{SO}_4(\text{aq})$ as a function of time, parameter: $c = c''(\text{BaCl}_2) = c''(\text{Na}_2\text{SO}_4)$

by Equation (1) (see Section 4). Figure 3 shows the concentration dependence of the delay time Δt_{del} for tow systems

$$\text{BaCl}_2(\text{aq})/m(\text{BaSO}_4)/\text{Na}_2\text{SO}_4(\text{aq}) \text{ and}$$

$$\text{Ba(OH)}_2(\text{aq})/m(\text{BaSO}_4)/\text{H}_2\text{SO}_4(\text{aq}).$$

For each concentration c ($c'\text{Ba}^{++}) = c''(\text{SO}_4^{--})$) a new cellophane skin is used. For both systems, Δt_{del}

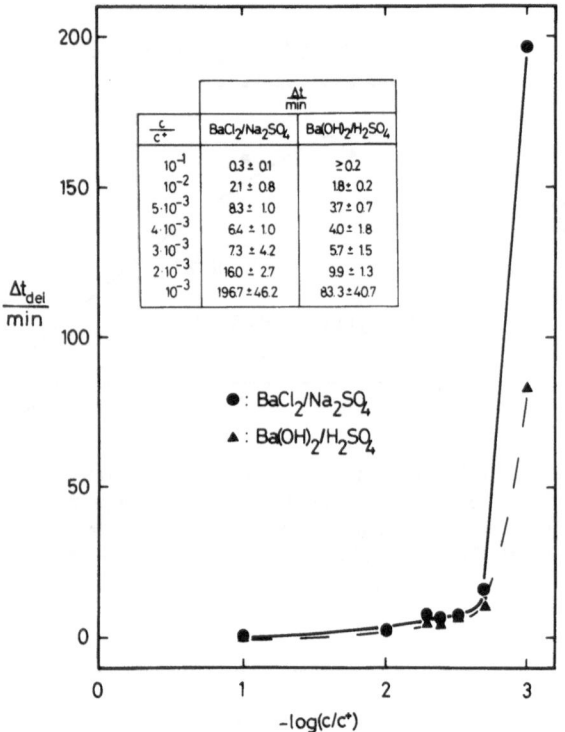

Fig. 3. Concentration dependence of delay time Δt_{del} as a function of logarithm of the concentration c; $c = c'(Ba^{++}) = c''(SO_4^{--})$ system 1: $Ba(OH)_2(aq)/m(BaSO_4)/H_2SO_4(aq)$; system 2: $BaCl_2(aq)/m(BaSO_4)/Na_2SO_4(aq)$

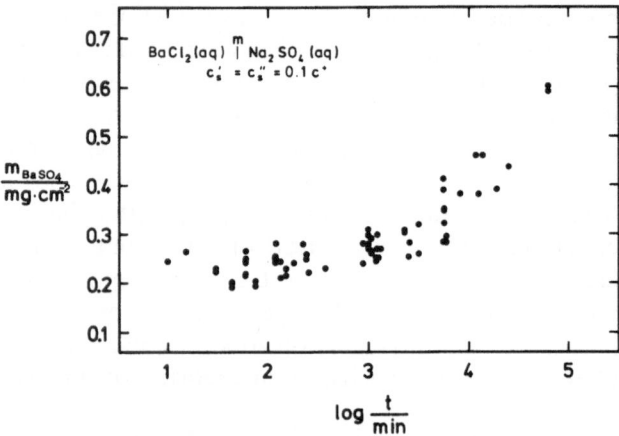

Fig. 4. Mass of $BaSO_4$ (m_{BaSO_4}) per unit area of cellophane skin as a function of time. $c'(BaCl_2) = 0.094$ c^+/cellophane skin/$c''(Na_2SO_4)$ = 0.1 c^+ (iso-osmotic concentrations)

increases with decreasing values of c. The shortest delay time is larger by a factor of about 10 than the characteristic diffusion time to obtain stationary concentration profiles within a cellophane skin used in this study ($t = \delta^2/D$; δ: thickness of the swollen cellophane skin, $\delta = 50$ µm; effective diffusion coefficient, $D \simeq 5 \cdot 10^{-6}$ cm^2 s^{-1}).

Therefore, the delay time must be related to the precipitation process taking place within the cellophane skin. Furthermore, there exists a characteristic concentration c^{**} (which is different from the characteristic concentration c^*) below which a $BaSO_4$ precipitation zone cannot acquire membrane properties and below which it never stops to grow. c^{**} has a value of the order $c^{**} \simeq 10^{-3}$ c^+. What is the cause of the existence of the characteristic concentration c^{**}? Measurements of the mass of $BaSO_4$ forming a precipitation zone with membrane properties (determination of the mass of $BaSO_4$ within a cellophane skin of known area after destroying the cellophane matrix by combustion) demonstrate that such zones contain about 0.3 mg

$BaSO_4/cm^2$ (see Fig. 4). Scanning electron microscope pictures indicate that typical precipitation zones have a thickness of about 10 µm [3]. From these data a concentration of $BaSO_4$ within the precipitation zone of the order of 10^{-3} mol cm^{-3} is calculated. A simple model proposed by Helfferich and Katchalsky [11] can be used to calculate the production rate j of a precipitate AB (solubility product K_s) from uncharged species A and B in a convection free layer. The concentration of A and B in the bulk phases is one of the parameters of the model

$$(c'(A) = c''(B) = c \text{ and}$$

$$D_A = D_B = D = 5 \cdot 10^{-6} \text{ cm}^2 \text{ s}^{-1}).$$

Results of such a calculation (with $K_s = 10^{-10}$ $(c^+)^2$; $K_s = K_s(BaSO_4)$) are shown in Figure 5.

The maximal values of the production rate j_{max} of precipitate calculated on the basis of this model are: $c = 10^{-1}$ c^+, $j_{max} = 2 \cdot 10^{-3}$ mol cm^{-3} s^{-1}; $c = 10^{-2}$ c^+, $j_{max} = 2 \cdot 10^{-5}$ mol cm^{-3} s^{-1}; $c = 10^{-3}$ c^+, $j_{max} = 2 \cdot 10^{-7}$ mol cm^{-3} s^{-1}; $c = 10^{-4}$ c^+, $j_{max} = 2 \cdot 10^{-9}$ mol cm^{-3} s^{-1}.

On the basis of these j_{max} values it is possible to estimate the time, Δt_{del}(calc) necessary to produce a $BaSO_4$ concentration of 10^{-3} mol cm^{-3}. It is found: $c = 10^{-2}$ c^+, Δt_{del}(calc) = 2 min; $c = 10^{-3}$ c^+, Δt_{del}(calc) = $1.5 \cdot 10^2$ min; $c = 10^{-4}$ c^+, Δt_{del}(calc) = $1.5 \cdot 10^4$ min. For $c = 10^{-2}$ c^+ and $c = 10^{-3}$ c^+ the values of Δt_{del}(calc) are same of the order of magnitude as those experimentally determined. For $c = 10^{-4}$ c^+ the delay time

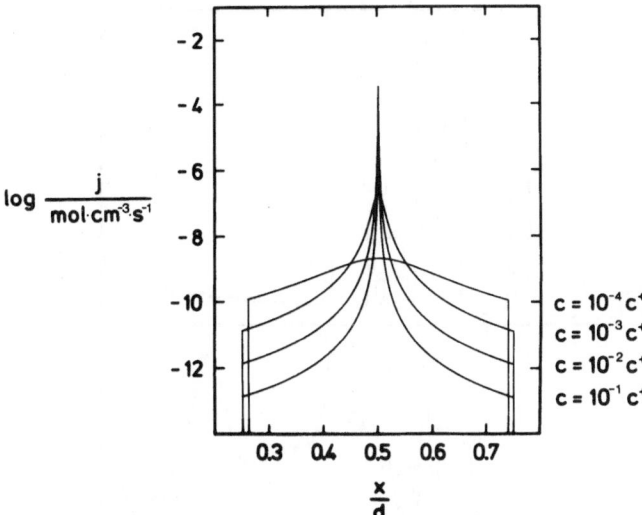

Fig. 5. Stationary production rate j of precipitate AB(s) (A(aq) + B(aq) AB(s)) within a convection free layer as function of the dimensionless space variable x/δ (δ: thickness of the layer). The curves are calculated on the basis of a model proposed by Helfferich and Katchalsky [11]. Solubility product $K_s = 10^{-10}$ (c$^+$)^2 ($K_s = K_s$(BaSO$_4$)). The concentration of the uncharged species A and B in the bulk phases has the same value ($c''(A) = c''(B) = c$). Diffusion coefficients $D_A = D_B = 5 \cdot 10^{-6}$ cm^2 s^{-1}. The maximal values j_{max} of the production rate of precipitate at $x/\delta = 0.5$ are given in text

Δt_{del}(calc) is too big to be observed experimentally. This is in agreement with the experimental facts.

From these considerations it is concluded that the delay time Δt_{del} is the time necessary to generate a BaSO$_4$ concentration of about 10^{-3} mol cm^{-3} within the precipitation zone. When the BaSO$_4$ concentration has reached this value the counter ion clouds of the charges adsorbed on the surface of the crystals forming the precipitation zone overlap effectively. Consequently, the production of BaSO$_4$ practically ceases. For times $\Delta t < \Delta t_{del}$ the membrane potential can be interpreted as a diffusion potential. The characteristic membrane potential of a precipitation membrane (see Eq. (1)) is built up after adsorption charges of the BaSO$_4$ crystals begin to overlap effectively. The characteristic concentration c^{**}, below which the experimentally determined delay time Δt(del) diverges, is different from the characteristic concentration c^* ($c^{**} < c^*$) introduced in connection with the study of maintainance of membrane properties (see Section 2). This is not surprising. c^* refers to a characteristic charge density of adsorption charges at the surface of BaSO$_4$ crystallites forming an existing BaSO$_4$ membrane. c^{**} is related to the production rate necessary to obtain a

tight packing of BaSO$_4$ crystals within the precipitation zone within minutes.

A closer analysis of the data shown in Figure 4 reveals that the formation of BaSO$_4$ does not stop completely after the precipitation zone has acquired membrane properties. The mass of BaSO$_4$ forming the precipitation membrane increases slowly. This is interpreted in the following way. The crystallites forming an intact BaSO$_4$ membrane are not in thermodynamic equilibrium as far as their size and shape are concerned. They form a disperse phase (see Fig. 6). In principle, the system is not in thermodynamic equilibrium until all the precipitate is gathered in one single crystal after an infinitely long time. Since the BaSO$_4$ crystals are immobilized within the cellophane skin, Ostwald ripening is probably the dominating process leading to this state. The driving force of Ostwald ripening is the difference in solubility between small and large crystallites.

These ripening processes take place only very slowly within a precipitation zone with membrane properties [10]. The distribution of positive and negative adsorption charges impedes the electrolyte transport. The ripening processes produce imperfections in the overlap of the counter ion clouds. Such imperfections can lead to the formation of channels of imperfections allowing the diffusion of BaCl$_2$ and Na$_2$SO$_4$. If the concentration of the Ba^{++} and the SO$_4^{--}$ ions in the bulk phases is sufficiently high, these channels are plugged up again immediately, by renewed precipitation of BaSO$_4$ (self healing process). In view of these findings it seems reasonable to assume that the characteristic concentration c^* introduced in Section 2 does not only reflect a lower limit of charge density of adsorption charges necessary to maintain the membrane properties of a BaSO$_4$ precipitation zone. Since Ostwald ripening takes place all the time, generating imperfections of overlap of the counterion cloud, the self healing process can take place effectively only if the concentration of Ba^{++} and SO$_4^{--}$ ions in the bulk phases is above a certain critical value.

4. Electrochemical experiments

Based on the assumed layered structure of the BaSO$_4$ precipitation membrane (see Section 1) it can be shown that the membrane potential of a precipitation membrane is given by [10]:

$$\Delta\varphi = \varphi' - \varphi'' = RT/F \ln (c_i''/c_i')^{1/z_i} \qquad (1)$$

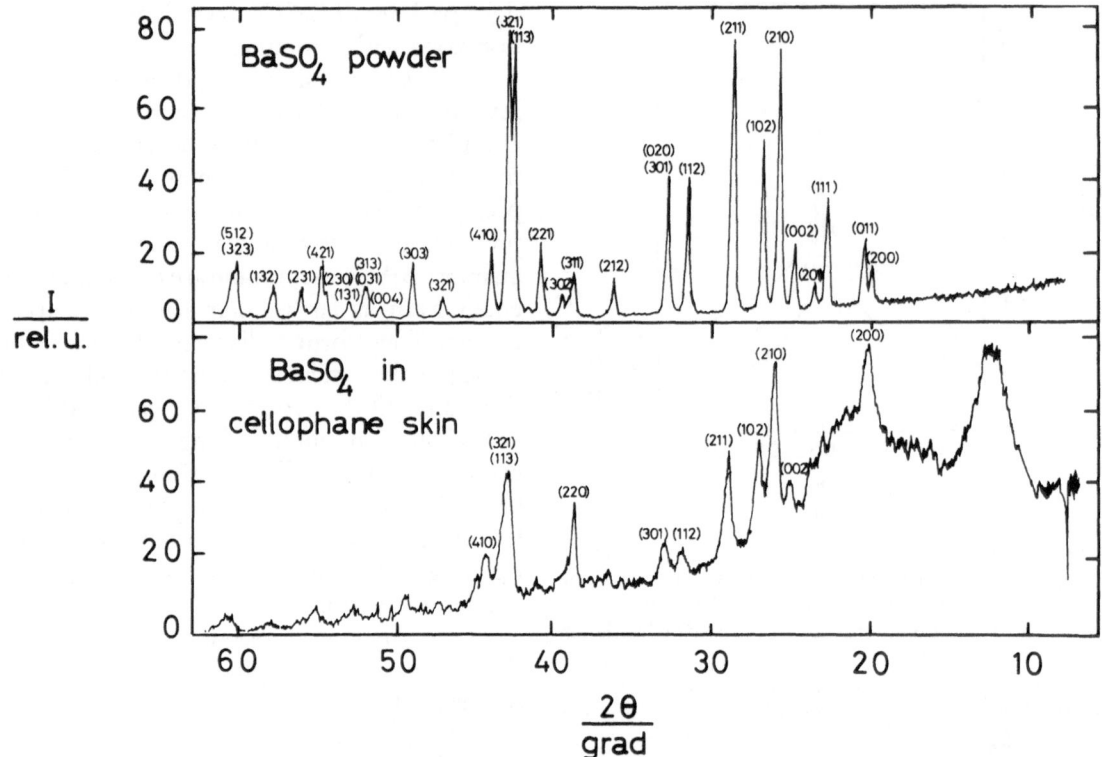

Fig. 6. X-ray diffraction pattern (Guinier method) of $BaSO_4$ powder and a cellophane skin (water content: 61%) containing a $BaSO_4$ precipitation zone with membrane properties. The precipitation zone is formed by interdiffusion of aqueous solutions of $BaCl_2$ and Na_2SO_4 of $c'(BaCl_2) = c''(Na_2SO_4) = 5 \cdot 10^{-2}\,c^+$. This corresponds to a supersaturation S (c/c_{stat}) of $S = 5 \cdot 10^{-3}$. The difference between the two diffraction patterns indicates that the $BaSO_4$ membrane is formed by crystallites with structural imperfections

φ: electrical potential; R: universal gas constant; T: thermodynamic temperature; F: Faraday number; c_i: molar volume concentration of the permeable species i; z_i: valency of ion species i including its sign; ($'$), ($''$): indices of the left and right bulk phase, respectively.

In deriving Equation (1), it is assumed that (a) the flow of the permeable species i across both layers can be described by the Nernst-Planck equation neglecting the contribution of convection to the transport; (b) the flow of the permeable species vanishes in the equilibrium state of the membrane system, after the precipitation zone has acquired membrane properties.

The validity of Equation (1) is checked by the following experiments:

(a) A $BaSO_4$ precipitation membrane is formed by interdiffusion of two equal, concentrated aqueous solutions of $BaCl_2$ and Na_2SO_4 across a cellophane skin:

$$c'(BaCl_2) = 10^{-2}\,c^+/m(BaSO_4)/c''(Na_2SO_4) = 10^{-2}\,c^+.$$

The electrical potential difference, the pH-difference and the concentration difference of sodium ions between the bulk phases are measured as a function of time. The results of the experiments are shown in Figure 7. It can be seen that the $BaSO_4$ precipitation membrane is permeable to 1-valent ions e. g. H^+, Na^+, OH^-, Cl^-. The concentration of the permeable ions in the bulk phase changes with time. In the equilibrium state the relation between the membrane potential and the concentration ratio $(c''_i/c'_i)^{1/z_i}$ of the permeable ions is given by Eq. (1)).

(b) $BaSO_4$ precipitation membranes are formed in the above mentioned way:

$$(c'(BaCl_2) = 5 \cdot 10^{-2}\,c^+/m(BaSO_4)/c''(Na_2SO_4)$$
$$= 5 \cdot 10^{-2}\,c^+;$$

$$c'(Ba(NO_3)_2) = 5 \cdot 10^{-2}\,c^+/m(BaSO_4)/c''(Ba(NO_3)_2)$$
$$= 5 \cdot 10^{-2}\,c^+.$$

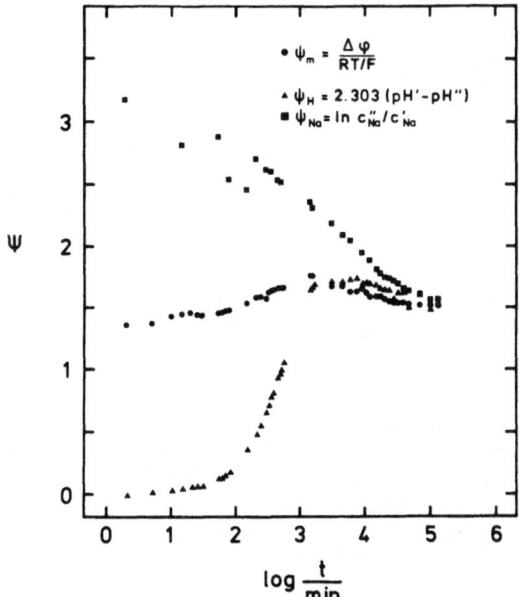

Fig. 7. Establishment of the equilibrium state of the system $c'(BaCl_2) = 10^{-2}\ c^+/m(BaSO_4)/c''(NaSO_4) = 10^{-2}\ c^+$. $t = 0$; pH' = pH'' = 4.00; ($\bullet$) $\psi_m = \Delta\varphi F/RT$; ($\blacktriangle$) $\psi_H = 2.303$ (pH' − pH''); ($\blacksquare$) $\psi_{Na} = \ln(c''_{Na}/c'_{Na})$; $t =$ time

Thereafter, the composition of the bulk phases is changed by adding known amounts of Na_2SO_4 to the left and right bulk phase. In this way defined concentration ratios $c''_{Na}/c'_{Na} = c'_{Cl}/c''_{Cl}$ and $c''_{Na}/c'_{Na} = c'_{NO_3}/c''_{NO_3}$ are produced. These concentration ratios are changed systematically. For each concentration ratio c'_{Cl}/c''_{Cl} and c'_{NO_3}/c''_{NO_3} the membrane potential is measured. The results of the experiments are shown in Figure 8.

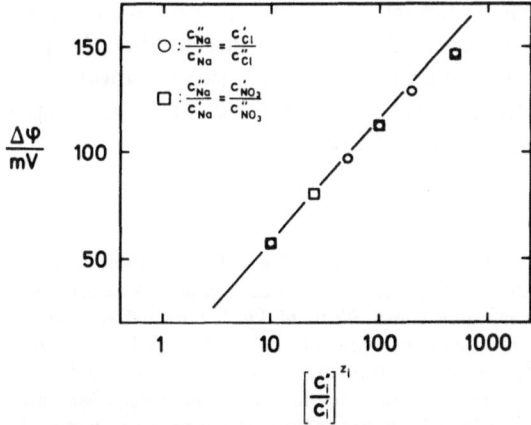

Fig. 8. Membrane potential $\Delta\varphi$ ($\varphi' - \varphi''$) of a $BaSO_4$ precipitation membrane as function of c'_{Cl}/c''_{Cl} ($\square$) and c'_{NO_3}/c''_{NO_3} ($\bigcirc$). The drawn out curve represents Equation (1)

The drawn out curve represents a graphical representation of Equation (1). It can be seen that the experimental data are described by Equation (1) to a good approximation.

Equation (1) can be used to explain the observation that the membrane potential of the system

(a) $Ba(OH)_2(aq)/m(BaSO_4)/H_2SO_4(aq)$ and

(b) $BaCl_2(aq)/m(BaSO_4/Na_2SO_4(aq))$

depends on the concentration $c'(Ba^{++}) = c''(SO_4^{--}) = c$ (see Figs. 2a and b). In system (a) the membrane potential is determined by the concentration difference of H^+ ions (or OH^- ions) between the bulk phases. In this case no additional electrolyte is generated by the precipitation process

$$(Ba(OH)_2(aq) + H_2SO_4(aq) = BaSO_4(f) + H_2O).$$

In system (b) the situation is differnt; the membrane potential is determined by the concentration difference of Cl^- and Na^+ ions between the bulk phases. This concentration difference is not determined by the stoichoimetric concentrations c. At the beginning of the experiment the right bulk phase (phase ('')) does not contain Cl^- ions and the left bulk phase (phase (')) does not contain Na^+ ions. But NaCl is generated with the cellophane skin by the precipitation process

$$(BaCl_2(aq) + Na_2SO_4(aq) = BaSO_4(f) + NaCl(aq)).$$

NaCl diffuses from there into the bulk phases in an more or less uncontrolled way. Even after the precipitation zone has attained membrane properties, NaCl is generated slowly by renewed precipitation of $BaSO_4$ to plug up imperfection within the precipitation zone generated by Ostwald ripening. This is reason for the general observation that the reproducibility of the membrane potential in systems of type (b) is not as good as in systems of type (a).

Is the presence of Ba^{++} ions (in the left compartement of the experimental set up used in this study) necessary to maintain the membrane properties of a $BaSO_4$ precipitation zone or can the Ba^{++} ions be replaced by other 2- or 3-valent cation species? To answer this question the following experiments are carried out.

A $BaSO_4$ membrane is generated in a cellophane skin by interdiffusion of aqueous solutions of $Ba(NO_3)_2$ and Na_2SO_4 ($c''(Ba(NO_3)_2) = c'(Na_2SO_4) =$

$5 \cdot 10^{-2}$ c$^+$). Thereafter, the Ba^{++} ion-containing solution is replaced by an aqueous solution ($c' = 5 \cdot 10^{-2}$ c$^+$) of NaNO$_3$, Mg(NO$_3$)$_2$ and La(NO$_3$)$_3$ respectively, and the membrane potential is measured as a function of time using two calomel electrodes. The results of these experiments are shown in Figure 9. In all cases the membrane potential starts at a value of about $\Delta\varphi = 150$ mV which is typical for an intact BaSO$_4$ membrane in contact with a Ba(NO$_3$)$_2$ and a Na$_2$SO$_4$ solution ($c' = c'' = 5 \cdot 10^{-2}$ c$^+$) and decreases with a time constant τ. τ decreases in the order $\tau_{(La^{+++})} > \tau_{(Mg^{++})} > \tau_{(Na^+)}$. In the system NaNO$_3$(aq)/$m$(BaSO$_4$)/Na$_2SO_4$(aq) the membrane potential reaches quickly values typical for a diffusion potential in that system if the BaSO$_4$ precipitation zone is absent.

This behaviour is interpreted as evidence of a loss of effective overlap of the negative charges of the counterion clouds surrounding the BaSO$_4$ crystals in layer 1 (see Section 1) and the effect of Ostwald ripening. A regeneration of the membrane properties by renewed precipitation of BaSO$_4$ is not possible because the left bulk phase no longer contains Ba^{++} ions. Preliminary results of measurements of the electrophoretic mobility of BaSO$_4$ crystals suspended in Mg(NO$_3$)$_2$ solutions and La(NO$_3$)$_3$ solutions respectively ($c = 10^{-3}$ c$^+$) indicate that the adsorpiton charges have a positive sign. The zeta potential of BaSO$_4$ crystals suspended in a solution of La(NO$_3$)$_3$ has about the same value as in an equal concentrated solution of BaCl$_2$ (about

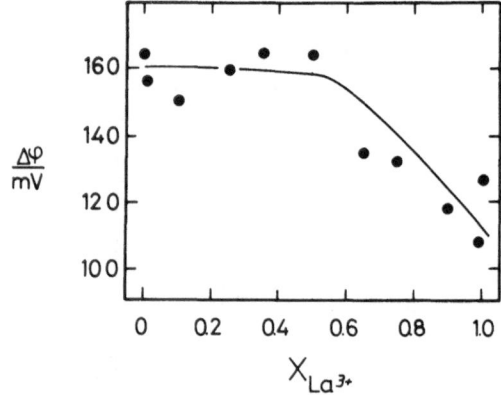

Fig. 10. Membrane potential ($\Delta\varphi = \varphi' - \varphi''$) of a BaSO$_4$ precipitation membrane in the presence of La(NO$_3$)$_3$. Generation of BaSO$_4$ membrane: $c'(Ba(NO_3)_2) = c''(Na_2SO_4 = 5 \cdot 10^{-2}$ c$^+$. The data refer to the system: $c'(La(NO_3)_3) + c'(BaCl_2) = 5 \cdot 10^{-2}$ c$^+$, m(BaSO$_4$/$c''(Na_2SO_4 = 5 \cdot 10^{-2}$ c$^+$. $x(La^{+++}) = c(La(NO_3)_3/(c(La(NO_3)_3 + c(Ba(NO_3)_2)$. At $x(La^{+++}) > 0.4$ the membrane slowly decreases with time (see Fig. 9)

30 mV). The zeta potential of BaSO$_4$ crystals suspended in a solution of Mg(NO$_3$)$_2$ and NaNO$_3$, is smaller (about 10 mV and 20 mV respectively). A preferential adsorpiton of Ba^{++} ions over Na$^+$ ions could be a reason for the unexpectedly high value of the zeta potential of BaSO$_4$ suspended in a solution of NaNO$_3$.

Similar conclusions can be drawn from measurements of membrane potential and measurements of electric current-voltage characteristics in systems in which the left bulk phase contains Ba(NO$_3$)$_2$ and La(NO$_3$)$_3$ in different proportions (see Figs. 10 and 11). Up to a mole fraction of about $x(La^{+++}) < 0.4$ the rectifying properties of a BaSO$_4$ membrane, as well as the membrane potential, is not much influenced by the presence of La(NO$_3$)$_3$. This indicates that the proposed layer structure of a BaSO$_4$ precipitation membrane remains intact in the presence of La(NO$_3$)$_3$ up to a certain mole fraction of $x(La^{+++})$.

Acknowledgements

We thank Prof. S. Haussühl for his willingness to measure and analyze Guinier diffraction patterns of cellophane skins containing BaSO$_4$ precipitation zones. We thank Dr. P. Hirsch-Ayalon for his continued interest in this work. The help of Petra Schwahn and M. Schwarz during stages of this study is gratefully acknowledged. D. Pfennig participated in many helpful discussions. Ohage Mohamed thanks the DAAD (Bonn) for its financial support which made his stay in Köln possible.

The study was supported by the Minister für Wissenschaft und Forschung des Landes Nordrhein-Westfalen.

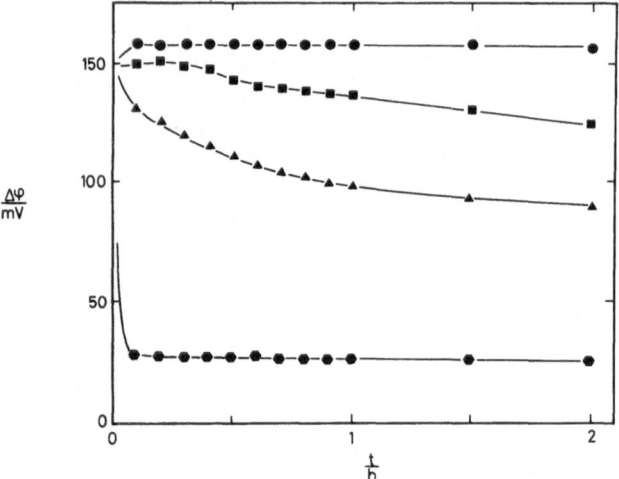

Fig. 9. Membrane potential $\Delta\varphi$ ($\varphi' - \varphi''$) of a BaSO$_4$ precipitation membrane in the presence of Na$^+$, Mg^{++} and La^{+++} ions as a function of time. (●) $c'(Ba(NO_3)_2) = c''(Na_2SO_4) = 5 \cdot 10^{-2}$ c$^+$; (○) $c'(NaNO_3) = c''(Na_2So_4) = 5 \cdot 10^{-2}$ c$^+$; (▲) $c'(Mg(NO_3)_2) = c''(Na_2SO_4 = 5 \cdot 10^{-2}$ c$^+$; (■) $c'(La(NO_3)_3) = c''(Na_2SO_4 = 5 \cdot 10^{-2}$ c$^+$

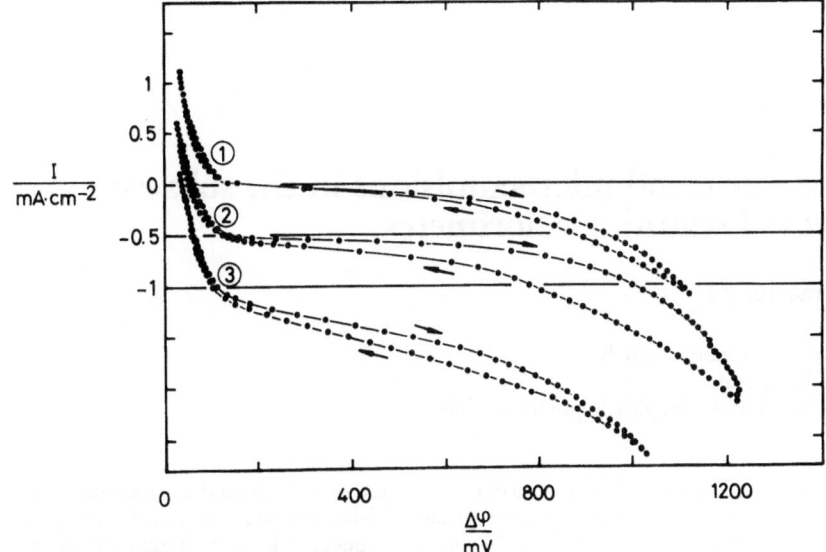

Fig. 11. Current-voltage characteristics of $BaSO_4$ precipitation membrane in the presence of $La(NO_3)_3$ measured under galvanostatic conditions. Curve (1) $c'(Ba(NO_3)_2 = c''(Na_2SO_4 = 5 \cdot 10^{-2} c^+$; curve (2) $c'(Ba(NO_3)_2 + c'(La(NO_3)_3) = 5 \cdot 10^{-2} c^+$; $x(La^{+++} = 0.25$; $c''(Na_2SO_4 = 5 \cdot 10^{-2} c^+$; curve (3) $c'(Ba(NO_3)_2 + c'(La(NO_3)_3) = 5 \cdot 10^{-2} c^+$; $x(La^{+++}) = 0.5$; $c''(Na_2SO_4 = 5 \cdot 10^{-2} c^+$. I = Electric current density; $\Delta\varphi$ = electric potential difference ($\Delta\varphi = \varphi' - \varphi''$). $x(La^{+++})$ = mole fraction of La^{+++} ion left bulk phase, $x(La^{+++}) = c(La(NO_3)_3)/(c(La(NO_3)_3) + c(Ba(NO_3)_2))$. The arrows indicate the direction in which $\Delta\varphi$ is varied. The hysteresis reflects concentration polarisation and pH-changes

References

1. Terazawa T (1954) Jpn J Med Sci, Tokyo 3:139
2. Hirsch-Ayalon P (1956) Chemisch Weekblad 52:557
3. Woermann D (1984) Z Phys Chem Neue Folge 142:117
4. Ayalon A (1984) J Membr Sci 20:93
5. van Oss CJ (1984) In: Matijevic E (ed) Surface and Colloid Science, Vol 13, Plenum Press
6. Ortmann R, Woermann D (1986) Z Phys Chem Neue Folge 148:231
7. Buchanan AS, Heymann E (1949) J Coll Sci 4:137
8. Morimato T (1964) Bull Chem Soc Japan 37:386
9. Hirsch-Ayalon P (1979) J Membr Biol 51:1
10. Ortmann R, Woermann D (1984) React Polym 2:133
11. Helfferich F, Katchalsky A (1970) J Phys Chem 74:308

Received December 24, 1986;
accepted January 23, 1987

Authors' address:

Prof. Dr. Dietrich Woermann
Institut für Physikalische Chemie
Universität Köln
Luxemburger Straße 116
D-5000 Köln, F.R.G.

Progress in Colloid & Polymer Science

Progr Colloid & Polymer Sci 73:66–75 (1987)

A study of the properties of water-in-oil microemulsions in the subzero temperature range by differential scanning calorimetry

D. Senatra[1][2]), Z. Zhou[1][3]), and L. Pieraccini[1][2])

[1]) Department of Physics, University of Florence, 50125 Florence, Italy
[2]) CISM (of the M.P.I.) and GNSM (of the C.N.R.) groups
[3]) Department of Physics, Hubei University, Wuhan, Hubei, People's Republic of China

Abstract: The thermal properties of two w/o microemulsions and one micellar solution were systematically investigated by means of differential scanning calorimetry (DSC). H_2O/hexadecane/K-oleate/hexanol and H_2O/dodecane/K-oleate/hexanol microemulsions with the mass ratios K-Oleate/hexanol = 0.6 and (K-oleate + Hexanol)/oil = 0.4 and a H_2O/AOT/isooctane 3 % micellar solution were studied as a function of increasing water concentration. Both DSC endothermal and DSC exothermal analyses were performed in the temperature interval 123–303 K. The dependence of the thermal properties upon both the particular temperature rate (1–2–4–6–8 K/min), and thermal cycling with repeated freezing/thawing runs, was studied. The freezing behavior of the water dispersed phase was analyzed in detail. Preliminary results are also reported about low temperature storage effects.

The analysis of endothermic processes due to the melting of water has shown that this component may exist into two configurations, namely, free and interphasal distinguished by the melting temperatures of 273 K and 263 K respectively. The exothermic analysis has proved very effective in the identification of the structural evolution of the systems as a function of water addition. The freezing temperature of the overcooled water fraction was found to exhibit a stepwise trend upon water addition, each "step" corresponding to specific and well defined both DSC-ENDO and DSC-EXO characteristics.

Key words: Microemulsions, free water, interphasal water, DSC study, overcooled water.

Introduction

The study of the low-temperature behavior of aqueous solubilized systems such as water-in-oil microemulsions, in which water droplets of diameter less than 100 nm are stably dispersed, offers a challenging research perspective because of its potential implication in the field of the properties of water both in the liquid and in the supercooled state.

The purpose of this work is to present the results of a systematic investigation of the thermal properties of oil external microemulsions by means of Differential Scanning Calorimetry (DSC) in the temperature interval 123–303 K.

The dependence of DSC analysis upon the temperature rate applied during the steady cooling of the liquid microemulsion samples, (exo-thermal DSC) and the continuous heating of previously frozen ones, (endothermal DSC), is examined. In particular the freezing behavior of the dispersed water phase is analyzed together with the effect that thermal cycling may have on the reproducibility of the DSC measurements. Some preliminary results are also reported about the effect that the storage of microemulsion samples at subambient temperature, has on both the freezing and the melting behavior of these highly disperse systems.

These studies are of interest not only from a theoretical point of view but also because of their possible implications in biology and in any field where fluid performance plays an important role.

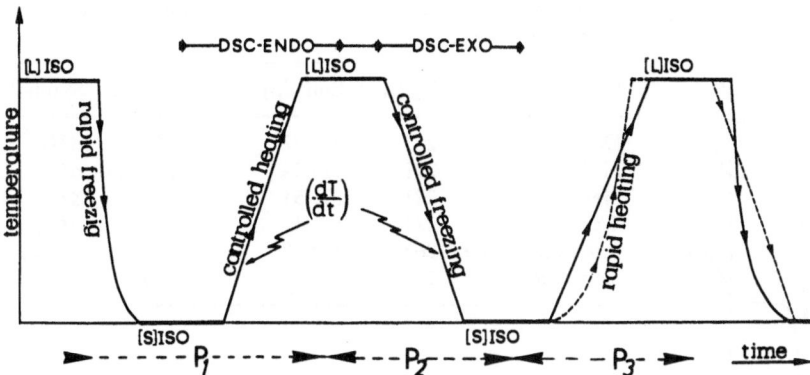

Fig. 1. Thermal cycles followed in both the DSC-ENDO and the DSC-EXO experiments. P_1–P_2–P_3: sequence of the temperature programs applied. See text for further explanation

Experimental

Materials

Water/dodecane and Water/hexadecane systems with potassium oleate and n-hexanol combined in the 3/5 mass ratio were used. In the pseudoternary phase diagram both systems exhibit a large solubility area corresponding to fluid, transparent and isotropic w/o type phases which are disjoined from the o/w ones by a region where highly viscous, turbid and birefringent phases are encountered [1–6]. Water-in-oil microemulsion samples were produced by adding water to a mixture with the fixed mass ratio of surface active agent to oil equal 0.4. The sample water content was expressed by the mass fraction C = water/total sample.

Microemulsion samples with concentration in the intervals $0.0245 \leq C_1 < 0.38$ (w/dodecane) $0.0298 \leq C_2 < 0.36$ (w/hexadecane) stored at room temperature for 1 year did not exhibit any noticeable change upon visual, calorimetric and dielectric observation.

Since the limits of the monophasic w/o domains depend on the temperature, all the samples studied were observed both, by quenching them in liquid nitrogen and, by keeping them in a thermostatic bath. From visual observation no phase separation could be observed. After several cycles of freezing and thawing, once melted, the samples that frozen were marble white, appeared newly isotropic and transparent. The same results did not apply to the samples belonging to the border line of the monophasic domains with $C_1 > 0.38$ and $C_2 > 0.36$.

A solution of inverted micelles was also investigated. The water/AOT (di(2-ethylhexyl) sodium sulfoccinate)/ isooctane system was chosen. Samples were obtained by solubilizing water in the AOT/alkane solution. The AOT concentration was maintained at 3 % [7–9].

All the experiments were performed with a maximum of 1 year and a minimum of 3-days-old samples. In order to avoid confusion, the w/dodecane and the w/hexadecane microemulsions will be referred from here onwards as "system-1" and "system-2".

Methods

The thermal analysis of the microemulsions was carried out with Differential Scanning Calorimetry (DSC) by means of a Mettler TA 3000 equipment with a low temperature DSC-30 measuring cell.

The heat flow rate (dQ/dt) as a function of temperature was recorded both during the controlled heating of solid (S), frozen microemulsions (DSC-ENDO), and, during the controlled freezing of the liquid (L) samples, (DSC-EXO). The main temperature programs adopted in the DSC study are plotted in Figure 1. The starting as well as the final temperatures of the samples were always preceded and/or followed by a 10 min isothermal period (L-ISO, S-ISO). The rapid freezing of the L-samples was obtained by applying a 100 K/min temperature rate.

The sequence of the temperature programs (P_1–P_2–P_3) was interchanged (see dotted lines in Fig. 1). A total of four complete DSC runs was made for each sample tested: two DSC-ENDO and two DSC-EXO. The standard temperature program sequences followed were: P_1–P_2–P_3 and P_2–P_3–P_1, respectively. Five different scan speeds were used, namely: 1–2–4–8 and 10 K/min.

Although small samples of about 2 mg would decrease the difference between the sample and the reference-crucible temperatures, mostly at the melting/crystallization points ($T_{m, fz}$ in Fig. 2), a sample weight varying between 3 and 4 mg was chosen in order not

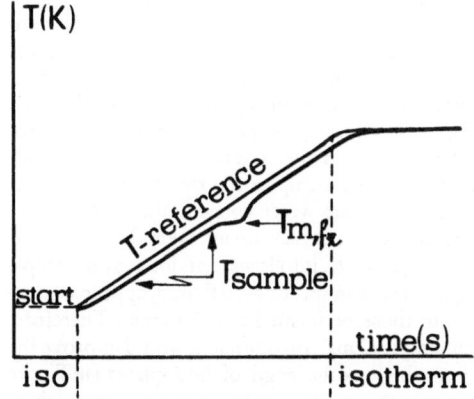

Fig. 2. Temperature difference (ΔT) between the sample and the reference pans in the DSC measuring cell. The larger ΔT occur at the melting and/or freezing points ($T_{m, fz}$). The plateaux at the thermal transitions depend on both the order of the transition involved and the amount (mg) of a given component in the sample. In multicomponent microemulsion systems several plateaux occur, therefore a careful choice of the sample weight for a reasonably good temperature accuracy is of paramount importance.

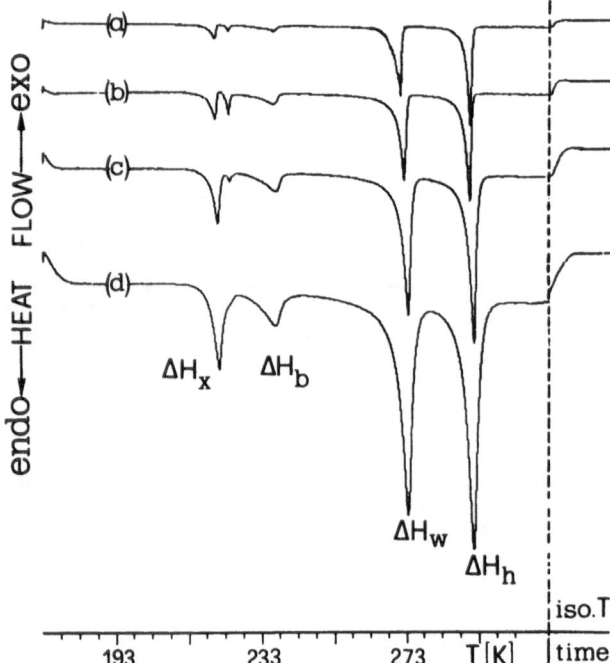

Fig. 3. Temperature rate dependence of the DSC-ENDO measures of a water/hexadecane sample with $C = 0.25$. From curve (a) to curve (d), the T-rates are as follows: 1, 2, 4 and 8 (K/min). The same sample was always measured. The final isothermal part $(Iso-T)$ of the recording is also shown. The smaller the temperature "jump" between the dynamic and the isothermal part of the DSC thermogram, the smaller the difference ΔT between the sample and the reference pan temperatures

Table 1. Reference melting temperatures, enthalpy values and freezing temperatures

Component	T_m (K)	Enthalpy (J/g)	T_{fz} (K)	Symbol
Dodecane	263	216.27	259	ΔH_d
Hexadecane	291	215.34	287	ΔH_h
Hexanol	225	150.54	209	ΔH_x
Water	273	333.42	252	ΔH_w
Interphasal water	263	312.38	—	$(\Delta H_w)_{263}$
Isooctane	166	89.73	145	ΔH_i

Purified and dried nitrogen gas with a flow of 30 ml/min was used to purge the DSC-30 cell during the measurements.

The reference values adopted throughout this paper and given in Table 1, were obtained by measuring both the melting (T_m) and the crystallization (T_{fz}) temperatures as welll as the enthalpy values (ΔH) of the components used·to formulate the three systems.

The effect that the storage of microemulsion samples at subambient temperatures in the range 278–233 K, may have on the freezing behavior of water was studied by keeping the samples for one hour at three different temperatures, namely: 278–263 and 233 K. After the storage under strictly controlled isothermal conditions, measurements were performed as follows: a DSC-EXO run from the storage temperature down to the normal final temperature; a DSC-ENDO run and another DSC-EXO run as a further control. Both measurements were repeated 4 days later, on the same samples maintained at room temperature. The latter research is still in progress.

to miss small thermal contributions of the order of 0.5 mW or less (Fig. 3) and to optimize the evaluation of the enthalpy changes (ΔH) associated with the recorded thermal events. At the present stage of the research, the medium sensitivity (20 $\mu V/K$) of the standard DSC sensor of the measuring cell was used to have a complete picture of the microemulsion thermal behavior; this is mainly because the amounts of heat associated with the thermal processes observed in these multi-component systems, are spread over a broad range of values that depend on both the nature and the amount of any component in the sample as well as, upon the structural transitions linked with the increase of the water concentration. Moreover, since the calorimeter at the selected sensitivity, takes the fixed number of 2400 points per degree K, it follows that, for a given sample the DSC recordings, at the scan speeds 4–10 K/min, appear as magnified with respect to those performed at 1–2 K/min. Therefore, while using a small sample (2 mg), on one hand, would improve the temperature accuracy, because the length of the T-plateau in Figure 2 depends on the mass of the component besides on the type of thermal process involved in the given transition; on the other hand it would decrease the accuracy in the evaluation of the enthalpy data when very small amounts of heat are either absorbed or released.

The sample weight of 3–4 mg was found experimentally to represent a reasonable compromise between the requirements of temperature accuracy, enthalpy accuracy and a good thermal peak resolution at all the different temperature rates employed in both the DSC-ENDO and the DSC-EXO analysis.

Results and discussion

DSC-ENDO

In this paper we will confine ourselves to the analysis and discussion of the results obtained mainly with the DSC study of the w/hexadecane microemulsion because, as reported elsewere [10–12], in the latter system the melting of the oil at 291 K was found not to interfere with any of the so far identified thermal events due to water. However, for sake of completeness, some information concerning the w/dodecane microemulsion and the water/AOT/isooctane system will be also given together with relevant reference indication.

As shown in Figure 4 where the DSC thermograms of system-1 and system-2 are plotted, in both microemulsions, independently of the particular hydrocarbon used as dispersing medium, it is possible to distinguish between samples that possess a "free water fraction", melting at 273 K, and those that do not (see Fig. 4 of Réf. [13]). However, as depicted in Figure 5, in the water/hexadecane system, in the concentration range 0.105–0.13, also an endothermic contribution asso-

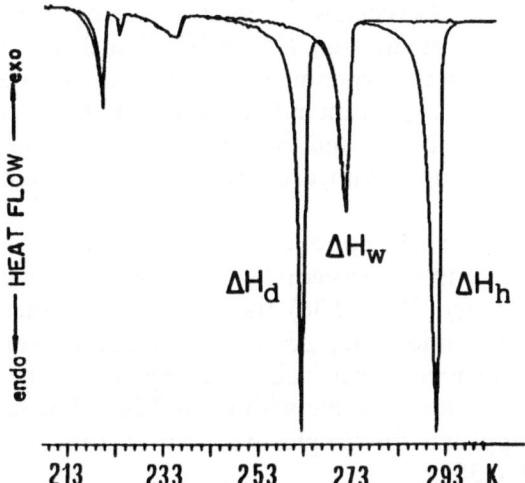

Fig. 4. Comparison between the DSC-ENDO recordings of a W/dodecane and a W/hexadecane microemulsion with the same water content. Both samples exhibit an endothermic contribution due to water melting at 273 K (Free water fraction). The two DSC curves differ only by the thermal peak associated with the melting of the hydrocarbon. dT/dt = 4 K/min; C_1 = 0.195; C_2 = 0.197

ciated with water melting at 263 K, "interphasal water", can be directly observed. The latter process, in the water/dodecane microemulsion, is sheltered by the melting of the oil (see Table 1) [10].

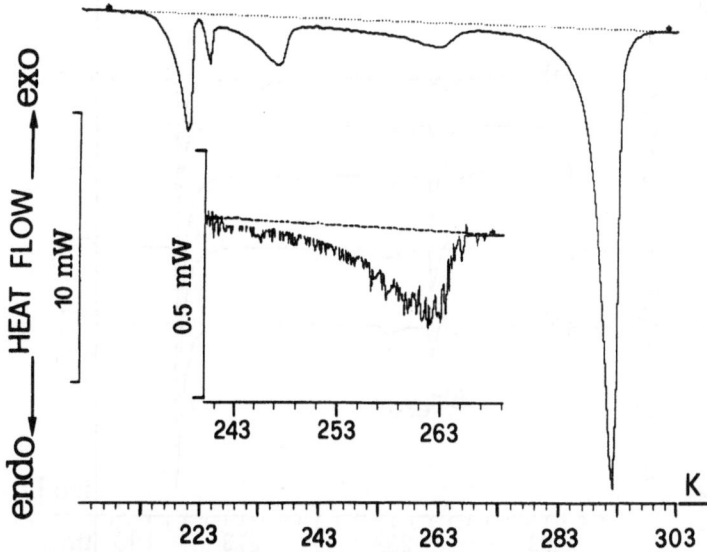

Fig. 5. "Interphasal water" melting at 263 K in a water/hexadecane microemulsion sample with C = 0.122. Complete DSC curve: dT/dt = 8 K/min. Enlargement: interphasal water peak at 2 K/min. The latter peak is not observable as an isolated contribution in samples of system-1

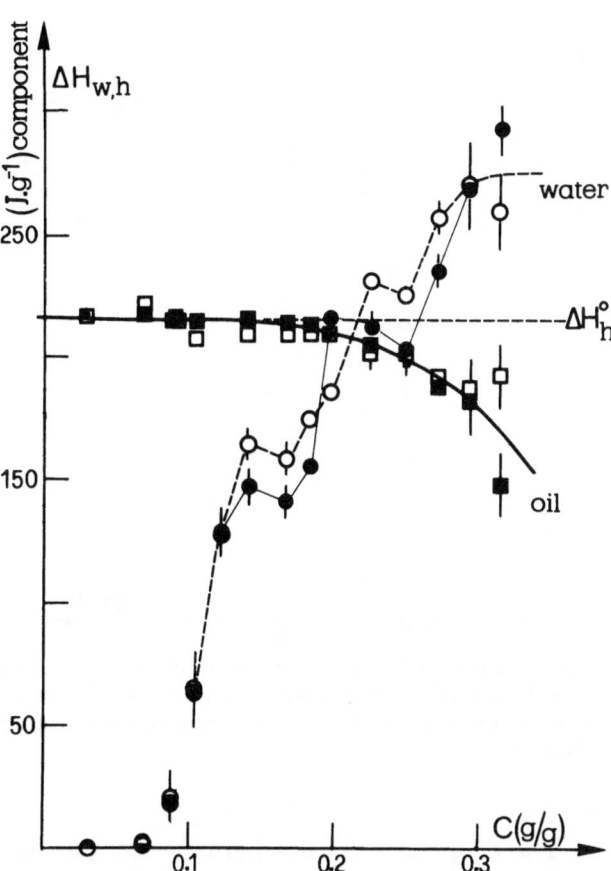

Fig. 6. Enthalpic changes associated with the melting of water (ΔH_w: ○-●) and hexadecane (ΔH_h: □-■) against increasing concentration. ΔH expressed in J/g of the given component in the sample. Open symbols: first experimental run; solid symbols: second run. ΔH_h° measured enthalpy of an hexadecane bulk sample. Singularities can be observed around: C = 0.122; C = 0.198; C = 0.273; C = 0.36. See also Figure 11 for comparison

The behavior of the enthalpy changes associated with the melting of, respectively, the water (ΔH_w) and the hexadecane (ΔH_h), is plotted against increasing concentration in Figure 6. The $\Delta H_{w,h}$ data are expressed in J/g of the given component in the sample. The enthalpy values reported in Figure 6 regard two different and independent experimental runs. In the first as well as in the second set of measurements, the DSC analysis was performed at the temperature rates of 2–4–8 (K/min). The thermal analysis was made both by studying the same sample at the different T-rates and, by using different samples with the same concentration, at any scan speed employed. For the corresponding diagram of the water/dodecane microemulsion (see Fig. 7 of Ref. [11]).

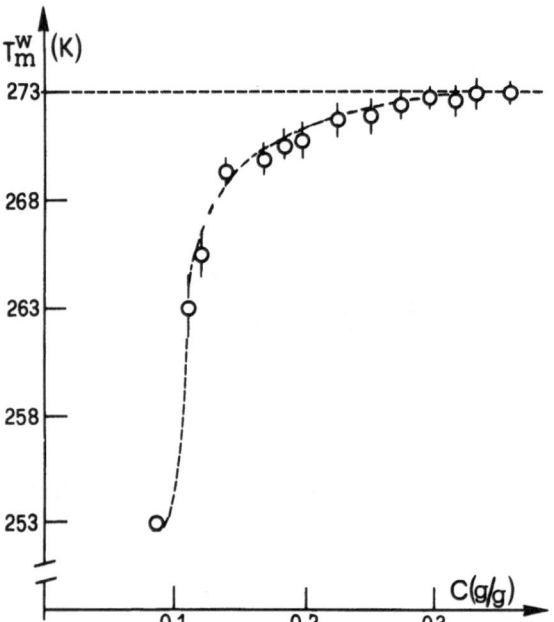

Fig. 7. Behavior of the melting temperature of water vs. concentration; the points are mean values calculated on the melting temperatures measured at the different T-rates. Bars are standard errors of the mean

The temperature of the thermal event due to the melting of the water measured at the aforementioned T-rates, is reported in Figure 7, as a function of increasing concentration. It should be noted that the latter values are "melting" temperatures, therefore they do not exactly correspond to the DSC recorder peak temperatures, because the Mettler TA 3000 equipment is, basically, conceived for differential thermal analysis (DTA).

The trends found for both T_m and ΔH_w against concentration for system-2, are in good accordance with the ones reported by other authors, for the same quantities in quite different disperse systems as emulsions [14], biologic tissues [15], lyotropic mesophases [16] and water adsorbed on porous materials and clays [17].

Temperature-rate dependence study

The results of a careful analysis of the dependence of the melting behavior of the samples upon the particular temperature rate applied are as follows:

1. General features of the three systems studied:
i) The overall behavior of the DSC-ENDO processes does not depend on the particular T-rate employed to freeze the liquid samples

ii) Thermal cycling with repeated freezing/thawing runs, does not affect the DSC-ENDO analysis

iii) Within the experimental accuracy of our measurements no significant differences could be observed between the 2-day- and the 1-year-old stable microemulsion samples with, of course, the same concentration.

2. Specific behavior of the recorded thermal events
i) For samples of system-2 with water concentration in the range (0.24–0.36) the different endotherms recorded, namely $\Delta H_x, \Delta H_b, \Delta H_w$ and ΔH_h, are quite independent on the particular T-rate applied, as shown in Figure 3 for a sample with $C_2 = 0.249$. The same result applies to the samples of system-1 in the C-range (0.222–0.38).

ii) For samples with concentration in the interval 0.0298–0.24 for system-2 and 0.024–0.2 for system-1, however, the peculiar thermal event depicted in Figure 8 and labeled (ΔH_{exo}), can be observed in the low temperature part of the DSC recordings where, the onset of the very first melting process is abruptly followed by the occurrence of a large exotherm. The (ΔH_{exo}) process, at T-rates higher than 4 K/min., looses its characteristic shape and appears as a well defined peak followed by a very small one. The latter is nothing else than the end of the (ΔH_{exo}) contribution whose com-

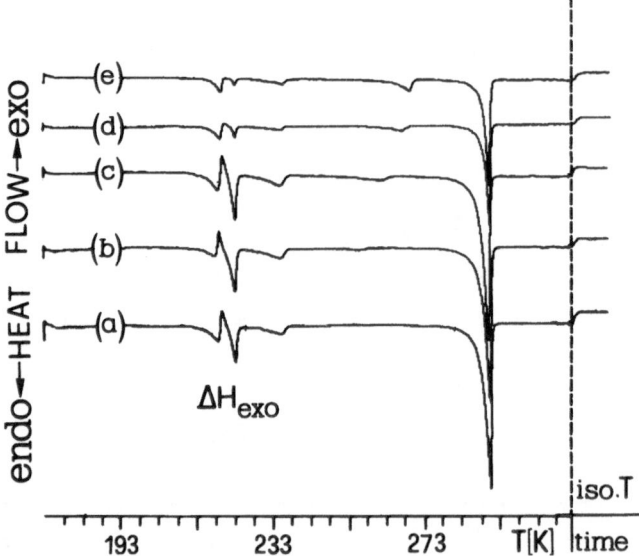

Fig. 8. DSC-ENDO recording of samples with increasing water content. T-rate 2 K/min. The low temperature exotherm ΔH_{exo} that abruptly interrupts the very first melting process of the samples is quite evident in curves (a), (b) and (c). Concentrations: $C_a = 0.069$; C_b 0.087; $C_c = 0.105$; $C_d = 0.138$; $C_e = 0.169$

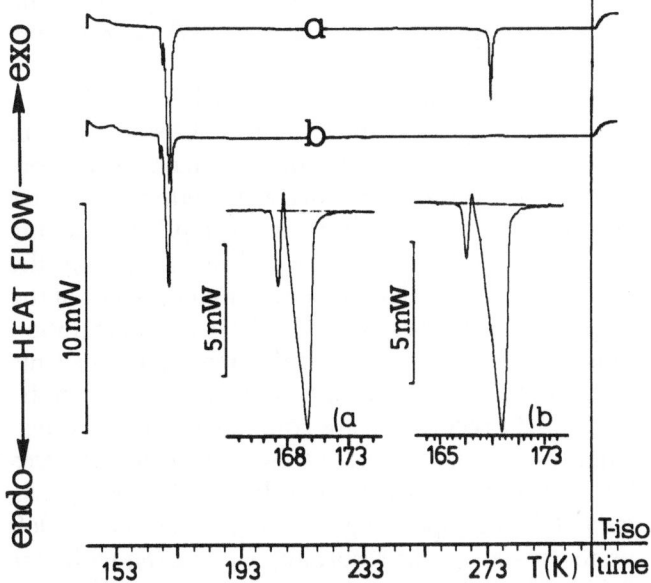

Fig. 9. DSC-endotherms of the water/AOT/isooctane system at 4 K/min. For molar fraction (H₂O/AOT) > 10 the contribution due to water melting at 273 K becomes evident, curve (a) [18]. For molar fraction < 10 no free water could be detected, curve (b). Note that the low temperature exotherm occurs also in this system at the beginning of the very first melting process, immediately before the thermal peak due to the fusion of the component with the lowest melting temperature (isooctane). The ΔH_{exo} processes of curves (a) and (b) are also shown in detail in the enlargement

plete development is "cut-off" because of the higher scan speed imposed. The low temperature exotherm follows a decreasing trend upon water addition and, eventually is no longer present in the upper isotropic and transparent microemulsion phase of the biphasic system that one obtains at $C > 0.36$, when the border of the monophasic domain is crossed. In the latter case the lower phase is an o/w type of lyotropic mesophase.

iii) The results of the DSC study of the water/AOT/isooctane system, are in good agreement with those reported by Boned et al. [18] mostly for what concerns the melting of the water component. However, as shown in Figure 9, the low temperature exotherm occurs also in this system at the beginning of the very first melting process immediately before the endotherm due to the fusion of the isooctane at 166 K.

iv) A specific investigation was performed with the aim of checking whether the ΔH_{exo} process were dependent on the particular cooling rate adopted to freeze the samples.

Several runs were made by applying different cooling rates from 100 K/min down to 0.5 K/min. The samples were thereafter reheated with firstly, the same scan speed used during the freezing part of the measurement and secondly, with much lower T-rates.

The DSC crucibles with the microemulsion samples were directly quenched in liquid nitrogen and, immediately after put into the DSC-30 measuring cell preset at 87 K. After an isothermal period of ≈ 20 min. at the above temperature, the DSC-ENDO data were taken by applying the heating rates of 8-4-2 and 0.5 (K/min). The low temperature of the DSC-30 cell was reached and maintained during the isothermal period with the help of a flux of nitrogen gas cooled by allowing the latter to flow through a copper coil immersed into a dewar filled with liquid nitrogen.

The low temperature exotherm was found in both w/o microemulsions and the water/AOT/isooctane system, not to depend on the freezing or the heating rates applied. Such a rather surprising finding seems to exclude the ΔH_{exo} process as interpretable in terms of an overcooled state of that component in the systems with the lower melting temperature (n-Hexanol for systems-1,2; isooctane for system-3). In fact, in this case we would have expected the recrystallization-like process to disappear or, at least to significantly decrease upon steady cooling with lower and lower temperature rates. In conclusion, the observed phenomenon is temperature rate independent, does not depend on the particular hydrocarbon used as dispersing medium or on the presence in the system of a cosurfactant. The experimental evidence here reported supports the following tentative explanation: a kinetically unstable phase develops upon steady cooling of the samples. Such a metastable phase, upon reheating, relaxes by crystallizing (large exotherm), as soon as the temperature has become high enough to allow the interphase tenants to acquire the necessary mobility to rearrange themselves into an equilibrium state. Structural studies are actually in progress to get a better insight into this type of thermal transition.

DSC-EXO

The results of the DSC-EXO study of the thermal behavior of microemulsion samples as a function of decreasing temperature, can be summarized as follows:

1. General features

i) Because of the liquid nature of the system as well as of their components, overcooling occurs upon freezing in all the samples. As shown for the water/

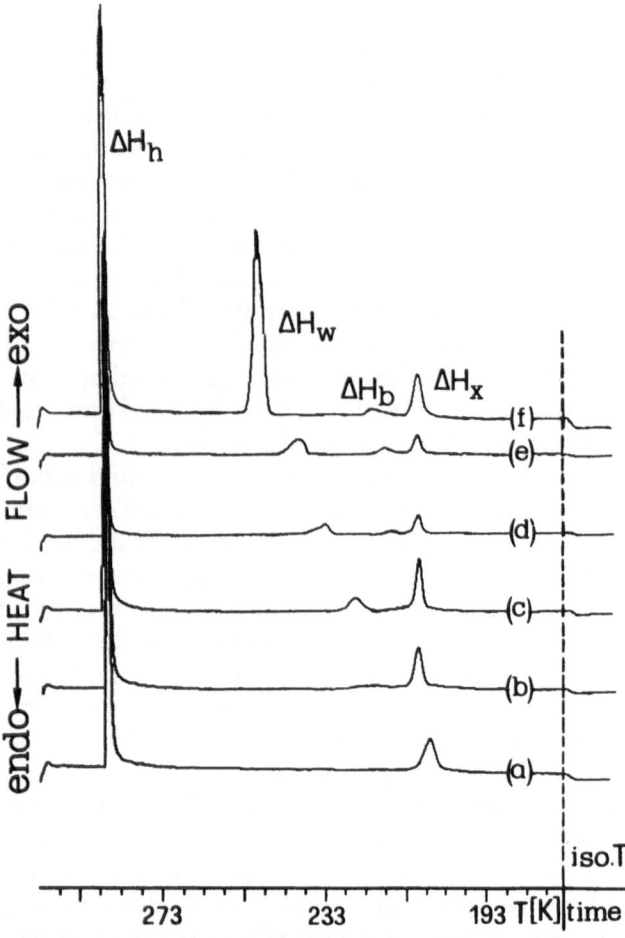

Fig. 10. DSC-EXO recordings of the W/hexadecane microemulsion samples with increasing water content. Concentrations are as in Figure 8. $C_f = 0.25$; $dT/dt = 2$ K/min. Both dynamic and isothermal parts of the DSC recordings are plotted. The onset of the thermal processes associated with the water component is quite well observable. For details see text

hexadecane system in Figure 10, curve (*f*), the different exotherms ΔH_h, ΔH_w, ΔH_b and ΔH_x, are all shifted towards lower temperatures even at the 2 K/min rate applied.

ii) The DSC-EXO recordings, under continuous cooling exhibit a temperature rate dependence and this, mostly for what concerns the exotherm due to the water component.

iii) The temperature cycling between 303 K and 123 K, with repeated freezing/thawing runs, does not affect the DSC-EXO results.

iv) As far as the DSC analysis is concerned, both microemulsions follow a quite well reproducible trend

upon decreasing and/or increasing temperatures. The same result applies to the water/AOT/Isooctane system.

2. Specific behavior of the thermal events

i) As in the case of the DSC-ENDO study, also in the freezing experiments measurements were performed by substituting normal water with heavy water [10–11]. The DSC analysis of the D_2O/hexadecane system has revealed very helpful in recognizing the thermal processes due to the water component of the microemulsion, mostly within the low concentration range where the onset of the different thermal events parallels the addition of water to the system. In fact the DSC-EXO recordings are complex spectra (see Fig. 10), in which the number of thermal events increases by increasing concentration from, the two-peaks curve (a), corresponding to a sample with $C = 0.068$ up to the four-peaks recording of curve (*f*), with $C = 0.249$. The latter, even if because of overcooling, the peak temperatures are shifted towards lower values, is practically the mirror image of the DSC-ENDO thermograms, as far as both the number and the shape of the thermal events is concerned. (see also Fig. 3).

ii) The DSC-EXO study of the water/dodecane system has shown that, the freezing of the dodecane at 259 K pilots in some way the exothermic "answer" of the microemulsion mainly for what concerns the freezing of overcooled water [19]. Thus the DSC-EXO curves of the water/dodecane system, exhibit an "apparent" reversible trend mainly at scan speeds > 4 K/min as reported in previous papers [11–13].

iii) The DSC-EXO study seems to provide some information about the structural modifications occurring in the system as a function of water addition. Such information is better evidenced by the plot of the freezing temperature (T_{fz}^w) of the ΔH_w process against concentration, given in Figure 11. The data were taken by applying a rate of 2 K/min. The application of higher scan speeds results in an enhancement of the stepwise trend of T_{fz}^w but, does not add any amount of information or change the pattern of the DSC recordings. Thus, as far as the type of exothermal DSC spectra found at any concentration tested, it emerges that the DSC-EXO measurements are indeed "temperature rate independent" in the sense that, more than upon the particular T-rate, the freezing behavior of the microemulsion appears rather to depend on the water content of the particular sample studied.

On the basis of the data collected for system-2 and represented in Figures 6–8 and 10–11, the concentration interval investigated, ($0.0298 \leq C < 0.4$) may be

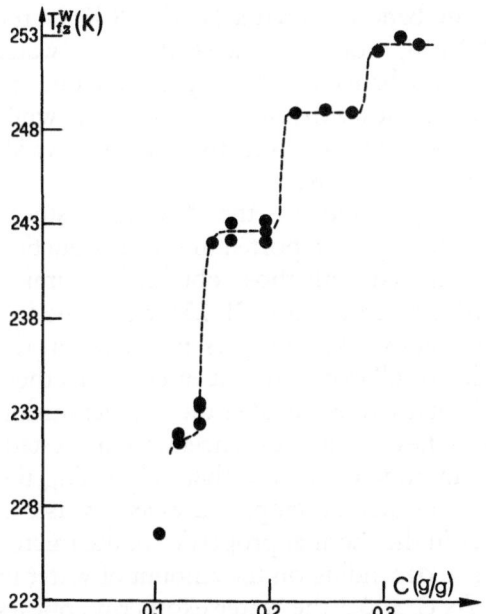

Fig. 11. Water freezing temperature against increasing concentration. Values taken from the DSC recordings performed at 2 K/min. The temperature steps fall within the C-subintervals where the discontinuities in the enthalpy data plotted in Figure 7 were found to occur. The first experimental point reported corresponds to the very first exotherm observed by adding water to the samples. (see curve (b) in Fig. 10)

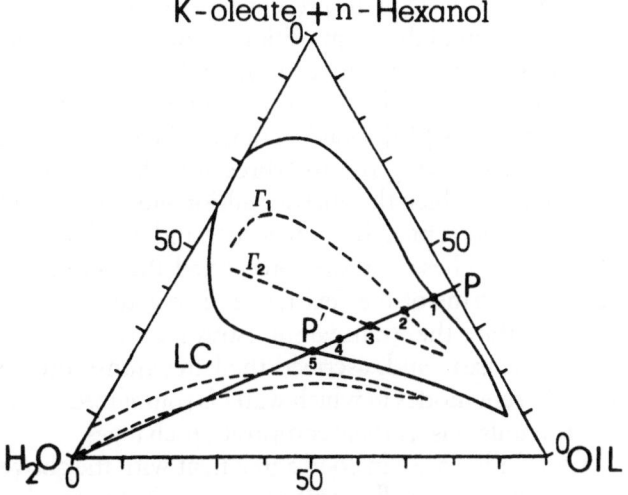

Fig. 12. Pseudoternary phase diagram of the water/hexadecane system and experimental path followed during the present study (PP'). Dots on the PP' line correspond to the composition of the samples at the intersection points between the PP' line and the w/o microemulsion monophasic domain (dots 1 and 5) as well as between the PP' line and Clausse and coworkers [20] Γ_1 and Γ_2 curves. See text for further explanation

devided in subintervals ΔC_i, each characterized by both specific DSC thermograms and typical thermal events for what concerns the water component.

The main distinguishing features of the subintervals are summarized in Table 2.

The above ΔC_i are represented in the pseudoternary phase diagram of the system, plotted in Figure 12, by the numbered dots [1–5] located on the line PP' defined by the given and fixed surface active agent to

Table 2. Thermal characteristics of w/o microemulsions in the different concentration subintervals

	DSC-ENDO Study				DSC-EXO Study		
ΔC	Free H_2O	Interphasal	Hexadecane	Enthalpy (J/g) from Figure 6	$(\Delta H_w)_{fz}$	T_{fz}^w	Type of structure
$0.03 < C < 0.105$	–		$\Delta H_h = \Delta H_h^\circ$	–		–	From hydrated soap aggregates to the
$0.105 < C < 0.122$	–	$(\Delta H_w)_{263}$	$\Delta H_h = \Delta H_h^\circ$	small ΔH_w contribution	1st ΔH_w exotherm	234	onset of w/o microemulsions
$0.122 < C < 0.198$	$\neq 0$		$\Delta H_h = \Delta H_h^\circ$	1st plateau	$\Delta H_w \neq 0$	242	From monodisperse w/o droplets to the
$0.198 < C < 0.273$	$\neq 0$		$\Delta H_h \simeq \Delta H_h^\circ$	2nd plateau	$\Delta H_w \neq 0$	249	appearance of larger droplets:prolate elipsoids
$0.273 < C < 0.355$	$\neq 0$		$\Delta H_h << \Delta H_h^\circ$	3rd plateau	$\Delta H_w \neq 0$	252	(onset bicontinuous structure?)

oil mass ratio used in the present experiment. Each dot on the line gives the composition of the samples at the intersection points between the PP' line and the dashed curves Γ_1 (dot 2) and Γ_2 (dot 3), that Clausse and coworkers [20] found to devide the monophasic w/o domain giving rise to "three well defined subareas" within which the microemulsion samples exhibit different and typical both dielectric and conductivity properties. These authors interpreted the above behavior in terms of the "existence of structural transitions within the transparent isotropic water-in-oil solubility area" and, ascribed the latter transitions to the different modes in which water is solubilized in the microemulsions as their composition changes.

Our findings are in good agreement with the above interpretation as well as with recent results reported in literature on the same system, by small angle neutron scattering techniques [5].

Storage effect

Microemulsion samples with concentration in the intervals $0.12 < C < 0.22$ and $0.22 < C < 0.35$ were investigated. Samples with water content in the first C-interval, stored for 1 h at 278 K and 263 K and, thereafter frozen by steady cooling, did not show up in the DSC-EXO recordings the exothermal peak due to the exadecane. All the other thermal events were found to occur unchanged in shape as well as unshifted in temperature. The second experiment, by remelting the whole samples and repeating the normal DSC sequence, confirmed that the microemulsions had not undergone any structural modification. The same results apply also to the samples stored at 233 K. In this case, the DSC-EXO recordings showed only the thermal processes occuring at temperatures below 233 K. Measurements performed after the complete melting of the 233 K-samples, gave both DSC-EXO and DSC-ENDO unmodified thermograms, proving that the storage and the thermal cycling had not affected the microemulsion integrity and structure.

Samples with concentration in the second interval, exhibited, upon storage at 278 K, 263 K and 233 K a DSC behavior alike the one observed in those with lower concentration. However, in the samples stored at 233 K, during the second measuring run, different results were obtained for what concerns the water endotherm as well as the water exotherm. In fact, the ΔH_w process in the DSC-EXO experiment, was divided into different thermal contributions with a sharp peak at a higher temperature (around 253 K), followed

by a composite band centered around 248 K. In the DSC-ENDO run, moreover, the shape of the water peak appeared unchanged with respect to the one exhibited by the samples with lower water content, while the peak temperature was slightly shifted towards higher values (273–274) K.

The results gathered in the "storage" study, although preliminary, as reported in the introduction section, if compared with those obtained in similar studies in ordinary emulsions [21–23], allow the following conclusions. At the present stage of the research which is still in progress, it emerges that thermal cycling by itself, does not alter the microemulsion thermal properties at all the concentrations tested. However, if in addition to the thermal cycling the samples are also stored at temperatures as low as 233 K, differences in the thermal properties of the microemulsions arise depending on the amount of water in the system. As $C > 0.3$ the water exotherm appears devided into small freezing peaks. The latter result may reflect a change of the droplets to a larger size, as confirmed by the shift to higher temperatures of the DSC-ENDO water peak with respect to the temperature at which the ΔH_w process occurred before storing the samples at 233 K.

Conclusions

The structural evolution of microemulsion systems upon water addition was probed by means of differential scanning calorimetry, (DSC).

The analysis of the endothermic processes associated with the melting of water has shown that the latter component may exist into two configurations namely, free and interphasal, distinguished by the melting temperatures of respectively 273 K and 263 K.

The temperature-rate dependence study has evidenced that the DSC-ENDO properties do not depend on the particular temperature rate applied during both the cooling and the heating part of the measure.

The study of the overcooling processes associated with the freezing of the liquid samples as well as components has evidenced, once more, the leading role of water in governing the behavior and stability of microemulsion systems. The DSC exothermal behavior of the samples, although temperature-rate dependent, was found mainly to reflect the system's structural configuration. In fact, depending on the water content, the concentration interval investigated results divided into adjacent subintervals, each being characterized by specific and well-defined thermal properties for what concerns the number of thermal events recorded and the water freezing temperatures.

A relation emerges between the droplet size distribution and the overcooling "ability" of the different water domains.

References

1. Bellocq AM, Fourche G (1980) J Coll Interf Sci 78:275
2. Bellocq AM, Fourche G, Chabrat P, Letamendia L, Rouch J, Vaucamps C (1980) Optica Acta 27:1629
3. Shah DO, Hamlin RM (1971) Science 171:483
4. Boned C, Clausse M, Lagourette B, Peyrelasse J, McClean VER, Sheffard RJ (1980) J Phys Chem 84:1520
5. Caponetti E, Magid LJ, Hayter JB, Johnson Jr JS (1986) Langmuir 2:722
6. Senatra D, Gambi CMC, Neri AP (1981) J Coll Interf Sci 79:433
7. Wong M, Thomas JK, Grätzel M (1976) J Am Chem Soc 98, 9:2391
8. Zulauf M, Eiche HF (1979) J Phys Chem 83:480
9. Kotlarchyk M, Huang JS, Sow-Hsin Chen (1985) J Phys Chem 89:4382
10. Senatra D, Gabrielli G, Guarini GGT (1986) Europhys Lett 2:455
11. Senatra D, Guarini GGT, Gabrielli G, Zoppi M (1984) J Phys, Paris 45:1159
12. Senatra D, Guarini GGT, Gabrielli G, Zoppi M (1985) In: Shah DO (ed) Macro-Microemulsions. Theory and Applications, ACS Symposium Series, Book 272, p 133
13. Senatra D, Gabrielli G, Guarini GGT (1987) In: Martellucci S, Chester AN (eds) Progress in Microemulsions, Plenum, New York, London, in press
14. Francks F (ed) (1975) Water a Comprehensive Treatise, Vol 5 and (1979) In: Water A comprehensive treatise, Vol 7
15. Mazur P (1965) Ann New York Acad Sci 125:658
16. Le Moigne J, Pouyet G, Francois J (1980) J Coll Polym Sci 258:1383
17. Banin A, Anderson PM (1975) Nature 255:261
18. Boned C, Peyrelasse J, Moha-Quchane M (1986) J Phys Chem 90:634
19. Senatra D, Gabrielli G, Guarini GGT (1986) Annali di Chimica, Italian Chemical Soc, in press
20. Clausse M, Boned C, Peyrelasse J, Lagourette B, McClean VER, Sheppard RJ (1981) In: Shah DO (ed) Surface Phenomena in Enhanced Oil Recovery, Plenum Pree New York, London 199
21. Morgan JoL, Drake NE, Mraw SC (1985) Proc A.F.C.A.T. and A.I.C.A.T. Journées de Calorimetrie at Analyse Thermiquie, 53
22. Rassmussen DA, Mac Kenzie AP (1972) In: Jellinek HHG (ed) Water Structure at the Water-Polymer Interface
23. Broto F, Clausse D (1976) J Phys C Solid State Phys 9:4251

Received February 19, 1987;
accepted March 2, 1987

Authors' address:

Prof. D. Senatra
Universita di Firenze
Dipartimento di Fisica
Largo Enrico Fermi, 2 (Arcetri)
I-50125 Firenze, Italy

Progress in Colloid & Polymer Science   Progr Colloid & Polymer Sci 73:76–80 (1987)

Electric birefringence of nonionic micellar solutions near the cloud point

V. Degiorgio and R. Piazza

Dipartimento di Elettronica-Sezione die Fisica Applicata, Università di Pavia, Pavia, Italy

Abstract: The Kerr coefficient B of aqueous solutions of the nonionic amphiphiles C_6E_3, C_8E_4, $C_{10}E_5$ and $C_{12}E_6$ is measured as a function of the temperature T in a temperature region close to the cloud point, and at the concentration corresponding to the minimum of the cloud curve. In all the investigated systems B is found to increase as T approaches the critical temperature T_c following a power-law behavior, $B \approx (T_c - T)^{-\psi}$, with an exponent $\psi \approx 0.9$. In the case of the system $C_{12}E_6$–H_2O which was studied in a wide temperature range, the data are consistent with a moderate micellar growth as a function of T.

Key words: Electric birefringence, micelles, critical phenomena.

Introduction

The measurement of the birefringence induced by an electric field on a suspension of particles in Brownian motion represents a very useful technique for studying the geometric anisotropy of the particles. Several investigations concerning the application of electric birefringence to inorganic colloids, polymers, and biological macromolecules have been reported in the literature [1], but only recently some studies referring to micellar solutions have appeared [2–7]. The experiments of References [3] and [4] have been performed with ionic amphiphiles forming rod-shaped micelles with a length of many hundred Angströms. These solutions present rather large Kerr coefficients. References [5–7] deal with solutions of nonionic amphiphiles from the family $C_iH_{2i+1}(OCH_2CH_2)_i$-OH, hereafter called C_iE_j. The system C_iE_j–H_2O is known to present a lower consolute curve (called cloud curve in the literature) with a minimum at a critical temperature T_c and a critical concentration c_c. The value of T_c depends on the hydrophilic-lyophilic balance of the amphiphile. The value of c_c is usually low (below 5 %) except for short-chain amphiphiles. Hoffmann et al. [5] have studied solutions of C_8E_4, $C_{10}E_4$ and $C_{12}E_4$ at low concentration below T_c. They found no electric birefringence for C_8E_4 and $C_{10}E_4$, and some

for $C_{12}E_4$ solutions. Neeson et al. [6] have studied $C_{12}E_6$ and $C_{12}E_8$ solutions at high concentration near the isotropic-hexagonal phase boundary. They attributed the observed electric birefringence to the existence of nonspherical micelles, which grow in size, near the boundary of the liquid crystalline phase. In the experiment of Reference [7] the Kerr coefficient of $C_{12}E_6$ and $C_{12}E_8$ solutions at the critical concentration was measured as a function of the temperature T as T approaches T_c. It was found for both systems that B grows considerably as the temperature distance $T_c - T$ is reduced. The experimental data are consistent with a power-law behavior of the type $B \approx (T_c - T)^{-\psi}$, where $\psi \approx 0.6$. We suggested in Reference [7] that the effects observed near T_c are connected with the existence of critical concentration fluctuations and do not imply the formation of very elongated micelles.

We present in this paper new and more detailed electric birefringence measurements performed on solutions of the nonionic amphiphiles C_6E_3, C_8E_4, $C_{10}E_5$ and $C_{12}E_6$. All systems have been studied as a function of the temperature at fixed concentration. We find indeed that B shows a power-law divergence in all systems, in agreement with the behavior expected for a critical binary mixture. The measured exponent ψ is however larger than predicted by existing theories. In the case of the system $C_{12}E_6$–H_2O which was studied in a wide temperature region our data are consistent with the hypothesis of a moderate micellar growth.

Experimental

The nonionic amphiphiles used are high-purity products prepared by the group of Dr. Platone (Eniricerche, S. Donato, Milano, Italy). In order to lower the sample conductivity, the compounds were further purified by means of repeated extraction with organic solvents to reduce the amount of residual ionic impurities. Typically, the resistivity of our samples was 300 kΩ cm at 20 °C.

A detailed description of our experimental apparatus can be found in Reference [8]. The set-up used in this work includes a quarter-wave plate inserted between the Kerr cell and the analyzer. Such a configuration allows us to derive the sign of B and to reduce the amplitude of the applied electric field as much as possible, thus minimizing residual heating. Another advantage of the insertion of the quarter-wave plate is that a good compensation of the effects due to the stress-induced birefringence of the cell windows can be obtained, by using the correction method discussed in Reference [9].

The Kerr cell was made from an optical glass cuvette which was specially built by Hellma with low residual stress birefringence. The cell has two square windows of 45 x 45 mm. The opticall path length is 60 mm. Two parallel massive stainless-steel electrodes are inserted into the cell. The electrodes are separated by a PTFE spacer having a thickness of 2.5 mm. The sample volume was chosen to be rather large (about 20 cm³) in order to reduce the effects of sample degradation due to the release of metallic ions from the electrodes. It is known that such a degradation can become appreciable when the sample is kept for many hours in the cell at high temperature. The cell temperature was controlled within 0.01 °C. Voltage pulses had height of 0.3–1 kV and duration of 10–300 µs. The birefringence pulses at the output of the detector are sent to a transient digitizer (Data 6000, Data Precision) which performs the average over a prescribed number of runs (typically 30).

The transition temperature of the solution is carefully determined in the cell itself by monitoring the cell turbidity as the temperature is increased in steps of 0.01 °C, allowing the sample a sufficiently long equilibration time between one step and another.

Results

The steady-state value of the birefringence pulse is proportional to the difference Δn between the index of refraction $n_{\parallel}$ (polarization parallel to the applied electric field) and $n_{\perp}$ (polarization perpendicular to the electric field). We have verified that, in the range of applied electric fields, Δn is indeed proportional to the square of the electric field E. The Kerr coefficient B is derived from the expression $B = \Delta n / \lambda E^2$. We always found positive values for B.

We have measured B as function of the temperature distance from the critical point for aqueous solutions of C_6E_3, C_8E_4, $C_{10}E_5$, and $C_{12}E_6$ (the latter in both H_2O

Table 1. Critical temperature and concentration of nonionic amphiphile solutions.

System	T_c (°C)	c_c (%)
C_6E_3–H_2O	46.05	13
C_8E_4–H_2O	40.57	7.5
$C_{10}E_5$–H_2O	37.81	3.6
$C_{12}E_6$–H_2O	51.49	2.2
$C_{12}E_6$–D_2O	48.78	1.6

and D_2O), all prepared at the critical concentration. The critical temperature and concentration of the used solutions are reported in Table 1. Note that the critical temperatures are similar, because amphiphiles with the same ratio i/j present substantially the same hydrophilic-lypophilic balance. On the contrary, the critical concentration varies considerably by changing the chain length, even at constant i/j ratio. This is probably due to the fact that c_c depends strongly on the micelle size.

The results obtained for B are shown in Figures 1–5. All the investigated solutions present, far from the cloud point, a very low Kerr coefficient, comparable to the value expected for pure water, $B \approx 3 \times 10^{-14}$ m/V². All systems show a considerable increase of B as the temperature is raised toward the critical point. In a double logarithmic plot B is seen to depend linearly on the temperature distance $T_c - T$ when the data are

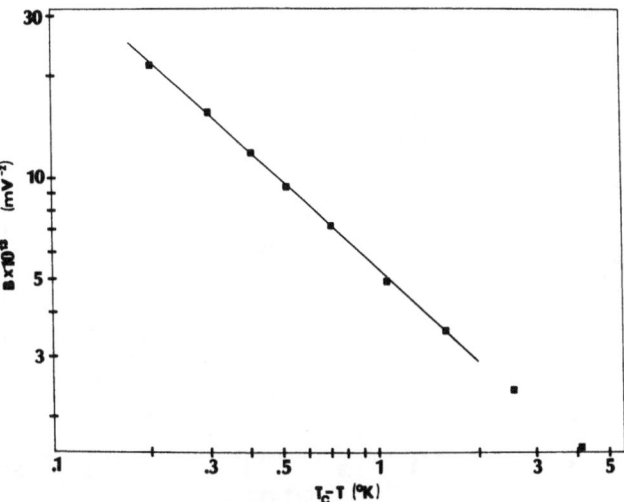

Fig. 1. Kerr coefficient B vs. the temperature distance from the critical point for a 13 % solution of C_6E_3 in H_2O

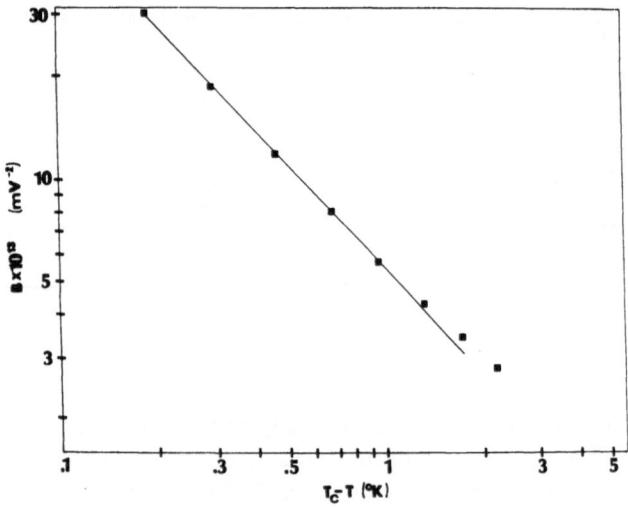

Fig. 2. *B* vs. temperature distance (as in Fig. 1) for a 7.5 % solution of C_8E_4 in H_2O

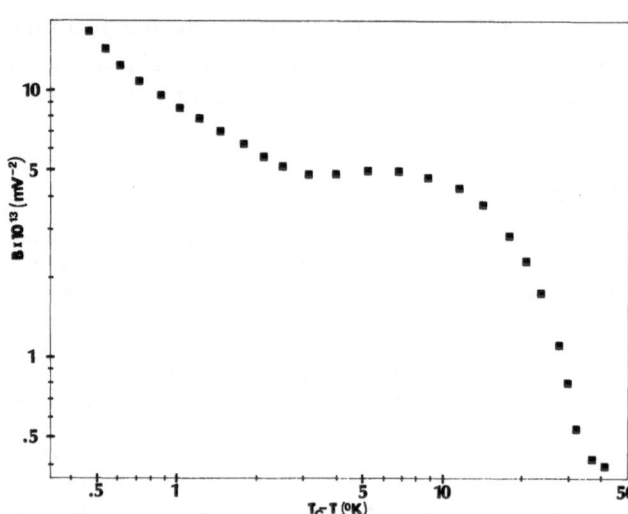

Fig. 4. *B* vs. temperature distance (as in Fig. 1) for a 2.2 % solution of $C_{12}E_6$ in H_2O

taken sufficiently close to T_c. Far from T_c the behavior of *B* may be more complex as shown in Figure 4 for the system $C_{12}E_6$–H_2O which was studied in a wide range of temperatures (8–50 °C).

In order to compare results obtained with different amphiphiles at distinct volume fractions ϕ, it is better to define a specific Kerr coefficient as $B' = B/\phi$. Since the density of the used amphiphiles is very close to the density of H_2O, we can take volume fractions to coincide with weight fractions. The values of B' (in units of 10^{-13} m/V^2) as calculated, for instance, at $T_c - T =$

0.5 °C are the following: 69 for C_6E_3, 120 for C_8E_4, 250 for $C_{10}E_5$, 590 for $C_{12}E_6$ in H_2O and 1750 for $C_{12}E_6$ in D_2O.

The temperature dependence of *B* is reported in Figure 6 for two cases in which the concentration does not correspond to the critical concentration. In this plot T^* is taken as the temperature at which phase separation is observed to occur in the scattering cell.

It should be mentioned that our experiment also gives information about the build up and the decay of birefringence. Since the analysis of the dynamic data is

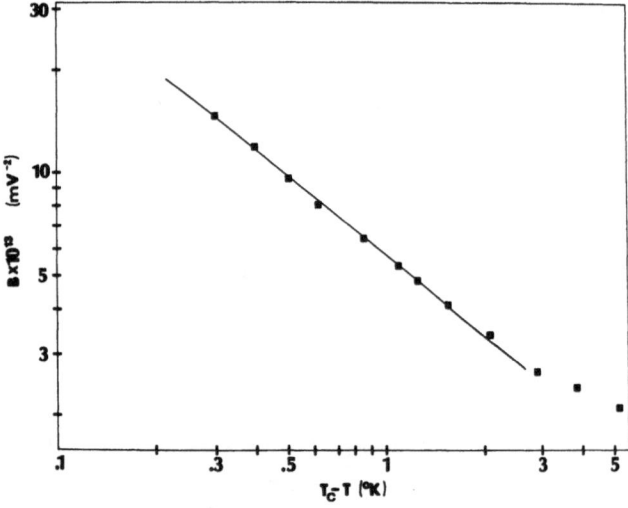

Fig. 3. *B* vs. temperature distance (as in Fig. 1) for a 3.6 % solution of $C_{10}E_5$ in H_2O

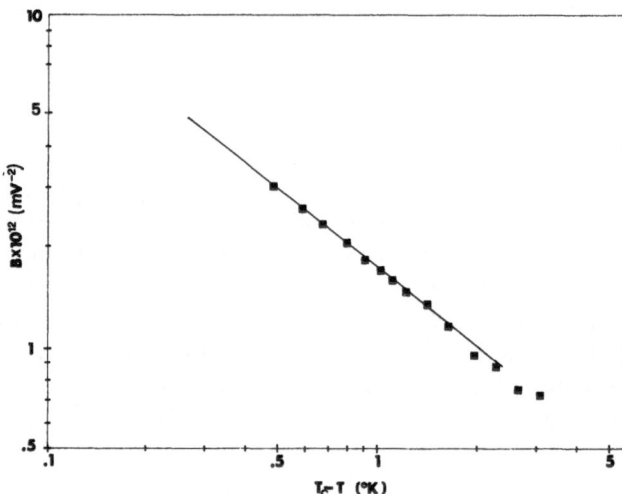

Fig. 5. *B* vs. temperature distance (as in Fig. 1) for a 1.6 % solution of $C_{12}E_6$ in D_2O

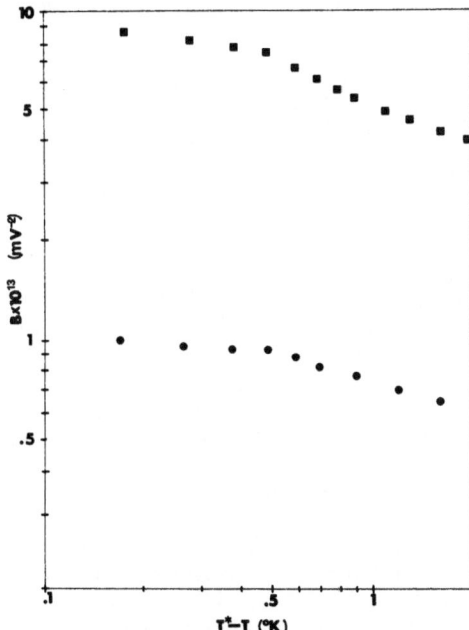

Fig. 6. B vs. temperature distance from the cloud temperature for a 7 % solution of $C_{10}E_5$ in H_2O (■, $T^* = 38.21$) and for a 6 % solution of $C_{10}E_5$ in H_2O (●, $T^* = 46.70$)

rather complex [7] and is not yet completed, we have chosen to concentrate in this paper only on the static behavior.

Discussion

Electric birefringence data on colloidal solutions can be simply interpreted in terms of the optical anisotropy of the individual colloidal particles only if the solution is sufficiently dilute to allow the neglection of interparticle interactions. It is important to stress that interactions may lead to a Kerr effect even if the colloidal particles are spherical. A clear example of this effect is represented by the experiments performed on microemulsions consisting of a dispersion of water-in-oil droplets [10]. In such a system electric birefringence may arise because the attractive interactions amoung the droplets generate statistical clusters which have a nonspherical shape. Of course, this effect becomes particularly significant near phase transition lines, such as the percolation threshold. A case which is particularly interesting for our discussion is that of a binary liquid mixture near a critical consolute point. It has been shown indeed that B may become very large in the critical region, even for liquid mixtures which pres-

ent negligible electric birefringence far from the critical point [8, 11]. The origin of the effect observed in critical systems is not in the anisotropy of individual molecules but is rather in the anisotropy of the spontaneous concentration fluctuations which are oriented (or deformed?) by the applied electric field. Theory [12] predicts that B should grow as the critical point is approached, following a power-law behavior as a function of the reduced temperature with a critical exponent that is close to the one associated with the divergence of the correlation range ($\psi \approx 0.6$).

The value of ψ for the system C_6E_3–H_2O, as derived from the data of Figure 1, is 0.88 ± 0.04. In the case of C_8E_4–H_2O the data shown in Figure 2 do not follow a straight line. If we take the limiting slope (straight line drawn in the figure), we find ≈ 1.05. It is possible, however, that for some unknown reason the critical temperature of this system had a drift during the measurement or was measured with an error of a few hundredths of a degree. If this latter hypothesis is correct, it is more indicative to measure the slope not very close to T_c. This would give $\psi \approx 0.9$. The best fit exponents for $C_{10}E_5$–H_2O and for $C_{12}E_6$–D_2O are, respectively, 0.75 ± 0.06 and 0.80 ± 0.06. In the case of $C_{12}E_6$–H_2O, it is not possible to assign a critical exponent. We can only say that the closest points to T_c are compatible with an asymptotic slope $\psi \approx 0.8$–0.9.

The data shown in Figure 6 confirm that the divergence of B is a critical effect, since it disappears when the temperature dependence is studied at a concentration which is very different from the critical one.

Aqueous solutions of nonionic amphiphiles have been studied in the last few years by a variety of techniques. The view [13] now almost generally accepted is that the cloud curve is a consolute curve at which the micellar solution separates into two micellar solutions with distinct concentrations, and the phase separation is driven by intermicellar interactions which become increasingly attractive as the temperature is raised. Critical concentration fluctuations will dominate the behavior of the micellar solution in a wide region around the cloud curve. It remains to be completely understood whether the micelle size and shape are changing when the temperature is increased to approach the cloud curve [14]. The data of Figure 4 suggest that the micelle shape is changing with T in $C_{12}E_6$–H_2O solutions. Otherwise it would be rather difficult to explain the very steep growth of B above 15 °C. If we fully attribute the variation of B between 15 ° and 40 °C to micellar growth, and if we assume that the micelle grows as a prolate ellipsoid with fixed semiminor axis, we can give an upper estimate of the increase of the aggregation number. The expression for the Kerr coefficient of a solution of (non-interacting) ellipsoids was given some years ago by Peterlin and Stuart [15]. The calculation of B requires knowledge of the axial ratio of the ellipsoidal micelle, of the index of refraction and the static dielectric constant of the amphiphilic material. We have measured for pure $C_{12}E_6$ a relative dielectric constant $\epsilon \approx 6$. The index of refraction is known from previous data [13], $n = 1.47$. The axial ratio below 15 °C must be very close to 2, if we consider that the aggregation number is somewhat above 100 [16, 17] and that the aggregation number of the minimal spherical micelle is around 60 for a dodecyl chain [18]. The result of our calculation is that the experimentally observed increase of B by a factor of 10 when the temperature goes from 8 °C to the critical region would correspond, if totally attributed to micellar growth, to an increase of the aggregation number by a factor of 3. Note that this result is consistent with recent data by Zana and Weill [17].

As a conclusion, we have shown in this paper that the Kerr coefficient of nonionic micellar solutions diverges on approaching the minimum of the cloud curve, according to a power-law behavior as a function of the reduced temperature. This effect is present in any critical binary mixture, and is due to the divergence of the size of critical concentration fluctuations. The measurement of B performed in a wide temperature range for the system $C_{12}E_6$–H_2O allows us to assign an upper limit to the increase of the aggregation number m with T. We find that m can increase by at most a factor of 3 which means that no very large micelles are formed in the cloud point region.

Acknowledgements

We thank E. Platone for the preparation of the nonionic amphiphiles, and T. Bellini for help in the measurements. This work was supported by grants from the Italian Ministry for Public Education.

References

1. Fredericq E, Houssier C (1973) Electric Dichroism and Electric Birefringence, Clarendon Press, Oxford; see also O'Konski CT (ed) (1976) Molecular Electro-Optics, Dekker, New York; Jennigs BR (ed) (1979) Electro-Optics and Dielectrics of Macromolecules and Colloids, Plenum, New York

2. Hoffmann H (1985) In: Dgiorgio V, Corti M (eds) Physics of Amphiphiles: Micelles, Vesicles and Microemulsions, North-Holland, Amsterdam, p 160
3. Schorr W, Hoffmann H (1981) J Phys Chem 85:3160
4. Nicoli DF, Elias JG, Eden D (1981) J Phys Chem 85:2866
5. Hoffmann H, Kielman HS, Pavlovic D, Platz G, Ulbricht W (1981) J Colloid Interface Sci 80:237
6. Neeson PG, Jennings BR, Tiddy GJT (1983) Faraday Disc Chem Soc 76:353
7. Degiorgio V, Piazza R (1985) Phys Rev Lett 55:288
8. Piazza R, Degiorgio V, Bellini T (1986) J Opt Soc Am B 3:1642
9. Piazza R, Degiorgio V, Bellini T (1986) Opt Commun 58:400
10. Guering P, Cazabat AM (1983) J Phys Lett, Paris 44:601; Eicke H-F, Hilfiker R, Thomas H (1985) Chem Phys Lett 120:272
11. Pyzuk W (1980) Chem Phys 50:281; Pyzuk W, Majgier-Baranowska H, Ziolo J (1981) Chem Phys 59:111
12. Goulon J, Greffe JL, Oxtoby DW (1979) J Chem Phys 70:4742; Hoye JS, Stell G (1984) J Chem Phys 81:3200
13. Corti M, Degiorgio V (1981) J Phys Chem 85:1442; Corti M, Minero C, Degiorgio V (1984) J Phys Chem 88:309
14. Nilsson PG, Wennerström H, Lindman B (1983) J Phys Chem 87:1377; Cebula DJ, Ottewill RH (1982) Coll & Polym Sci 260:1118; Triolo R, Magid LJ, Johnson JS, Child HR (1982) J Phys Chem 86:3689; Almgren M, Löfroth J (1984) In: Mittal KL, Lindman B (eds) Surfactant in Solution, Plenum, New York, p 627; Strey R, Pakusch A (1987) In: Mittal KL, Bothorel P (eds) Surfactants in Solution, Plenum, New York, to be published; Kato T, Seimiya T (1986) J Phys Chem 90:3159
15. Peterlin V, Stuart HA (1939) Z Physik 112:129; O'Konski CT, Krause S (1976) In: O'Konski CT (ed) Electro-Optics, Dekker, New York, p 63
16. Zulauf M, Weckström K, Hayter JB, Degiorgio V, Corti M (1985) J Phys Chem 89:3411
17. Zana R, Weill C (1985) J Phys Lett, Paris 46:953
18. Israelachhvili JN, Mitchell DJ, Ninham BW (1976) J Chem Soc Faraday Trans II, 72:1525

Received December 24, 1986;
accepted January 20, 1987

Authors' address:

Prof. V. Degiorgio
Dipartimento di Elettronica
Universita di Pavia
Via Abbiategrasso 209
I-27900 Pavia, Italy

Progress in Colloid & Polymer Science   Progr Colloid & Polymer Sci 73:81–89 (1987)

Electric birefringence and elastic and quasi-elastic light scattering investigation of the critical behavior of Triton X-100 in aqueous solution

Y. Dormoy, E. Hirsch[1]), S. J. Candau, and R. Zana[2])

Laboratoire de Spectrométrie et d'Imagerie Ultrasonores, Unité associée au C.N.R.S., Université Louis Pasteur, Strasbourg, France
[1]) Ecole Nationale Supérieure de Physique de Strasbourg, Laboratoire des Sciences de l'Image et de la Télédétection, Strasbourg, France
[2]) Institut Charles Sadron (C.R.M.), C.N.R.S., Strasbourg, France

Abstract: Aqueous solutions of Triton X-100 have been investigated by means of elastic and quasi-elastic light scattering, viscosity, and electric birefringence in the temperature range between room temperature and critical temperature T_c. The intensity of scattered light and the correlation length ξ have been found to follow power laws of $(T_c - T)/T_c$ with exponents equal to those predicted by the renormalisation group theory. Nevertheless some deviations from the Kawasaki-Ferrell universal plot are noted when the correlation range increases much, close to T_c. The decay and the rise of the electric birefringence show the presence of two relaxation processes. The fast relaxation process has been attributed to the individual micelles and its analysis has yielded information on the shape and dimension of the Triton X-100 micelles. The slow process which becomes predominant close to T_c appears to be due to the micelles clusters, present at these temperature. It yields values of the correlation range in good agreement with those obtained from light scattering. The results show that the micelles are anisodiametric and that fluctuations of micelle concentration are anisotropic.

Key words: Electric birefringence, light scattering, critical phenomena, non-ionic surfactant.

Introduction

Low concentration aqueous solutions of many nonionic surfactants, upon heating, become cloudy at a well defined temperature which depends on surfactant concentration [1]. The so-called cloud point curve is a lower consolution curve above which the micellar solution separates into two isotropic micellar solutions of different concentrations. The cloud point curve presents, in the concentration-temperature plane, a minimum at a temperature T_c and a concentration C_c. A general feature of the nonionic surfactants investigated up to now is that the coexistence curve is strongly asymmetric around the minimum and the concentration C_c low [1]. Light scattering experiments performed along an isoconcentration path show an enhancement of the turbidity as the temperature is increased. Such a phenomenon was first interpreted as due to micellar growth upon heating [2]. However, static and dynamic light scattering studies have now

clearly demonstrated that, in the vicinity of the cloud point transition, the properties of micellar solutions are mainly determined by long-range concentration fluctuations similar to those occurring in binary liquid mixtures near a critical consolution point [1]. More specifically, both the osmotic compressibility and correlation range for concentration fluctuations diverge at T_c. However, often the nonionic surfactant solutions do not show the universal behavior observed for critical binary mixtures of two low molecular weight components. Indeed, the measured critical exponents have been found to depend on the surfactant and on the solvent [1, 3, 4]. This behavior for the surfactants of the $C_n E_m$ type (n = number carbon atoms of the hydrophobic chain and m = number of ethylenoxide groups E of the hydrophilic moiety) is far from being understood. A theoretical treatment based on decorated lattice Ising models [5, 6] including a directional interaction between the components of the system, shows that the temperature range of validity of the exponent depends strongly on the directionality of the intermolecular interaction and on the molar volume difference of the components. This treatment may explain the non-universality of the critical behavior of nonionic micellar solutions.

A further complication in the study of nonionic micelles comes from the fact that an increase of their aggregation number with temperature may be superimposed to the critical fluctuations. Notice that different studies using different techniques have led to contradictory results [7–12] concerning the change of micelle size with temperature.

The critical behavior of fluid mixtures has also been investigated by electric birefringence. A theory based on a droplet model [13] predicts a divergence of the Kerr constant at the critical temperature, associated with an induced anisotropy of the correlation length along the direction of the electric field. The predictions of the theory have been verified for mixtures of low molecular weight components [14] and also for nonionic micellar systems [15]. However the droplet model does not provide any description of the transient behavior of the electric birefringence. A recent experimental study of $C_{12}E_6$ and $C_{12}E_8$ micellar solutions [15] has shown a strong asymmetry between rise and decay of birefringence. The decay time was too short to be experimentally measured. The rise of the birefringence was strongly non exponential and followed, sufficiently close to the critical point, a universal function of scaled time.

In this paper, we report a study by static and dynamic light scattering and by transient electric birefringence of the critical behavior of aqueous solutions of Triton X-100. In contrast to the isomerically pure C_mE_m nonionic surfactants whose critical behavior has been previously investigated by light and neutron scattering and electric birefringence, Triton X-100, is a commercial product which refers to p-2,3,3-trimethylpentylphenoxy-polyoxyethylene, where the polyoxyethylene chain contains an average number of 10 ethoxy groups. The distribution of the lengths of the hydrophilic moiety is known [16].

It was therefore interesting to investigate how such a polydisperse surfactant, which is a good representative of commercial nonionic surfactants, would behave close to T_c, with respect to isomerically pure ones. As will be shown below, the power law divergences for the scattered intensity and the correlation range ξ of concentration fluctuations were found to be consistent with the theoretical predictions of the renormalization group theory.

The rise and decay curves of the electric birefringence are described by two relaxation processes that we have attributed to the rotational diffusion of an isolated micelle and the reorientation of a fluctuation of size ξ respectively.

Material and methods

The sample of Triton X-100 (Rohm and Haas, USA) contained 2.30% H_2O. It was dissolved without further purification in distilled water at room temperature and carefully degassed before use.

Light scattering experiments

Elastic and quasi-elastic light scattering experiments were performed using the same apparatus as in previous studies [17]. The measurements involved the intensity I_s of the light scattered by the micellar solutions and the autocorrelation function of this intensity. This function was analyzed by means of the cumulant method [18] to yield the variance v and the first reduced cumulant $\langle \Gamma \rangle / 2K^2$ where $\langle \Gamma \rangle$ is the average decay rate of the autocorrelation function and K the magnitude of the scattering wavevector:

$$K = [4\pi n \sin (\theta/2)]/\lambda \qquad (1)$$

where θ is the scattering angle, n the refractive index of the scattering medium and λ the wavelength of the incident light in vacuo. At a temperature T far from T_c, $\langle \Gamma \rangle / 2K^2$ provides a measurement of the mutual diffusion coefficient D of the micelles.

Electric birefringence measurements

The electric birefringence apparatus was similar to that described elsewhere [19]. A 5 nw He-Ne laser beam ($\lambda = 632.8$ nm), polarized at $3\pi/4$ to the vertical, propagates parallel to the surface of two flat stainless steel electrodes, which are set vertically, 2.5 mm apart in a quartz spectrophotometer cell of 4 cm path length filled by the solution under study. The cell temperature is held constant to better than 0.005 °C.

Due to the very low birefringence of the micellar solutions, rectangular pulses with the maximum available voltage have been applied to the electrodes. As a result it has not been possible to perform measurements over a wide range of electric fields, nor to study the birefringence behavior following the reversing pulse since our pulse generator is voltage-limited in this configuration. A quarter wave device was set before the analyzer. The signal detected by a photodiode was digitized by a transient recorder Datalab DL 922 (8 bits, 2 Ko memory, 20 MHz maximum sampling rate) together with the high voltage electric pulse applied to the electrodes. The digital signals were processed by a local computer. For each experimental condition, an average of several recordings was stored for further analysis.

The Kerr constant is obtained from the measurement of the steady-state birefringence Δn_o (difference between the refractive indices in directions parallel and perpendicular to the electric field).

Assuming that Δn_o is proportional to the square of the applied electric field E, the Kerr constant is given by:

$$K = \Delta n_o/nE^2 \qquad (2)$$

where n is the refractive index of the medium.

Viscosity

The viscosity was measured using an Oswald viscosimeter immersed in a water bath. The temperature stability was better than 0.005 °C. At all temperatures, the investigated systems showed a Newtonian behavior for the shear gradients available on the setup.

Cloud point measurements

A series of sealed-glass cells filled with solutions of different concentrations were immersed in a water bath whose temperature stability was better than 0.005 °C. The temperature was increased by steps of 0.01 °C near the transition at a very low rate. The phase-separation temperature was taken as that where the turbidity of the solution, illuminated by a white light beam, increased suddenly.

Results

Consolution curve and Kerr constant

Figure 1 shows the consolution curve of Triton X-100 obtained in this work. The curve shows a rather flat asymmetric minimum in the concentration range 0.02–0.04 g/cm³ at $T_c \simeq 62.8$ °C. This value is somewhat lower than that reported in a recent study [20]. The difference in T_c probably reflects the different amounts of impurities present in the two samples which were of different origins. This is a common feature even with supposedly isomerically pure anionic surfactants.

On Figure 1 is also plotted the concentration dependence of the Kerr constant at $T = T_c - 1$ °C. The corresponding curve exhibits a maximum at a concentration close to 4×10^{-2} g·cm⁻³. Most of the measurements described below have been performed at a surfactant concentration $C = 4.33 \times 10^{-2}$ g·cm⁻³. As part of this work some limited experiments have shown that the critical behavior observed was rather insensitive to C in the range $2 - 5 \times 10^{-2}$ g·cm⁻³. A similar observation has been reported by Corti et al. [21] for micellar solutions of isomerically pure C_nE_m surfactants.

It was also observed that the application of numerous high voltage electric pulses produces a slight decrease of the consolution temperature.

Viscosity

The kinematic viscosity η_s of Triton X-100 solution at $C = 4.33 \times 10^{-2}$ g·cm⁻³ is shown in Figure 2. The viscosity starts to increase with temperature at a temperature more than 20 °C below T_c. A similar behavior has also been reported by Corti et al. [21] for pure ionic surfactants. This effect might arise from the fact that far from the critical point a substantial micellar growth is induced by a temperature rise [12].

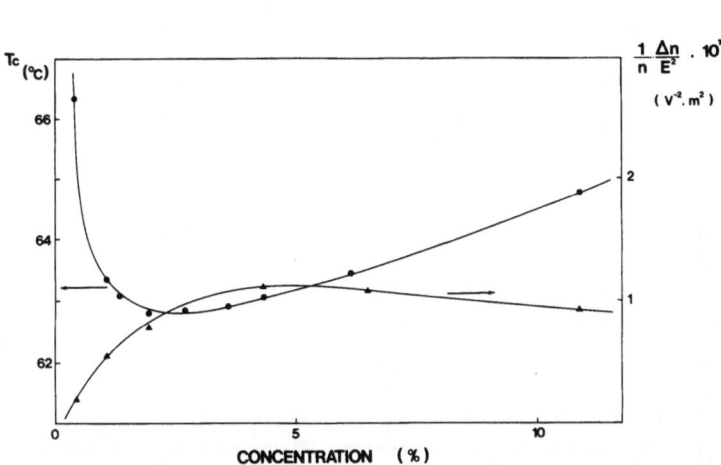

Fig. 1. Cloud curve (●) and variation of the Kerr constant (▲) with the Triton X-100 weight concentration in % at $T = T_c - 1$°C

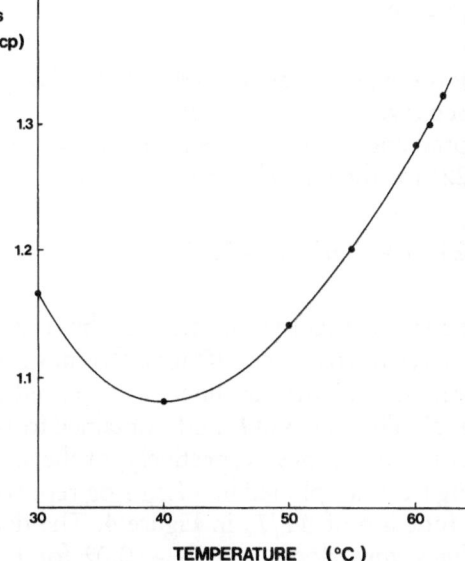

Fig. 2. Temperature dependence of the viscosity of a $4.33 \cdot 10^{-2}$ g/cm³ Triton X-100 solution

Light scattering

Near a critical consolution point, the intensity I scattered from a binary mixture depends on the scattering wavevector according to the Ornstein-Zernike relation:

$$I = I_o/(1 + K^2 \xi^2). \qquad (3)$$

The extrapolated scattered intensity at zero wavevector is related to the derivative of the osmotic pressure Π with respect to concentration, according to:

$$I_o = A C R T \, (dn/dC)^2 \, (\partial \Pi / \partial C)_{T,p}^{-1} \qquad (4)$$

where A is an instrumental constant, dn/dC the refractive index increment of the solution, R the gas constant and T is the absolute temperature. When the system approaches the critical point in the single-phase region at the critical concentration, both $(\partial \pi / \partial C)_{T,p}^{-1}$ and the correlation range ξ diverge according to the following power laws:

$$\left(\frac{\partial \pi}{\partial C}\right)_{T,p}^{-1} \alpha \left(\frac{\Delta T}{T_c}\right)^{-\gamma} \qquad (5)$$

$$\xi = \xi_o \left(\frac{\Delta T}{T_c}\right)^{-\nu}. \qquad (6)$$

The two critical exponents γ and ν are related through the relation:

$$\gamma = \nu(2 - \eta) \qquad (7)$$

where η is the exponent associated with the divergence of the viscosity.

The predictions of the renormalization group theory [22] for the critical exponents are:

$$\gamma = 1.24 \quad \nu \simeq 0.63 \quad \eta = 0.031. \qquad (8)$$

Figure 3 shows that the variations of the reciprocal of the scattered intensity with K^2 for different values of the reduced temperature are linear as expected from Equation (3). The values of I_o and ξ obtained from the intercepts and the slopes, respectively, of the straight lines of Figure 3 are plotted in a Log-Log representation as a function of $\Delta T/T_c$ in Figure 4. The data fit straight lines with slopes $- 1.24 \pm 0.02$ for I_o and $- 0.59 \pm 0.05$ for ξ, in good agreement with the theoretical predictions.

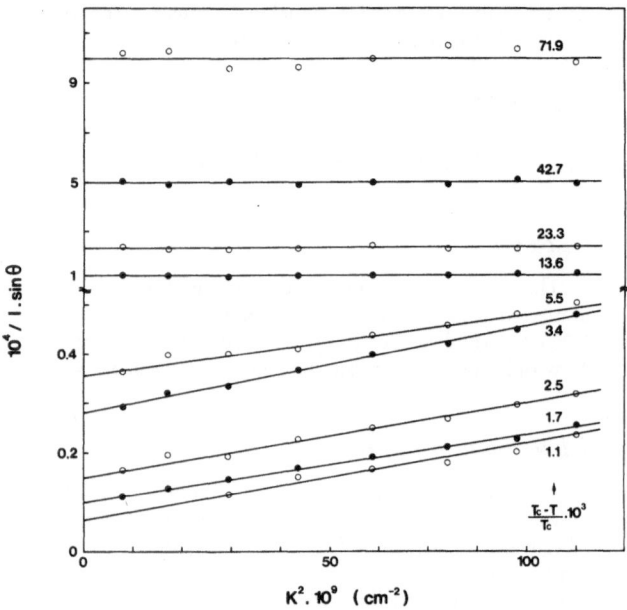

Fig. 3. Variation of the reciprocal of the intensity scattered by a $4.33 \cdot 10^{-2}$ g/cm³ Triton X-100 solution with the square of the wave vector, at different reduced temperatures, as indicated on the curves

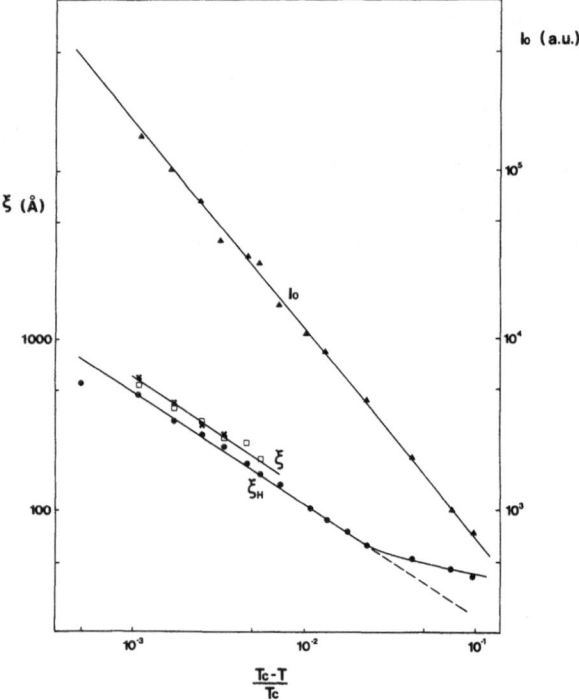

Fig. 4. Variations of the scattered intensity (in arbitrary units) and of the range of correlation fluctuations; ($\bullet$) values of ξ_H; ($\times$) values of ξ from the results of Figure 5 analyzed using Equation (9); ($\square$) values of ξ from the plots of Figure 3 with the reduced temperature

Turning to dynamic light scattering, we recall that the decay Γ of the critical concentration fluctuations is given by the Kawasaki-Ferrell equation [23–25]:

$$\Gamma = h \frac{k_B T}{6\pi \eta_s} \xi^{-3} H(K\xi) \qquad (9)$$

where h is a constant which has been found equal to 1.15 for both micellar and low molecular weight components critical mixtures [26], k_B is the Boltzmann constant and $H(K\xi)$ is given by:

$$H(x) = -\frac{3}{4}\left[1 + x^2 + (x^3 - x^{-1})\tan^{-1}(x)\right]. \qquad (10)$$

In the limit $K\xi \ll 1$, Equation (9) reduces to:

$$\frac{\Gamma}{K^2} = \frac{k_B T}{6\pi \eta_s \xi_H}\left(1 + \frac{3}{5} K^2 \xi^2\right) \qquad (11)$$

where the hydrodynamic correlation length ξ_H is related to ξ through the relationship

$$\xi_H = \xi/h. \qquad (12)$$

The analysis of the autocorrelation function of the photocurrent provides a measurement of $\langle\Gamma\rangle/2K^2$. From the intercept of the $\langle\Gamma\rangle/2K^2$ verus K^2 plots of

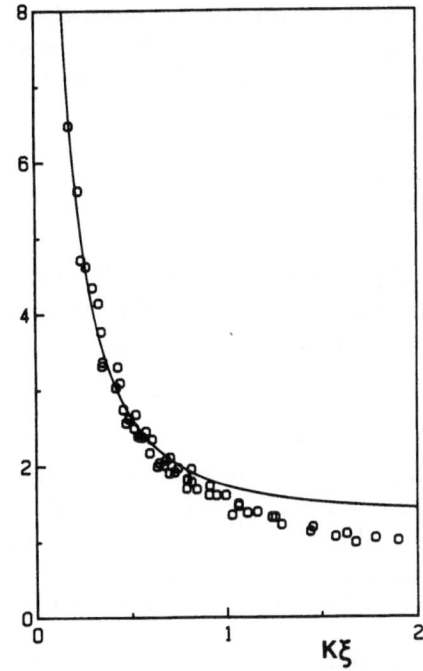

Fig. 6. Variation of the reduced linewidth Γ^* with $K\xi$. The curve going through the experimental data points at low $K\xi$, and deviating from these points at $K\xi \geq 1$ obeys Equation (13)

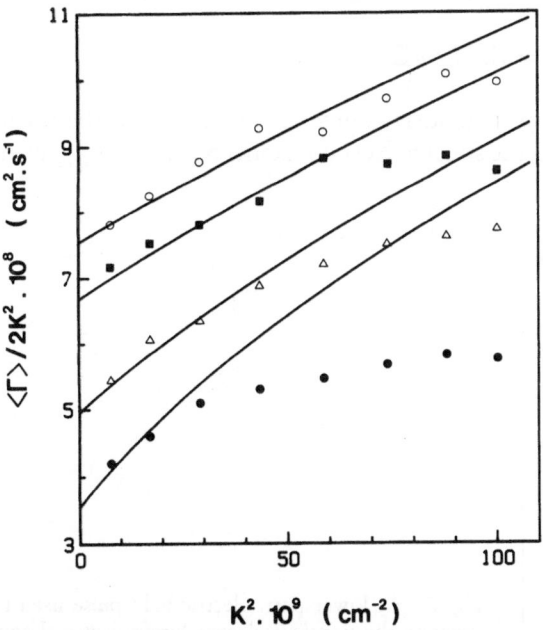

Fig. 5. Fitting of the experimental values of $\langle\Gamma\rangle/2K^2$ at different wave vectors and reduced temperatures to the Kawasaki-Ferrell equation (9) $\Delta T/T_c \cdot 10^3 =$: 1.1 (●); 1.7 (△); 2.5 (■); 3.4 (○)

Figure 5 and the fit of these experimental points to Equation (9), we have determined ξ_H and ξ as a function of $\Delta T/T_c$. The variations of these two parameters are plotted in Figure 4. The values of ξ so obtained are in good agreement with those obtained from the wavevector dependence of the scattered intensity. Also ξ_H obeys a power law of $\Delta T/T_c$ in a large reduced temperature range with an exponent $\nu = -0.66$ and a prefactor $\xi_{HO} = 5.4$ Å. We observe that the ratio ξ/ξ_H is close to 1.15, value reported by others. We also note that this value ξ_{HO} is close to that reported for C_6E_3 (which is characterized by the Ising values of γ and ν) and C_8E_4 for which values of γ and ν close to Ising ones have been found [21].

Therefore, both static data and dynamic data obtained in the low $K\xi$ range are consistent with the theoretical predictions for critical binary mixtures.

On the contrary, in the large K range and close to the critical point the behavior of the decay rate of concentration fluctuations deviates from that predicted in Equation (9). This is clearly seen in Figure 5 where the curves going through the experimental data have been calculated from Equation (9): the smaller $\Delta T/T_c$, the lower the K value at which occurs the deviation from the Kawasaki-Ferrell equation.

This anomalous behavior appears more clearly in the so-called Kawasaki plot [23–25] of Figure 6 which represents the $K\xi$ dependence of the reduced line-width Γ^* defined as:

$$\Gamma^* = h \frac{H(K\xi)}{(K\xi)^3}. \tag{13}$$

The curve calculated from Equation (13) with $h = 1.15$ goes well through the experimental data at small $K\xi$ but a large deviation appears for $K\xi > 1$.

Transient electric birefringence

The classical theory developped for non-interacting rigid particles in solution [27] predicts that the decay of the birefringence after removal of the electric field is exponential with a time constant $(6 D_r)^{-1}$, where D_r is the rotational diffusion constant of the particle:

$$\Delta n(t) = \Delta n_o e^{-6 D_r t}. \tag{14}$$

For spherical particles of radius R immersed in a solvent of viscosity η_o, D_r is given by:

$$D_r = \frac{k_B T}{8\pi \eta_o R^3}. \tag{15}$$

This equation can also be used for non-spherical particles. It then yields R_{eq}, radius of the equivalent spherical particle.

The rise of the birefringence at the onset of the field is also described by a single exponential process with a time constant $(6 D_r)^{-1}$ provided that the birefringence of the sample arises from a purely induced electric dipole.

Figure 7A illustrates the time dependence of the birefringence $\Delta n(t)$ induced by applying an electric field pulse to the Triton X-100 solution at $T = 50\,^\circ$C. The decay of the birefringence is enlarged in Figure 7B. Our results differ from those of Degiorgio and Piazza [15] for the isomerically pure $C_{12}E_6$ in two respects:

(i) The experimental rise and decay of the birefringence were always nearly symmetrical.

(ii) Both the rise and decay show two relaxation processes (1 and 2 on Fig. 7B) with widely separated time constants, in agreement with the preliminary observation of Wright [28] for Triton X-100.

Far from the critical temperature, both rise and decay curves are satisfactorily described by a sum of two exponentials, of comparable amplitudes. In the critical range ($\Delta T/T_c < 10^{-2}$), the birefringence is dominated by the slow process. The latter was analyzed by discarding the beginning of the curve (fast process) and fitting the long time range with a first order distribution of exponentials (cumulant method) to provide the average decay time and the variance of the distribution.

Temperature dependence of birefringence amplitude

Defining the specific Kerr constant as:

$$K_{sp} = \Delta n_o / n \varphi E^2 \tag{16}$$

where φ is the surfactant volume fraction, we have calculated the specific Kerr constants $K_{sp,1}$ and $K_{sp,2}$ from

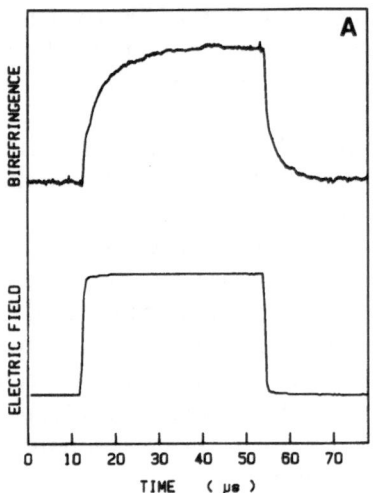

Fig. 7. (A) lower part: electric field pulse used to generate the transient electric birefringence shown on top. (B) Enlarged view of the decay of the birefringence. Temperature: 50 °C; Triton X-100 solution at 4.3×10^{-2} g/cm^{-3}

the amplitudes of the birefringence Δn_{o1} and Δn_{o2} of the fast and slow relaxation processes, respectively, obtained from the fitting of the decay of the electric birefringence. These quantities are plotted in Figure 8 as a function of the reduced temperature. The amplitudes of Δn_{o1} and Δn_{o2} are obtained from two exponential fits to the decay curves; the fast relaxation amplitude increases only little with T whereas the slow relaxation amplitude follows a power law:

$$K_{sp,2} = K^o_{sp,2} \left(\frac{T_c - T}{T_c} \right)^{-\psi} \qquad (17)$$

with the critical exponent $\psi = 0.55$, which indicates a critical behavior. This behavior of $K_{sp,2}$ is limited to the range $\Delta T / T_c < 10^{-1}$; then $K_{sp,2}$ becomes smaller than $K_{sp,1}$, to eventually vanish far from T_c. Notice that Degiorgio and Piazza reported values of 0.52 for $C_{12}E_6$ and 0.49 for $C_{12}E_8$ [15].

Temperature dependence of the relaxation time constants

a) Slow relaxation process

The analysis of the slow relaxation process of the birefringence decay curve using a distribution of exponentials yields a mean rotational diffusion constant $D^D_{r,2}$, and in turn a mean equivalent radius $R^D_{eq,2}$ using for η_o in Equation (15) the viscosity of the solution in order to permit a comparison of $R^D_{eq,2}$ to the correlation length ξ_H obtained by quasi-elastic light scattering. Notice that the reduced variance increases from 0.05 far from the critical temperature to 0.7 near the transition.

Figure 9 shows that the agreement between the mean equivalent radius $R^D_{eq,2}$ and the correlation length ξ_H is very good, for $\Delta T / \bar{T}_c < 10^{-2}$: the two quantities follow a power law with nearly the same critical exponent 0.65. In the same T-range the analysis of the slow relaxation process of the birefringence rise leads to a mean equivalent radius $R^R_{eq,2}$ equal to $R^D_{eq,2}$ within the experimental accuracy.

For $\Delta T / T_c > 10^{-2}$, $R^D_{eq,2}$ remains constant while the amplitude of this relaxation still continues to decrease (see Fig. 8). On the other hand, the ξ_H curve deviates from the straight line. The means that, even far from T_c, there is still a contribution to the electric birefringence associated with a slow process which is not significant in quasi-elastic light scattering experiments. Such a result could be explained for instance by the presence

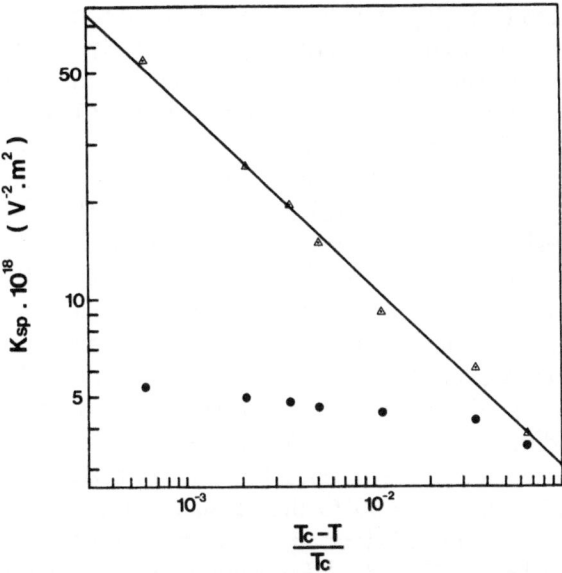

Fig. 8. Variation of the specific Kerr constants of the fast (●) and slow (△) relaxation processes of the electric birefringence, with the reduced temperature, for a 4.3% solution of Triton X-100

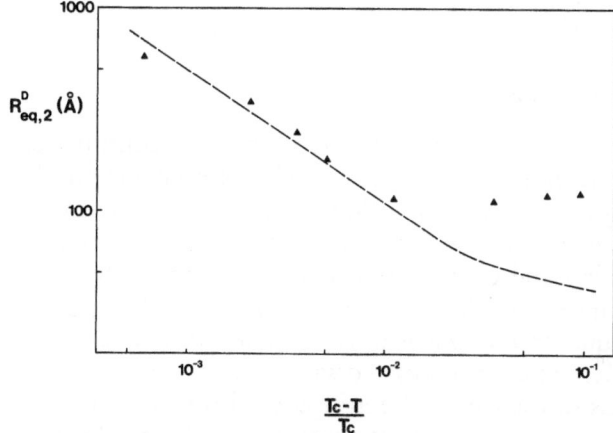

Fig. 9. Comparison between the values of the equivalent radius $R^D_{eq,2}$ obtained from the slow component of the birefringence decay (▲), and the correlation range ξ_H from quasi elastic light scattering (---)

in the solution of few large anisotropic micelles or groups of micelles. However, the observed effect might be more complex since the equivalent radius $R^R_{eq,2}$, obtained from the birefringence rise is found larger than $R^D_{eq,2}$, by a factor $\simeq 2$ for $\Delta T / T_c > 10^{-2}$.

b) Fast relaxation process; individual micelle characteristics

The analysis of the fast relaxation process of the birefringence rise and decay with a single exponential function (Eq. (14)), yields the values of the rotational diffusion constants D_{r1}^R and D_{r1}^D, respectively. These values are nearly equal whichever the temperature, the measurement of D_{r1}^R being accurate only for $\Delta T/T_c > 5 \cdot 10^{-3}$. The fast relaxation process may be attributed to the rotational diffusion of the isolated micelle. Therefore the micelle is oriented in the electric field by a purely induced electric dipole. Moreover, D_{r1}^D remains smaller than the translational diffusion coefficient $\Gamma/2K^2$ far from the critical temperature, indicating that the micelles are not spherical. Assuming that they are prolate ellipsoids, we can use inversion of the Perrin equations [29], relating translational and rotational diffusion constant to determine the geometrical characteristics of the micelles. For instance at $\Delta T/T_c = 9.5 \cdot 10^{-2}$, that is $T = 30\,°C$, the translational and rotational diffusion coefficients are respectively $4.65 \cdot 10^{-7}$ cm$^2 \cdot$ s^{-1} and $6.70 \cdot 10^5$ s^{-1}. Inserting these values in the Perring equation and taking for η_o the viscosity of the solvent leads to values of the micelle semi-minor and semi-major axis: 30 Å $\pm$ 5 Å and 100 Å $\pm$ 10 Å, respectively.

Discussion

The results presented above provide confirmatory evidence of the critical behavior of nonionic micellar systems close to the cloud point. Light scattering experiments have shown that both osmotic compressibility and correlation length follow power laws as a function of the reduced temperature with exponents equal to the universal Ising values [22]. The specific Kerr constant associated with the slow relaxation process of the electric birefringence also follows a power law with an exponent equal to the value predicted by the droplet model [13]. Notice that for the birefringence the correct exponent is observed even though the mechanism responsible for the occurrence of the birefringence is not the one for which the theory has been worked out (see below).

Thus the polydisperse Triton X-100 behaves like the isomerically pure C_6E_3 (normal Ising exponents and small ξ_{HO} value) whereas the other pure nonionic surfactants C_8E_4, $C_{12}E_6$, $C_{12}E_8$, $C_{14}E_7$ are characterized by exponents smaller than Ising ones and abnormaly large ξ_{HO} values (notice that for $C_{12}E_6$, Corti et al. [30] report low exponents, whereas Strey and Pakusch [31] report Ising values). It would appear as if the reported values of the exponents and of ξ_{HO} for all of the pure C_nE_m surfactants but C_6E_3, were mostly dependent on the surfactant hydrophobic moiety, rather than on the

hydrophobic head group, contrary to the theoretical interpretation which was briefly mentioned in the introduction. Indeed C_6E_3 and Triton X-100 have strongly differing polyoxyethylene chains, but their effective hydrophobic moieties contain 6 and 8.5 atoms respectively (indeed a phenoxy group is known to be equivalent to about 3.5 carbon atoms). Also C_8E_4 is characterized by exponent values close to Ising ones and a small ξ_{HO}. The differences between experimental and theoretical values are important only for the C_{12} or C_{14} nonionic surfactants. Clearly, additional work is required to understand the behavior of nonionic surfactants.

Our results show that the transient electric birefringence experiment probe reorientational processes with a characteristic range equal to the hydrodynamic correlation length as measured by quasielastic light scattering. Such a behavior already observed for critical water in oil microemulsions [32] indicates that the critical fluctuations are somewhat anisotropic.

This anisotropy is likely to originate from the non spherical shape of the isolated micelles which is evidenced at temperatures far from the critical temperature. Recall that a substantial micellar growth of the Triton X-100 micelles has also been inferred from fluorescence decay experiments in the same T-range [12]. Orientational short range correlations between the micelles would lead to anisotropic polarizability of the fluctuations and therefore birefringence under an electric field.

This anisotropic structure of the fluctuations is also likely to be at the origin of the deviation observed in the high $K\xi$ range of the Kawasaki plot (Fig. 6). When one probes fluctuations of size approaching the individual micellar size, the Kawasaki-Ferrell model developed for critical mixtures of low molecular weight component is no longer adequate.

Conclusion

Elastic and quasielastic light scattering and transient electric birefringence have been used to show that aqueous solutions of Triton X-100 close to the critical temperature and concentration obey the universal scaling laws predicted by the renormalization group theory and thus behave much like binary mixtures of low molecular weight compounds. In this respect the behavior of Triton X-100 and C_6E_3 differs significantly from other nonionic ethoxylated surfactants such as $C_{12}E_6$, $C_{12}E_8$ and $C_{14}E_7$. The latter are characterized by exponents lower than Ising values, and large ξ_{HO}

values. These differences appear to be mainly related to the length of the hydrophobic moiety rather than to the extent of the hydrophobic group of the surfactant molecule. More work is needed in order to elucidate the observed differences.

References

1. Degiorgio V (1985) In: Degiorgio V, Corti M (eds) Physics of Amphiphile: Micelles, Microemulsions and Vesicles, North Holland, Amsterdam, p 303; Corti M, Degiorgio V (1975) Opt Comm 14:274
2. Balmbra J, Clunie J, Corkill J, Goodman J (1962) Trans Faraday Soc 58:1661; (1964) 60:979
3. Corti M, Degiorgio V (1985) Phys Rev Lett 55:2005
4. Degiorgio V, Piazza R, Corti M, Minero C (1985) J Chem Phys 82:1025
5. Wheeler JC (1975) J Chem Phys 62:433
6. Reatto L, Tan M (1984) Chem Phys Lett 108:292
7. Cebula D, Ottewill R (1982) Coll & Polym Sci 260:1118
8. Brown W, Johnson R, Stilbs P, Lindman B (1983) J Phys Chem 87:4548
9. Ravey JC (1983) J Coll Interf Sci 94:289
10. Zulauf M, Weckstrom K, Hayter JB, Degiorgio V, Corti M (1985) J Phys Chem 89:3411
11. Kata T, Seimiya T (1986) J Phys Chem 90:3159
12. Zana R, Weil C (1985) J Phys Lett, Paris 46L:953
13. Goulon J, Greffe JL, Oxtoby DW (1979) J Chem Phys 70:4742
14. Pysuk W, Zboinsky K (1977) Chem Phys Lett 52:577
15. Degiorgio V, Piazza R (1985) Phys Rev Lett 55:288
16. El Seoud O, Vidotti G, Miranda O, Martins A (1980) J Coll Interf Sci 76:265
17. Candau SJ, Zana R (1981) J Coll Interf Sci 84:206
18. Koppel DE (1972) J Chem Phys 57:4814
19. Candau SJ, Dormoy Y, Mutin PH, Debeauvais F, Guenet JM (19??) Polymer, to be published
20. Valanlikar B, Manohar C (1985) J Coll Interf Sci 108:403
21. Corti M, Minero C, Degiorgio V (1984) J Phys Chem 88:309
22. Wilson KG, Kogut JB (1974) Phys Report 12c:75
23. Kawasaki K (1968) Phys Lett 26a:543; (1969) 30a:325
24. Kawasaki K (1970) Phys Rev a1:1750; Ann Phys N Y 61:1
25. Ferrell ? (1970) Phys Rev Lett 24:169
26. Beysens P (1982) NATO Adv Study Ins Ser, Ser B 82, 2, 72:25
27. Benoit H (1951) Ann Phys, Paris 6:561
28. Wright A (1976) J Coll Interf Sci 55:109
29. Perrin J (1936) J Phys Rad 6:1
30. Corti M, Degiorgio V (1981) J Phys Chem 85:1442
31. Strey R, Pakusch A (1987) In: Mittal K, Bothorel P (eds) Proceedings of the International Symposium on Surfactants in Solutions, Plenum Press, New York, in press
32. Guering P, Cazabat AM (1983) J Phys Lett 44:601

Received February 9, 1987;
accepted February 11, 1987

Authors' address:

Prof. Dr. S. Jean Candau
Laboratoire de Spectrométrie
Université Louis Pasteur
Unité Associée au CNRS no 851
4, rue Blaise Pascal
F-67070 Strasbourg Cédex, France

Progress in Colloid & Polymer Science

Progr Colloid & Polymer Sci 73:90–94 (1987)

The static structure factor of a system of fully aligned rod-like macroions

J. K. G. Dhont and R. Klein

Fakultät für Physik, Universität Konstanz, Konstanz, F.R.G.

Abstract: The static structure factor for fully aligned charged rod-like macromolecules is calculated from the Ornstein-Zernike equation in MSA-closure. The hard-core of the macroions is taken ellipsoidal. The reason for choosing ellipsoidal hard-cores is that the ellipsoid-of-nearest approach is conveniently mapped on the spheres by a simple contraction of the axis of alignment. The direct correlation function inside the contracted hard-core is expanded in a power series. The closure relation for the total correlation function is then satisfied if a certain non-linear functional in the expansion coefficients is zero. The numerical values for a finite set of expansion coefficients for which this functional is minimum are obtained by numerical methods.

Key words: Rod-like macromolecules, static structure factor, light scattering.

Introduction

For spherically symmetrical charged macromolecules a great deal has been learned from analysis of the Ornstein-Zernike equation with some appropriate closure relation [1]. Some work has been done in this respect for non-spherically symmetric macroions [2, 3].

In this paper the hard-core is taken into account in an exact way (for ellipsoids) and a realistic ansatz for the long-ranged interaction potential is made. Even for the MSA-closure we did not succeed in obtaining analytical results for the structure factor.

It must be kept in mind that the MSA-closure is appropriate if the charge on the macroions is not too large. For arbitrary charges the HNC-closure is appropriate.

Flexibility of the hard-core and fluctuations in the director are not considered. An estimate on the influence of both is extremely difficult to make.

Some of the detailed steps in the mathematics involved are left out in this paper. For these details the reader is referred to Reference [4].

Theory

The Ornstein-Zernike equation with mean-spherical approximation closure relation reads:

$$h(\vec{r}) = c(\vec{r}) + \varrho \int d\vec{r}' \, h(\vec{r} - \vec{r}') \, c(\vec{r}'), \tag{1a}$$

$$c(\vec{r}) = -\beta \, U(\vec{r}); \, \vec{r} \notin V, \tag{1b}$$

$$h(\vec{r}) = -1; \qquad \vec{r} \in V. \tag{1c}$$

$h(\vec{r})$ is the total correlation function, $c(\vec{r})$ is the direct-correlation function, ϱ is the macroion particle number density, $\beta = 1/k_B T$, and V is the volume occupied by the hard-core of an ellipsoid with its center of symmetry at the origin. $U(\vec{r})$ is the long-ranged interaction potential due to the charge on the surfaces of the macroions. More about $U(\vec{r})$ will be said later.

The ellipsoids are aligned in the z-direction of our coordinate system. The following transformation, a contraction of the z-axis relative to the x, y-axis, maps the ellipsoid-of-nearest approach which is twice as large as the hard-core onto spheres of unit radius:

$$A : (x, y, z) \rightarrow \left(\frac{x}{a}, \frac{y}{a}, \frac{z}{b} \right). \qquad (2)$$

$a(b)$ is the width (length) of the small (large) principal axis of the hard-core ellipsoid. Any function $f(\vec{r})$ is transformed accordingly: $F(\vec{R}) := f(A^{-1} \vec{R})$, with $\vec{R} = A \vec{r}$. Transformed functions are denoted by the corresponding capital letter. Equations (1) are thus transformed to:

$$H(\vec{R}) = C(\vec{R}) + \frac{6}{\pi} \phi \int d\vec{R}' \, H(\vec{R} - \vec{R}') \, C(\vec{R}'), \quad (3a)$$

$$C(\vec{R}) = - \beta \, U(\vec{R}) \quad R := |\vec{R}| > 1, \qquad (3b)$$

$$H(\vec{R}) = - 1; \qquad R \leq 1. \qquad (3c)$$

ϕ is the volume fraction of hard-core.

To solve this equation, $H(\vec{R})$ and $C(\vec{R})$ are expanded in a Legendre series:

$$H(\vec{R}) = \sum_{n=0}^{\infty} H_{2n}(R) \, P_{2n}(\cos \theta), \qquad (4a)$$

$$C(\vec{R}) = \sum_{n=0}^{\infty} C_{2n}(R) \, P_{2n}(\cos \theta), \qquad (4b)$$

where θ is the angle between $\vec{R}$ and the z-axis. Fourier transformation of Equation (1a) with respect to $\vec{R}$ gives:

$$H(\vec{K}) := \int d\vec{R} \, H(\vec{R}) \exp\{i \, \vec{K} \cdot \vec{R}\}$$

$$= C(\vec{K}) + \frac{6}{\pi} \phi \, H(\vec{K}) \, C(\vec{K}). \qquad (5)$$

Note that the physically relevant wave-vector $\vec{k}$ probed in light-scattering experiments is related to $\vec{K}$ by $\vec{k} \cdot \vec{r} = \vec{K} \cdot \vec{R}$, or:

$$\vec{k} = A \, \vec{K}. \qquad (6)$$

$\vec{K}$ is thus the probed wave-vector $\vec{k}$ rescaled with respect to the principle dimensions of the ellipsoidal hard-core.

Equations (4) yield upon Fourier transformation:

$$H(\vec{K}) = \sum_{n=0}^{\infty} H_{2n}(K) \, P_{2n}(\cos \theta'), \qquad (7a)$$

$$C(\vec{K}) = \sum_{n=0}^{\infty} C_{2n}(K) \, P_{2n}(\cos \theta'), \qquad (7b)$$

where θ' is the angle between $\vec{K}$ and the z-axis. The K-dependent expansion coefficients are Bessel transforms:

$$H_{2n}(K) = 4\pi(-1)^n \int_0^{\infty} dR \, R^2 \, j_{2n}(KR) \, H_{2n}(R), \quad (8a)$$

$$C_{2n}(K) = 4\pi(-1)^n \int_0^{\infty} dR \, R^2 \, j_{2n}(KR) \, C_{2n}(R). \quad (8b)$$

j_{2n} are spherical Bessel functions.

Next, each $C_{2n}(R)$ for $R \leq 1$ is expanded as:

$$C_{2n}(R) = \sum_{p=0}^{\infty} a_{2n, 2p} R^{2p}; \, R \leq 1; \, n = 0, 1, 2, \ldots \, (9)$$

The problem now is to determine the coefficients $\{a_{2n,2p}\}$ such that the closure relation (3c) is satisfied. The necessary and sufficient condition for this can be shown to be [4]:

$$f(\{a_{2n,2p}\}) := \sum_{n=0}^{\infty} \int_0^1 dR \left[\int_0^{\infty} dK \, K^2 \, j_{2n}(KR) \, H_{2n}(K) \right.$$

$$\left. + 2\pi^2 \, \delta_{n,0} \right]^2 = 0. \qquad (10)$$

In any practical numerical application of this scheme for solving the Ornstein-Zernike equation, the series expansions (7) are of course truncated. The level of truncation depends solely on the properties of the long-ranged interaction potential $U(\vec{R})$. This potential is discussed in the next section.

The long-ranged interaction potential

The following ansatz for a realistic interaction potential is chosen:

$$U(\vec{r}) = \frac{\sigma^2}{4\pi\varepsilon} \oint_{\partial V_1} ds_1 \oint_{\partial V_2} ds_2 \, \frac{\exp\{-\varkappa |r_1 - r_2|\}}{|r_1 - r_2|}. \quad (11)$$

In explanation: every pair of infinitesimal surface elements (ds_1 and ds_2) on the surfaces of two distinct ellipsoids (∂V_1 and ∂V_2) interact through a screened Coulomb potential with screening length $\varkappa^{-1} \cdot \varepsilon$ is the dielectric constant of the solvent and σ is the constant surface charge density. Using Equations (11), (9) and (3b) it can be shown that [4]:

$$C_{2n}(K) = 4\pi(-1)^n \sum_{p=0}^{\infty} a_{2n,2p} \int_0^1 dR\, R^{2+2p}\, j_{2n}(KR)$$
$$- \beta\, U_{2n}^o(K), \qquad (12a)$$

with:

$$U_{2n}^o(K) = \frac{2(4n+1)}{\pi\, \varepsilon\, a^2\, b} \int_1^{\infty} dR\, R^2 j_{2n}(KR) \int_0^{\infty} dq\, q^2 j_{2n}(qR)$$
$$\times \int_0^1 dx\, P_{2n}(x)\, \frac{g^2(x,q)}{x^2 + q^2 \left[\frac{1}{a^2} + \left\{\frac{1}{b^2} - \frac{1}{a^2}\right\} x^2\right]}, \quad (12b)$$

where:

$$g(x,q) = \sigma\, \pi\, a\, b \int_0^1 dz \left[1 + \left\{\left(\frac{a}{b}\right)^2 - 1\right\} z^2\right]^{1/2} \cos\left(\frac{qzx}{2}\right)$$
$$\times J_o\left(\frac{q}{2}(1-z^2)^{1/2}(1-x^2)^{1/2}\right), \qquad (12c)$$

where σ is the constant surface charge density.

The series expansions (7) are truncated at $n = N$, with N such that to a good approximation $U_{2M}^o(K) = 0$ for all $M > N$. As an example, for $a = 180$ Å, $b = 3000$ Å (Tobacco Mosaic Virus) and $\varkappa^{-1} = 30$ Å (corresponding the the Debye screening length in many experiments with TMV), the coefficients $U_{2n}^o(K)$, divided by Q^2, are given in Figure 1. Thus: $N = 2$ in this case. If $\varkappa^{-1}$ becomes larger more coefficients are needed: for $\varkappa^{-1} = 100$ Å it is found that $N = 5$. If $\varkappa^{-1}/b \gg 1$, the macroions will behave as spheres do, and clearly very many coefficients are needed. The given theory can be applied in numerical practice without problems for $\varkappa^{-1}/b \lesssim 0.05$. Otherwise too many coefficients have to be accounted for.

The flow-scheme for numerical analysis

To obtain the coefficients $H_{2n}(K)$, and from these the structure factor $S(\vec{K})$,

$$S(\vec{K}) = \sum_{n=0}^{N} S_{2n}(K)\, P_{2n}(\cos\theta'), \qquad (13a)$$

with,

$$S_{2n}(K) = \delta_{n,o} + \frac{6}{\pi}\, \phi\, H_{2n}(K), \qquad (13b)$$

the functional in Equation (10) must be minimized with respect to $\{a_{2n,2p}\}$. For this purpose it is convenient to define:

$$\tilde{H}_m(K) := \sum_{n=0}^{N} H_{2n}(K)\, P_{2n}(\cos\theta'_m), \qquad (14a)$$

$$\tilde{C}_m(K) := \sum_{n=0}^{N} C_{2n}(K)\, P_{2n}(\cos\theta'_m), \qquad (14b)$$

with $\{\theta_m\, m = 0, 1, \ldots, N\}$ a set of angles in $[0, \pi]$ such that the $(N+1) \times (N+1)$-matrix Equations (14) are invertible. From Equation (5) it follows immediately that:

$$\tilde{H}_m(K) = \tilde{C}_m(K) + \frac{6}{\pi}\, \phi\, \tilde{H}_m(K)\, \tilde{C}_m(K);$$
$$m = 0, 1, \ldots, N. \qquad (15)$$

The flow-scheme for the determination of $\{a_{2n,2p}\}$ is as follows: choose a starting set $\{a_{2n,2p}^{(0)}\}$; calculate for this set $C_{2n}(K)$, $n = 0, 1, \ldots, N$ from Equations (12); from Equation (14b) obtain $\tilde{C}_m(K)$ for all $m = 0, 1, \ldots, N$; claculate $\tilde{H}_m(K)$ from Equation (15); use these to get $H_{2n}(K)$ $n = 0, 1, \ldots, N$ by inverting Equation (14a); now the functional $f(\{a_{2n,2p}^{(0)}\})$, Equation (10) can be calculated. Choosing successively different coefficients and comparing the numerical values of the functional $f(\{a_{2n,2p}\})$ as obtained by the described flow-scheme, it is iterated down to its smallest value.

To stay on the "physical branch" of all possible solutions, we started at low concentrations with the known solution at infinite dilution, $a_{2n,2p} = -\delta_{n,o}\,\delta_{p,o}$, and used successively iterated coefficients as the starting set for a larger concentration.

An example

As an example we consider here $\varkappa^{-1} = 30$ Å and $\varkappa^{-1} = 100$ Å, with $a = 180$ Å, $b = 3000$ Å (TMV), and total charge of $Q = 100$ e. For $\varkappa^{-1} = 30$ Å the coefficients $U_{2n}^o(K)$ are plotted in Figure 1. In the iteration at most 12 coefficients $a_{2n,2p}$ for each n were needed, depending on the concentration. For $\varkappa^{-1} = 30$ Å the expansion coefficients $S_{2n}(K)$ for the structure factor $S(\vec{K})$ for two concentrations are shown in Figures 2 a and b; see Equation (13b). For this small

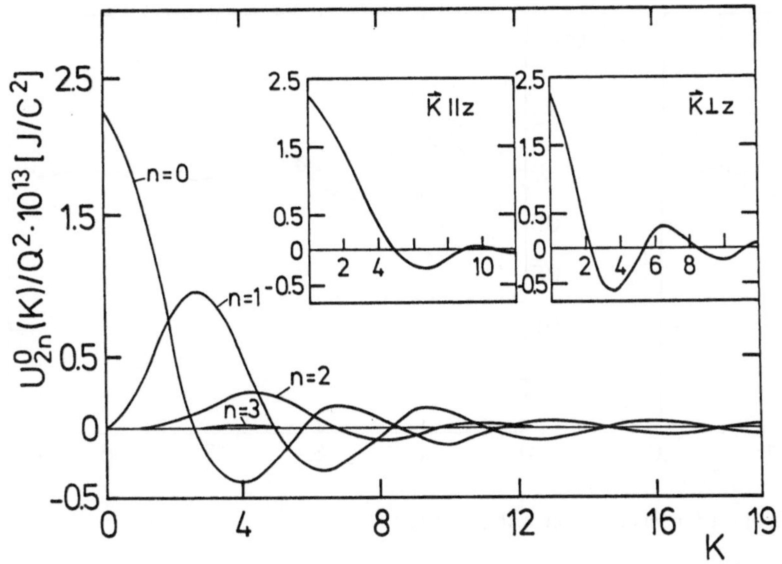

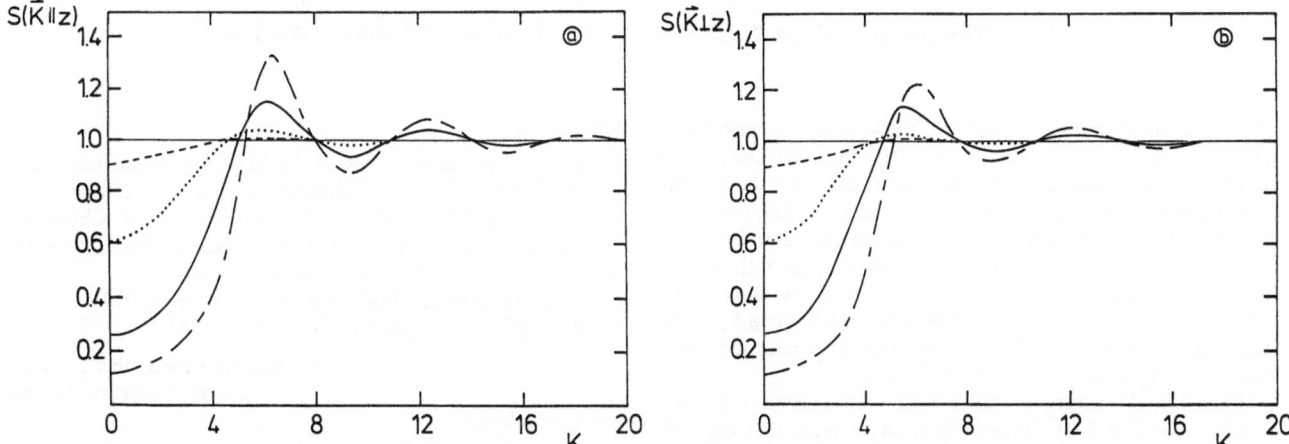

Fig. 1. The coefficient $U_{2n}^o(K)$ as defined in Equation (12b), normalized with respect to Q^2, for $\varkappa^{-1} = 30$ Å and the dimensions of TMV: $a = 180$ Å, $b = 3000$ Å. The dielectric constant is that of water at 25°C : $6.9 \cdot 10^{-10}$ C²/Nm². The inserted figures show $U^o(\vec{K}) := \sum_{n=0}^{\infty} U_{2n}(K) \, P_{2n}(\cos \theta')$ for $K \parallel z$ ($\cos \theta' = 1$) and $\vec{K} \perp z$ ($\cos \theta' = 0$)

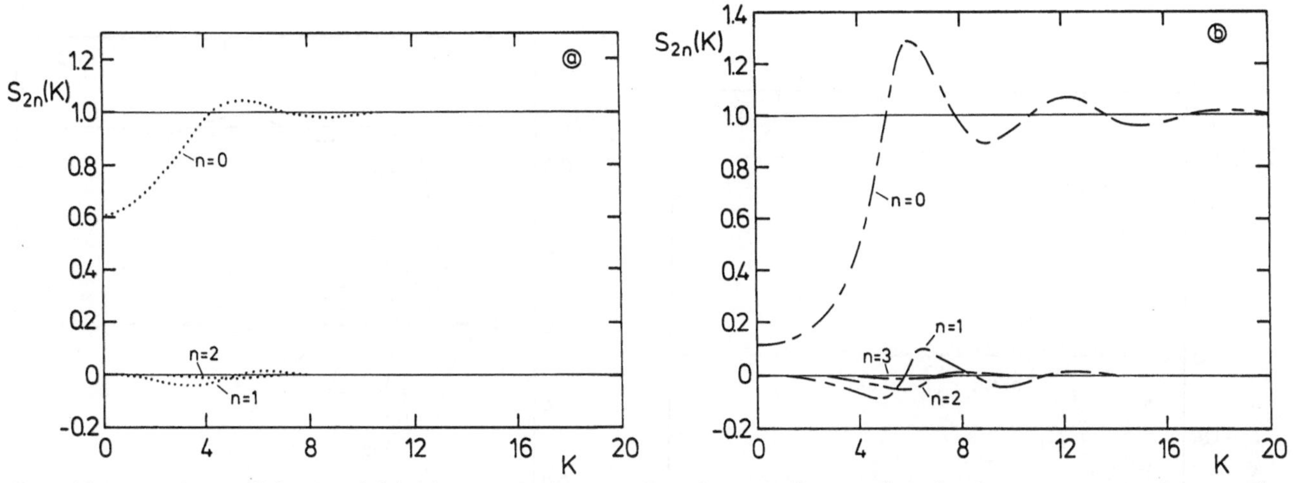

Fig. 2. The expansion coefficients $S_{2n}(K)$ for the structure factor, see Equations (13), for two volume fractions: $\phi = 0.05$ (a) and $\phi = 0.30$ (b). The screening length is $\varkappa^{-1} = 30$ Å, to total charge is $Q = 100$ e, and the temperature is 25°C

Fig. 3. The structure factor as a function of the rescaled wave-vector $\vec{K}$ for $\vec{K} \parallel z$ (a) and $\vec{K} \perp z$ (b), for various volume fractions ϕ: $---\phi = 0.01$; $\cdots \phi = 0.05$; —— $\phi = 0.15$; and $---\phi = 0.30$. The screening length is $\varkappa^{-1} = 30$ Å, the total charge is $Q = 100$ e and the temperature is 25°C

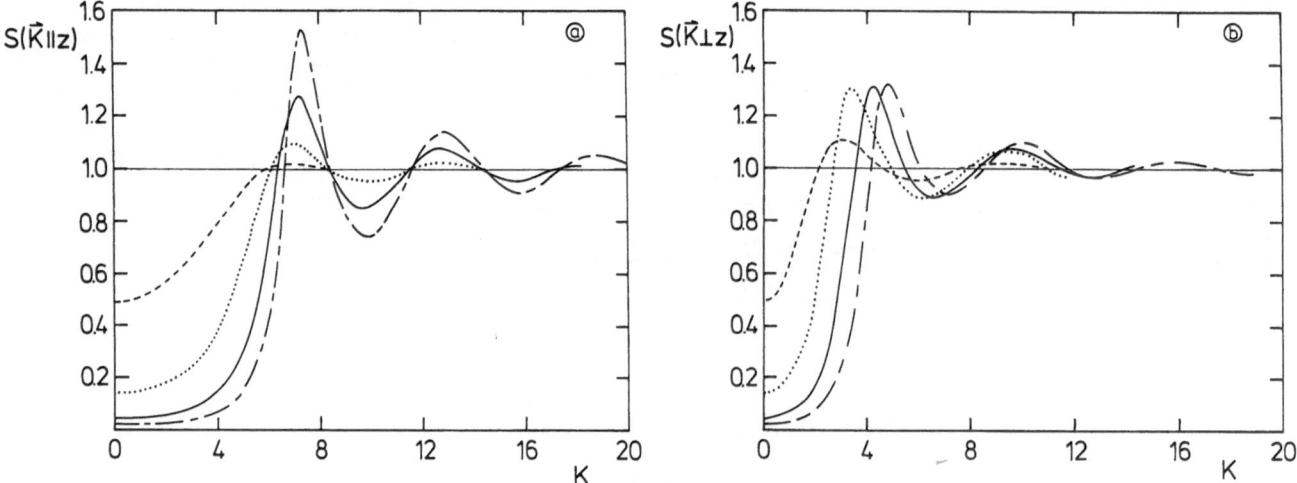

Fig. 4. The same as for Figure 3, but with $x^{-1} = 100$ Å

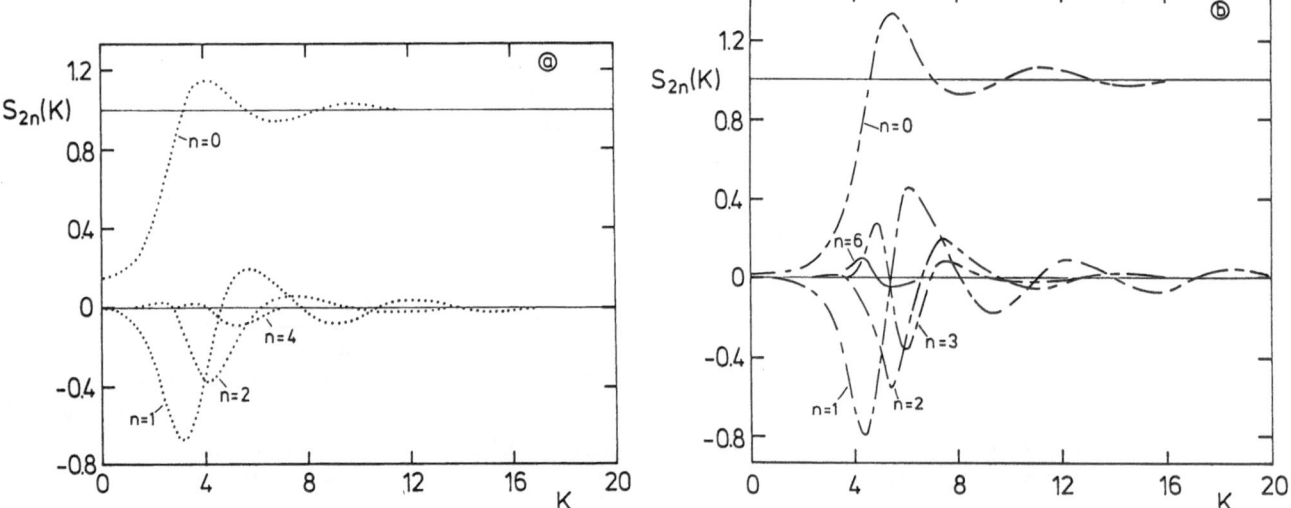

Fig. 5. The same as for Figure 2, but with $x^{-1} = 100$ Å. (a): $\phi = 0.05$ and (b): $\phi = 0.30$

charge and small screening length the "anisotropic" contributions to $S(\vec{K})$, that is, the contributions to $S(\vec{K})$ from the coefficients $S_{2n}(K)$ for $n > 0$ (remember: $P_o(\cos \theta') \equiv 1$), are not so large. This results in only minor differences in $S(\vec{K} \parallel z)$ and $S(\vec{K} \perp z)$; see Figure 3a and b. The "anisotropy" of $S(\vec{K})$ may be increased by either increasing the screening length x^{-1} or the charge Q, or both. Indeed as can be seen from Figure 4a and b, for $x^{-1} = 100$ Å, but still $Q = 100$ e, there is a considerable difference between $S(K \parallel z)$ and $S(K \perp z)$. This is of course due to the large contribution from the coefficients $S_{2n}(K)$, $n > 0$, which are shown in Figure 5.

Even if $S(\vec{K})$ is isotropic, $S(\vec{K}) \equiv S_o(K)$, the experimental structure factor $S(\vec{K}) \equiv S_o(A^{-1}\vec{k})$, see Equation (6), is strongly dependent on the direction of $\vec{k}$: in this simple case this anisotropy entirely comes from the trivial rescaling of the wave vector. In particular, the difference between $S(\vec{k} \parallel z)$ and $S(\vec{k} \perp z)$ is a difference in scaling of the k-axis with a factor of b/a.

References

1. For an extensive overview of the MSA-approximation and its relation to the Poisson-Boltzmann equation see: Blum L (1980) In: Eyring H, Henderson D (eds) Theoretical Chemistry: Advances and perspectives, Vol 5, Academic Press, pp 1–63
2. Pynn R (1975) J Phys Chem Solids 36:163
3. Hayter JB, Pynn R (1982) Phys Rev Lett 49(15):1103
4. Dhont JKG (1986) J Chem Phys 85(10):5983

Received December 24, 1986;
accepted January 20, 1987

Authors' address:

J. K. G. Dhont
Fakultät für Physik, Universität Konstanz
Bückestraße 13, Postfach 55 60
D-7750 Konstanz, F.R.G.

Progress in Colloid & Polymer Science   Progr Colloid & Polymer Sci 73:95–106 (1987)

The aggregation behaviour of tetradecyldimethylaminoxide

H. Hoffmann, G. Oetter, and B. Schwandner

Lehrstuhl für Physikalische Chemie I, Universität Bayreuth, Bayreuth, F.R.G.

Abstract: Static and dynamic light scattering, electrical birefringence, kinetic and rheological measurements were carried out on aqueous solutions of the zwitterionic surfactant tetradecyldimethylaminoxide ($C_{14}DMAO$) up to the liquid crystalline (l. c.) region. In the dilute concentration range the system forms globular micelles up to about 10 mM solutions. For higher concentrations, rodlike aggregates can be detected which increase in size with increasing concentration. At the overlap region of 100 mM solutions the forces between the micelles are repulsive up to the phase boundary and the light scattering decreases. With increasing temperature the system makes a sphere/rod transition which is indicated both by the temperature dependence of light scattering and electric birefringence. The experimental data show that the surfactant behaves like a nonionic system for low temperatures and like an ionic system for high temperatures. Also the cmc value, which is between a purely cationic and a nonionic system, shows the ambivalent nature of the surfactant. The system has no cloud point and forms a nematic phase as the first l. c. phase.

Key words: Micelles, electric birefringence, phase diagram, sphere/rod transition, light scattering.

Introduction

In many applications of surfactants, mixtures of nonionics with anionics are used to optimize certain solution properties such as interfacial tensions. Alkylpolyglykolethers are usually used as nonionics. The aggregation behaviour of these systems has been studied in detail [1, 2]. There are, however, other nonionics which are also of technical importance. One particular class are the glycosides, another the zwitterionic systems like alkyldimethylammoniumpropansulfonate and there are also the n-alkyldimethylaminoxides (C_nDMAO). Of particular interest is the dodecyldimethylaminoxide which has recently been studied. It was shown by Ikeda [3] that this system forms large rodlike micelles with lengths of several thousand Angström in the presence of salt. The system undergoes a sphere/rod transition when, for a given surfactant concentration, the salt concentration is increased from zero to moderate concentrations of 10^{-2} M. In many respects this system behaves as cetylpyridinium- or cetyltrimethylammoniumsalicylate which also form rather long rodlike micelles with persistence

lengths exceeding 1.000 Å [4]. The results by Ikeda on the dodecyldimethylaminoxide are in contrast with the results of Herrmann on the alkyldimethylaminoxides [5], who observed small globular micelles for surfactant solutions in the presence and in the absence of salt. He noted, however, a shift of the critical micellar concentrations to lower concentrations with increasing temperature, a behaviour which is most unusual. In a very recent NMR investigation by Rosano on the dodecyldimethylaminoxides [6] no observations were made which pointed to the existence of rodlike micelles. Both the studies by Rosano and Herrmann were carried out at rather low surfactant concentrations and it is possible that unisometric micelles would have been observed if the solutions had been studied at higher concentrations. Since the existence of rodlike micelles is of practical importance in surfactant chemistry we reinvestigated the aggregation behaviour of the alkyldimethylaminoxides as a function of the surfactant, the salt concentration and the temperature. Some of our results are reported in this paper.

Experimental results and their interpretation

The phase diagram of the system

The compound $C_{14}DMAO$ was obtained as a 25 % solution from the Hoechst Co. in Gendorf. The solution was freeze dried and the solid compound recrystallized several times from acetone. The crystalline product had a melting point of 131 °C. Surface tension measurements showed no minimum at the cmc; usually taken as evidence for purity. The cmc was $1.6 \cdot 10^{-4}$ what is somewhat lower than the value $2.7 \cdot 10^{-4}$ reported by Herrmann [5]. It is significant that the cmc of the system is about an order of magnitude higher than the cmc of a C_{14}polyglycolether of intermediate hydrophilic chain length: an indication that repulsive dipolar electrostatic contributions play an important part in the cmc. The cmc is somewhere between purely cationic and purely hydrophilic nonionic systems (Fig. 1). In contrast to the normal nonionics, the aminoxides, furthermore, have no cloud points, even in the presence of excess electrolyte. It is also noteworthy that the alkylaminoxides behave differently in

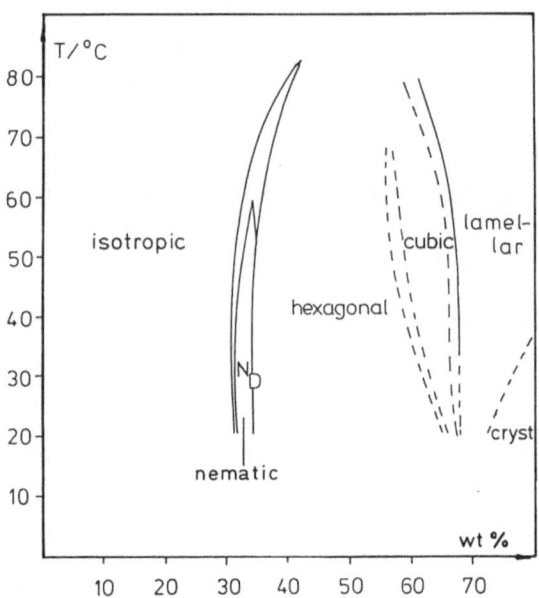

Fig. 2. Phase diagram of the system $C_{14}DMAO$-water

this respect when compared to the alkylphosphinoxides. These systems have a cloud point, as was recently shown by Pospischil [7] in a detailed study of the phase and solubilization behaviour of dodecyl- and tetradecyldimethylphosphinoxide. Both compounds had a cloud point and the first l. c. phase for higher concentration was a hexagonal phase.

Figure 2 gives the phase diagram for the $C_{14}DMAO$ system. The main differences with respect to the phosphinoxide are the lack of the cloud point and also the presence of a nematic lyotropic liquid crystalline phase

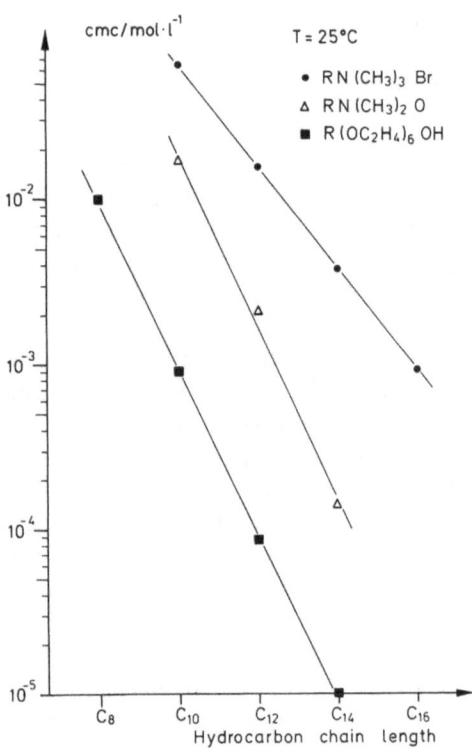

Fig. 1. The cmc values for different surfactant systems as a function of the chainlength of the alkyl group at 25 °C

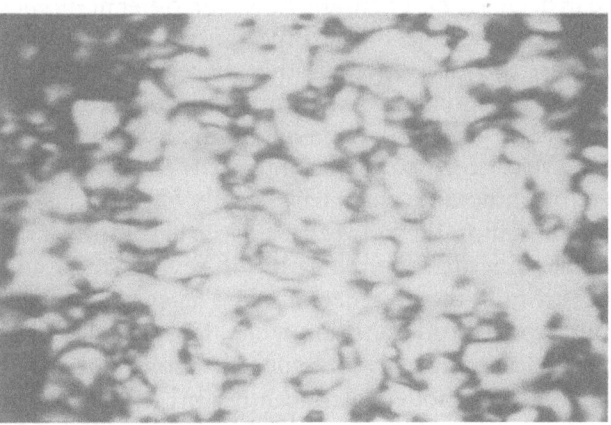

Fig. 3. Texture of the nematic phase of $C_{14}DMAO$ (34 wt %) in water at 25 °C

in a narrow concentration range, which is then followed by a hexagonal phase. The nematic phase can be recognized on its characteristic Schlieren texture under a polarization microscope (Fig. 3). While the microscopic structure of the nematic phase is not known at present, it is likely that the phase consists of small rodlike micelles, the orientation of which is correlated. The observed nematic phase is, as far as we know the first nematic phase which has been observed for a commercial nonionic surfactant. All the other nematic phases have been reported on ternary systems or on binary perfluoro-systems [8]. The aminoxides are coupled to an acid-base equilibrium and the possibility exists that the phase diagram is effected by this equilibrium.

$$
\begin{array}{ccc}
\text{CH}_3 & & \text{CH}_3 \\
| & & | \\
\text{C}_{14}\text{–N}^+\text{–O}^- + \text{H}_2\text{O} & \rightleftharpoons & \text{C}_{14}\text{–N}^+\text{–OH} + {}^-\text{OH} \\
| & & | \\
\text{CH}_3 & & \text{CH}_3
\end{array}
$$

The basicity of the oxide group is, however, rather weak and only a few tenths of a per cent of the compound are present in the protonated form. Furthermore, small additions of NaOH which would shift the equilibrium further to the right, did not effect the phase diagram. It is therefore very likely that the differences in the other diagrams between the aminoxides and the phosphinoxides are not due to the influence of the protonated form but rather reflect the stronger hydrophilicity of the aminoxide group with respect to the phosphinoxide group.

Light scattering

Concentration dependence

Figure 4a shows light scattering data which were made over a wide concentration range. For low concentrations the scattering intensity is somewhat higher in the presence of salt. The characteristic feature of the data is the strong maximum of the scattering intensity as a function of the concentration. The decrease of the scattering intensity reflects strong repulsive forces between the micelles which reduce the concentration fluctuations and hence reduce scattering. The effective molecular weights of the micelles which were determined from the light scattering measurements are given in Table 1. Data from electric birefringence measurements indicate that globular micelles could be pres-

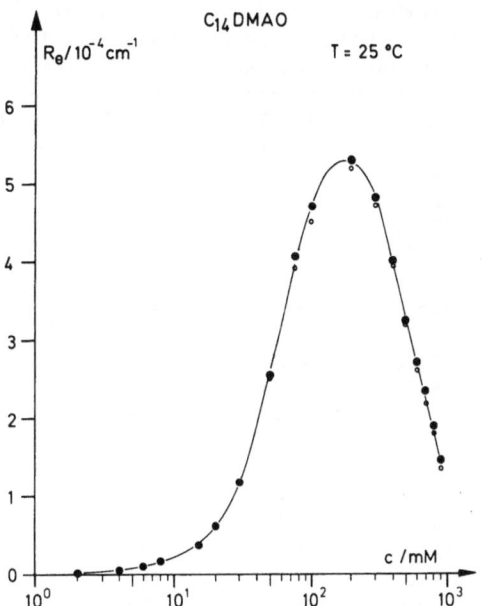

Fig. 4. a) The relative scattering intensity as a function of the concentration of C_{14}DMAO at 25 °C. (●) $R_{6°}$; (O) $R_{180°}$

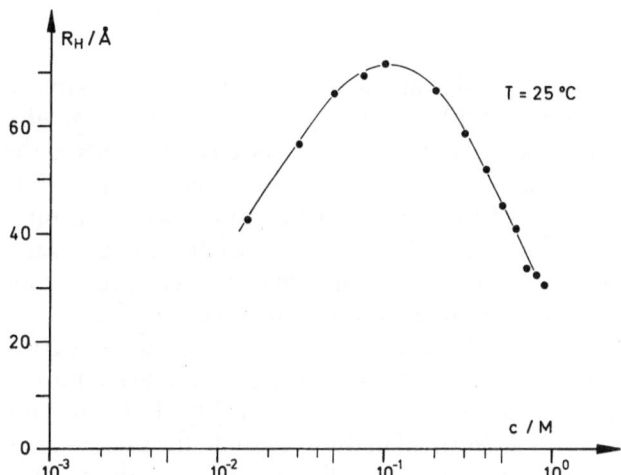

Fig. 4. b) The hydrodynamic radius as a function of the concentration of C_{14}DMAO at 25 °C

ent up to about 10 mM solutions while anisotropic micelles should be present at higher concentrations. The maximum in the scattering intensities under such conditions usually reflects the situation for which the large dimensions of the micelles are of about the same size as the mean distance between them. For the present system, this situation is reached at room temperature at around 100 mM. In agreement with these conclusions are the results from dynamic light scattering measurements. The hydrodynamic radius, which can

Table 1. Values for the effective molecular weight M_{eff}, the hydrodynamic radius R_H and the radius of gyration R_G, the lengths L_{eff} (calculated from M_{eff}, assuming cylindrical aggregates with a density of 0.9 g/cm³ and a short radius of 18 Å), L_{RG} (calculated from R_G) and L_D (calculated from R_H), the disorientation times τ_n (at 20 °C) and the kinetic times τ_2 for various aqueous concentrations of $C_{14}DMAO$ at 25 °C

c/mM	M_{eff}/gmol⁻¹	R_H/Å	R_G/Å	L_D/Å	L_{eff}/Å	L_{RG}/Å	τ_n/μs	τ_2/ms
4	34 300	—	—	—	—	—	—	—
6	42 500	—	—	—	—	—	—	—
8	50 700	—	—	—	—	—	—	—
15	62 900	42.4	—	110	114	—	0.7	—
20	76 600	—	—	—	139	—	0.8	7.0
30	98 500	56.3	—	172	179	—	1.0	—
40	—	—	—	—	—	—	—	5.0
50	127 400	65.8	58	230	232	—	1.1	4.5
75	135 400	69.3	122	256	246	—	2.0	3.5
100	117 500	71.4	133	—	—	460	2.8	2.8
150	—	—	—	—	—	—	—	2.0
200	73 900	66.4	121	—	—	450	8.2	1.6
300	40 000	58.4	95	—	—	327	22	1.2
400	24 800	52.1	57	—	—	195	50	0.95
500	16 100	45.0	73	—	—	251	95	—
600	11 100	40.9	114	—	—	394	200	—
700	8 260	33.5	180	—	—	625	400	—
800	5 820	32.6	137	—	—	474	850	—
900	4 040	29.9	183	—	—	635	1 700	—

be determined from the correlation time, passes over a maximum with increasing concentration (Fig. 4b). The position of the maximum also corresponds to the beginning overlap of the anisotropic micelles. The decrease of the hydrodynamic radius for concentrations above 100 mM may not signal that the aggregates become shorter again but rather that the repulsive interactions between the micelles increases. In many respects, both the dynamic and static light scattering data are very similar to the data on cetylpyridiniumsalicylate in the presence of salt [9]. Long rodlike micelles are formed in this system. In the present case, the micellar rods are smaller and therefore the point of beginning overlap is shifted to higher concentrations. The lengths of the rods which have been calculated from both the static and dynamic light scattering data are also given in Table 1. Below the overlap concentration they show a good agreement.

Temperature dependence

The results on the temperature dependence of the light scattering data are most interesting. In the concentration range between 10 and 200 mM in which the measurements were carried out, between 10° and 60 °C the scattering intensity increases with temperature and then decreases for higher temperature (Fig. 5).

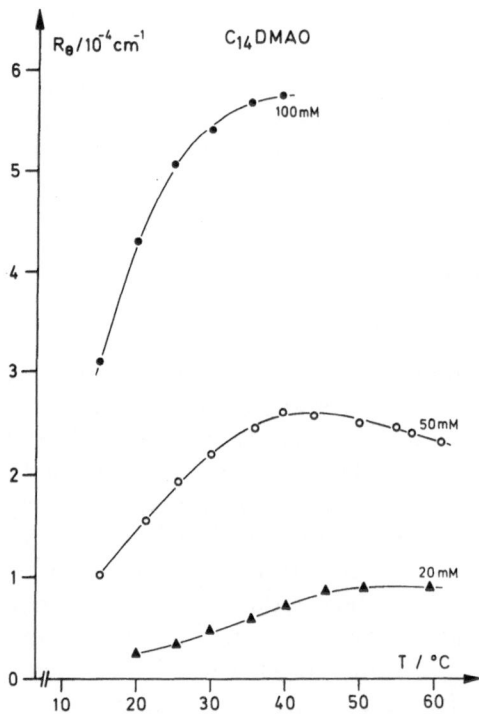

Fig. 5. The relative scattering intensity at an anlge of 6° as a function of the temperature for $C_{14}DMAO$ at various concentrations

In the low temperature range the system therefore behaves like the nonionic alkylpolyglycolethers, while at higher temperatures the system behaves like ionic systems which form rodlike micelles. This behaviour is a clear inidcation of the ambivalent nature of the system. It has the properties of both the ionic and nonionic surfactants. The unusual nature must be due to two competing effects. It is conceivable that the growth of the micelles is caused by the desolvation of the dipolar group with increasing temperature what results in a reduction of the interfacial area which the headgroup requires at the micellar interface. This effect is probably an energy term and leads to a stabilization of the micelle. It is this effect which is probably also responsible for the lowering of the cmc with increasing temperature. For higher temperatures this energy term is probably overcompensated by an entropic term which favours smaller micelles. The evaluated micellar parameters are given in Table 2.

Electric birefringence

Temperature dependence of the Kerr constant B

The unusual temperature dependence of light scattering data is also directly reflected in the electric birefringence behaviour. The Kerr constant passes over a maximum with increasing temperature (Fig. 6). It is noteworthy that the electric birefringence vanishes completely below a characteristic temperature, which indicates that the micelles undergo a sphere/rod transition with increasing temperature. With increasing concentration the transition temperature shifts somewhat to lower values.

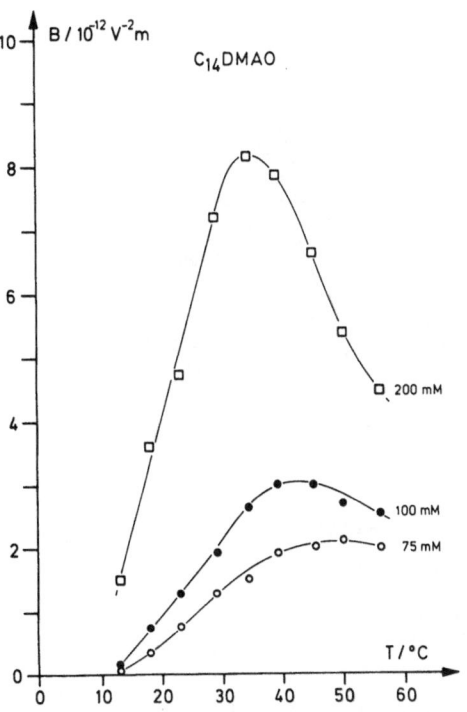

Fig. 6. The Kerr constant B as a function of the temperature for solutions of $C_{14}DMAO$ at various concentrations

Concentration dependence of the Kerr constant B

For a given temperature the amplitudes of the birefringence increase continuously with increasing concentration, up to the phase boundary of the l. c. region. The Kerr constants of the effects are shown in a double log plot in Figure 7. The diagram also shows the plot for the specific Kerr constant B_{sp}. In the dilute concen-

Table 2. Values for the short and the long disorientation time constants τ_{n1} and τ_{n2} (calculated from AC-electric-birefringence measurements) and the lengths L_n of the rods (calculated with Broersma's formula from τ_{n1}) as a function of the temperature at different concentrations of $C_{14}DMAO$

$T/°C$	$c = 75$ mM			$c = 100$ mM			$c = 200$ mM		
	$\tau_{n1}/\mu s$	$\tau_{n2}/\mu s$	$L_n/Å$	$\tau_{n1}/\mu s$	$\tau_{n2}/\mu s$	$L_n/Å$	$\tau_{n1}/\mu s$	$\tau_{n2}/\mu s$	$L_n/Å$
13	0.7	—	290	0.7	—	290	1.2	6.0	363
18	0.8	2.0	309	0.9	2.5	324	1.6	7.5	411
23	1.0	3.1	341	1.1	5.0	355	2.2	10.0	470
29	1.1	5.2	358	1.2	7.0	371	2.3	12.0	475
34	1.3	3.8	386	1.7	8.0	430	3.5	23.0	575
39	1.4	6.6	400	2.0	11.0	463	3.1	22.0	552
45	1.4	8.0	403	1.6	9.0	426	2.9	21.0	541
50	1.6	10.0	428	1.5	8.5	418	2.7	21.0	529
56	1.3	8.0	397	1.3	7.5	397	2.2	15.0	491

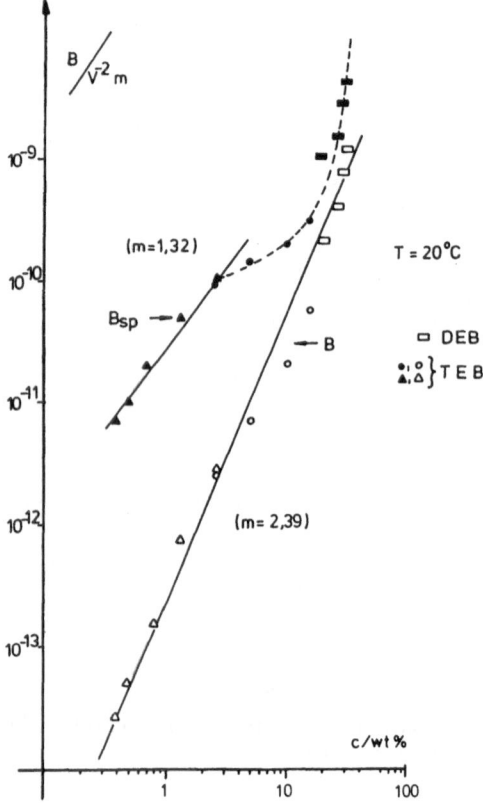

Fig. 7. The Kerr constant B and the specific Kerr constant B_{sp} as a function of the concentration for solutions of $C_{14}DMAO$ at 20 °C. DEB = dynamic electric birefringence; TEB = transient electric birefringence

exist. For small concentrations the build-up and the decay of the transient electric birefringence signal was below 1 μs and too fast to be measured accurately with the used instruments. For concentrations above 100 mM the decay time could be resolved, and a plot of the disorientation times against the concentration is given in Figure 8. In the concentration region between 2 % and 30 % of surfactant, the time constants increase from around 2 μs to 10 ms. A typical transient signal is shown in Figure 9a. The signal could not be fitted very well with a single time constant but rather two time constants which were about a factor 2–3 apart were necessary. This can be taken as evidence that not all rodlike micelles have the same length but show some polydispersity. The time constants in the plot are determined from the tail of the decay curves.

AC-measurements

For concentrations above 15 % of surfactant a steady-state value of birefringence could no longer be reached by applying rectangular field pulses. Furthermore, long rectangular pulses can lead to the decomposition of the solutions. Therefore the region of high surfactant concentrations was examined with the dynamic birefringence method using long electrical

tration region in which the rods do not overlap, the specific Kerr constant increases in the double scale more or less linearly with the concentration. In this concentration region, the specific Kerr constant obeys the equation $B_{sp} \sim (C/C_t)^x$ in which C_t is the transition concentration for the sphere/rod transition and x is a dimensionless scaling factor which has a value around 1.3. The increase of the specific Kerr constant with the concentration is a clear indication of the growth of the rods. For concentrations above the overlap, the specific Kerr constant follows a different law. This switch probably indicates that the process responsible for the birefringence is changing.

The time constants

In the dilute region the birefringence is determined by the orientation of the individual rods, while in the overlap the birefringence is a collective property in which correlation between the individual micelles can

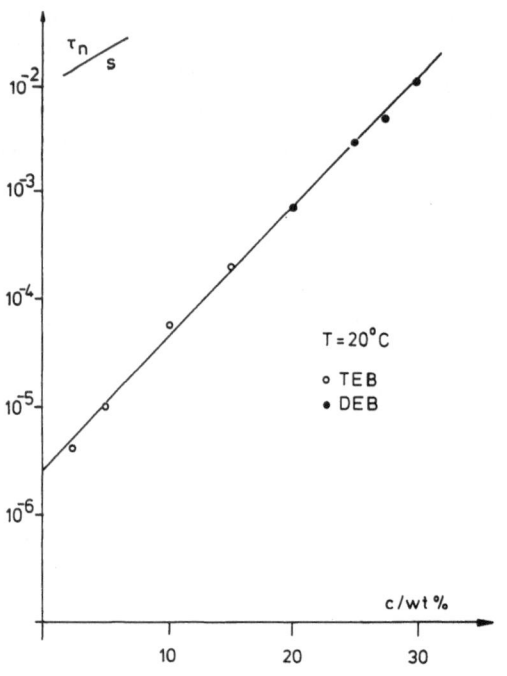

Fig. 8. The disorientation time τ_n as a function of the concentration for solutions of $C_{14}DMAO$ at 20 °C

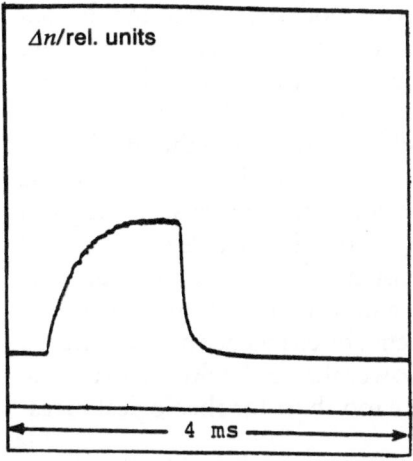

Fig. 9. a) Typical transient electric birefringence signal obtained for a 15 wt% solution of C_{14}DMAO at 25 °C. $E = 2.5$ kV/cm; pulse length = 1.3 ms

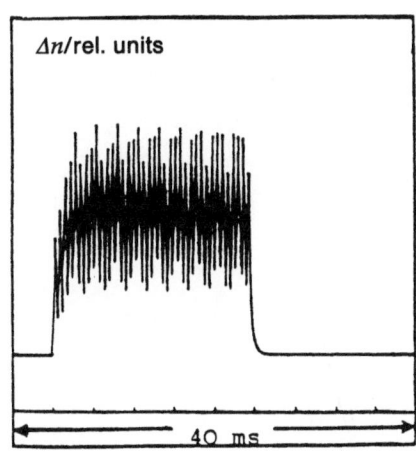

Fig. 9. b) Typical dynamic electric birefringence signal obtained for a 20 wt% solution of C_{14}DMAO at 20 °C. $E = 1.1 \cdot 10^3$ V/cm; $f =$ 250 Hz

sinusoidal pulses with varying frequencies. In this case the electrical field varies with the frequency

$$E = E_o \cdot \sin (\omega t) \tag{1}$$

and as a consequence, the birefringence $\Delta n(t)$ is given by

$$\Delta n(t) = \Delta n_{st} + \Delta n_{alt} \cos (2\omega t - \varphi) \tag{2}$$

where Δn_{st} and Δn_{alt} are the values for the stationary time independent birefringence and the alternating birefringence [10]. φ is the angle between Δn_{alt} and E. A typical signal is shown in Figure 9b.

For low frequencies Δn_{st} is identical with Δn_{alt}:

$$\Delta n_o = \lim_{\omega \to 0} \Delta n_{st}(\omega) = \lim_{\omega \to 0} \Delta n_{alt}(\omega) . \tag{3}$$

The Kerr constant B is then given by:

$$B = \Delta n_o / \lambda_o (E_o / \sqrt{2})^2 \tag{4}$$

where λ_o is the vacuum wavelength of light.

For concentrations above 15% of surfactant the Kerr constant and the disorientation times were determined by the dynamic elelctric birefringence method.

In a previous paper [11] we tried to explain the rather long time constants by coupling processes between rotational and translational motions, as suggested in the theories of Doi and Edwards [12]. In these theories it is assumed that slowing down of the rotation of rodlike molecules begins as soon as the rotational volumes begin to overlap. It was shown, however, both in theoretical and in experimental work [13, 14] that mutual hindrance does not begin at the overlap concentration but rather at much higher concentrations. For stiff rodlike Teflon fibers the rotational time constant is not affected at all when the concentration is increased far beyond the overlap concentration. The fact that slowing down does indeed occur at the overlap in the case of surfactants must therefore have a different origin. It is conceivable that the individual rods begin to form supermolecular network structures and the observed time constants are due to the relaxation process of the deformed network structures. In any case, whatever the origin of the rather long time constants is, it must be clear that any stress on the system which was applied either by electric field or by a shear stress relaxes with the observed time constant and the system is stress free and optically isotropic for times longer than these relaxation times. It is also clear that these time constants should be identical when a mechanical stress is applied to the solution and therefore they should also control the decay of the shear stress and the size of the zero-shear viscosities. This will be further elucidated in the next paragraph.

Rheology

Figure 10a shows the zero-shear viscosity as a function of the surfactant concentration. The viscosity

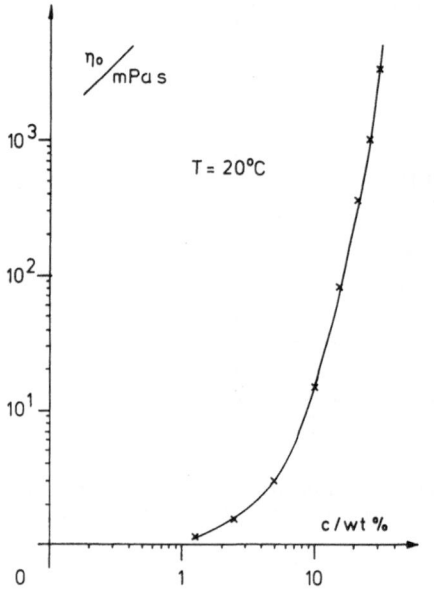

Fig. 10. a) The zero shear viscosity η_o as a function of the concentration for solutions of $C_{14}DMAO$ at 20 °C

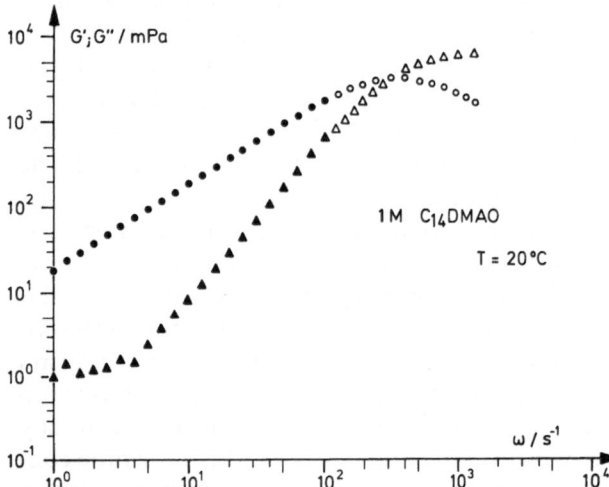

Fig. 10. b) The storage modulus G' and the loss modulus G'' as a function of the angular frequency ω for a solution of 1 M $C_{14}DMAO$ at 20 °C. (▲) G'; (●) G''; (△), (○) extrapolated values

begins to rise at the overlap and keeps rising up to the l. c. concentration range. In the simplest case, the zero-shear viscosity is given by a single relaxation time and a shear modulus:

$$\eta_o = G_o \cdot \tau . \qquad (5)$$

The rheological time constant τ can be determined from the shear-rate dependence of the viscosity or the frequency dependence of both the storage and the loss modulus. For high surfactant concentrations were were able to determine the time constant in this way (Fig. 10b). From the known G' and G'' values at low frequencies it is possible to extrapolate the angular frequency at which the storage modulus is identical to the loss modulus. At this point the time constant τ is given by $\tau = 1/\omega$ according to the simple case of a Maxwell model. The time constant of 2.8 ms turned out to be practically identical with one of 2.95 ms which was determined from electric birefringence measurements. For concentrations lower than 300 mM the structural relaxation times were too short for rheological measurements.

The structural relaxation time can also be determined from the extinction angle of flow birefringence. Data from such measurements are shown in Figure 11, where both the birefringence and the extinction angle are plotted against the shear rate. A relaxation time of 52 µs was calculated from the slope of the $\dot{y} - \chi$ plot. This time constant is again in good agreement with the relaxation time as determined from electric birefringence measurements.

Kinetic measurements

For an understanding of the macroscopic properties of micellar solutions, it is important to know how fast the micellar equilibration can adjust itself after the sys-

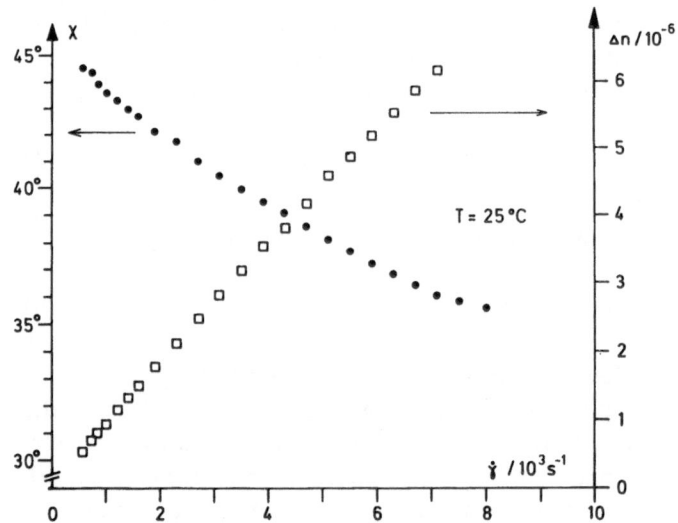

Fig. 11. The optical extinction angle χ and the birefringence Δn as a function of the shear rate $\dot{y}$ for a 300 mM solution of $C_{14}DMAO$ at 25 °C

tem is perturbed by a change of temperature, pressure or concentration. During this process, the number, density and size of the micelles changes and a redistribution of surfactant material between different micelles takes place. The reequilibration process can take place by surfactant monomer exchange but also by collision processes of micelles. The time constant for the equilibration can vary for the same surfactant concentration over many orders of magnitude from periods of hours to periods of microseconds, depending on the particular condition of the system, such as ionic strength, temperature, chainlength and concentration of additives like alcohols and so on. While it is possible to make predictions of the effects of those parameters on the kinetics of the total system, it is usually not possible to predict the absolute value of the time constant with any certainty. Therefore we measured the equilibration times experimentally. The experiments were carried out with a T-jump apparatus where the change of the system after a sudden T-jump could be monitored by following the change in the light scattering intensity. Some results are shown in Figure 12, where the reciprocal relaxation time is plotted against the surfactant concentration. With increasing concentration the relaxation times decrease from around 10 ms to 1 ms at 400 mM. For higher concentrations the scattering intensity of the solution was too small to detect. It is likely, however, that the plot of Figure 12 can also be extrapolated to higher concentrations and that at least approximate time constants can

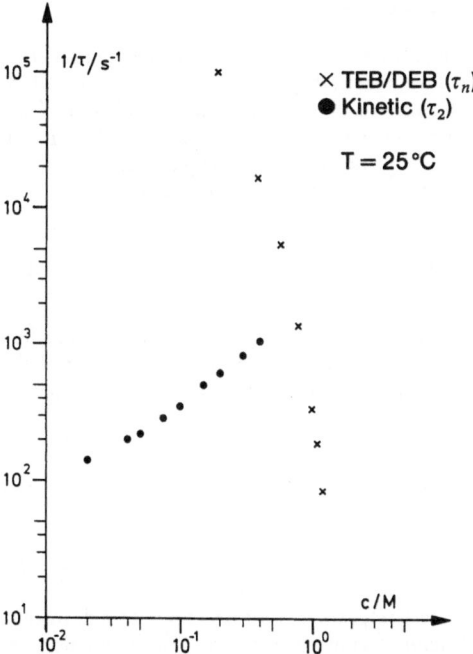

Fig. 13. The reciprocal kinetic relaxation time τ_2 and the reciprocal structural relaxation time τ_n as a function of the concentration of $C_{14}DMAO$ at 25 °C

be obtained. The time constants τ_2 for the chemical equilibration process are compared with the structural relaxation times τ_n in Figure 13. It is noteworthy that in the whole concentration range where both time constants could be measured, τ_2 is always much longer then τ_n. This clearly shows that the structural relaxation time is not kinetically controlled. During the time periods of τ_n the micelles can be treated as constant entities which do not change with time. Only for concentrations close to the phase boundary does τ_n become close to and even longer than τ_2. In this range it is therefore conceivable that the rheological properties are affected by the chemical equilibration processes.

The linear increase of $1/\tau_2$ with the surfactant concentration is very remarkable and significant for the equilibration mechanism. In systems where this mechanism proceeds by a step-wise process without contribution of collisions between micelles, it is usually observed that $1/\tau_2$ decreases with increasing concentration. The fact that the opposite is true for the present system indicates that the equilibration is collision controlled. Under such conditions, however, one usually observes a much faster increase of $1/\tau_2$ with concentration, as in the present case.

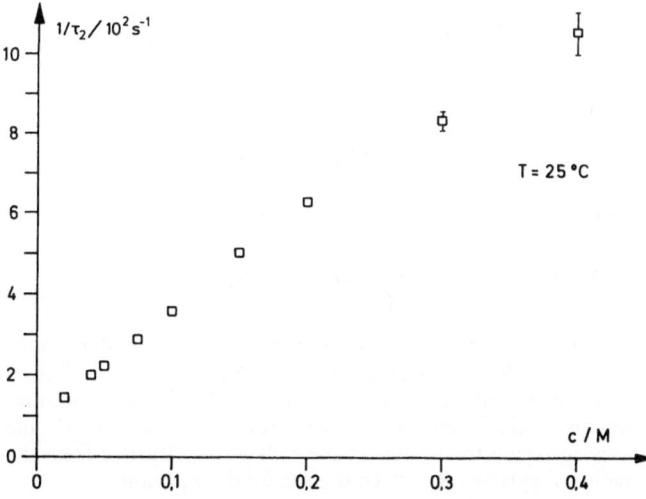

Fig. 12. The reciprocal kinetic relaxation time τ_2 as a function of the concentration of $C_{14}DMAO$ at 25 °C

Theoretical and experimental

Different techniques were used to investigate the micellar solutions and to determine the size, shape and dynamic properties of the micelles. The molecular weight of the micelles was obtained from light scattering data, obtained with a Chromatix KMX-6 instrument. Some essential equations should be explained:

The Rayleigh factor R_θ is defined by the expression

$$R_\theta = I_{sc} \cdot r^2 / I_o \cdot V_{sc} \qquad (6)$$

where I_{sc} and I_o are the intensities of the scattered and the incident light, r is the distance between the scattering volume v_{sc} and the detector and θ is the scattering angle. R_θ is also given by Equation (7):

$$R_\theta / K \cdot \Delta c = M_{eff} = M_w \cdot S(q) \cdot P(q) . \qquad (7)$$

Here Δc is the concentration of the scattering aggregates and defined by the expression $\Delta c = c_o - $ cmc. M_w is their molecular weight. $P(q)$ is the scattering factor or form factor of a single micelle and $S(q)$ is the structure factor which contains the interaction between the aggregates [15]. In the Rayleigh region where the size of the micelles is smaller than $\lambda_o/20$, $P(q)$ becomes equal to one. The scattering factor q is determined by

$$q = (4\pi n/\lambda_o) \cdot \sin(\theta/2) . \qquad (8)$$

Here n means the refractive index and λ_o is the vacuum wavelength of the incident light. The effective molecular weight M_{eff} can easily be calculated from Equation (7) if K is known, which is given by

$$K = (4\pi^2 n_o^2 / N_L \lambda_o^4) \cdot (dn/dc)^2 \qquad (9)$$

n_o is the refractive index of the solvent.

The necessary dn/dc values were obtained with a KMX-16 intrument also from Chromatix. The evaluation of the data was done in the same way as described in a previous study on the system dodecylpyridiniumsalicylate [4].

From M_{eff} the lengths of the micelles L_{eff} could be calculated under the assumption of the micellar density of 0.9 g/cm³. The L_{eff} values were calculated only for the dilute concentration range where the anisotropic micellar structures did not overlap. Here the evaluation was facilitated by the fact that the micelles practically carried no charge, so that the intermicellar interaction was very small and could be neglected.

In the theoretical equations for the light scattering, the structure factor S could therefore be set to unity. This was also true for the evaluation of the dynamic light scattering data which were carried out with a Malvern correlator.

According to the theory [16] the correlation function can be determined by the following equation:

$$G = \langle I(t) \cdot I(t + \Delta t) \rangle / \langle \bar{I} \rangle^2$$
$$= 1 + (\Sigma A_i \cdot \exp(-D_i q^2 \Delta t))^2 . \qquad (10)$$

For a system with a single scattering species the reciprocal correlation time τ_c is given by

$$1/\tau_c = D_c q^2 \qquad (11)$$

D_c is the collective diffusion coefficient of the scatterer. The correlation functions were fitted with a single correlation time τ_c. From the D_c values the hydrodynamic radius R_H can be calculated according to the equation

$$D_c = kT/6\pi \eta R_H \qquad (12)$$

where η is the viscosity of the solvent. This equation is only true for an ideal system without interacting particles. For interacting particles without hydrodynamic interaction, Pusey et al. [17] have shown that the measured diffusion coefficient D_c is related to the diffusion coefficient D_{iso} for noninteracting particles, by Equation (13)

$$D_c = D_{iso} / S(q) . \qquad (13)$$

If hydrodynamic interaction is taken into account, as shown by Ackerson [18], Equation (13) changes to

$$D_c = D_{iso} \cdot \frac{1 + H(q)}{S(q)} . \qquad (14)$$

In the hardsphere model, $H(q)$ is given by

$$H(q) = 0.75 (S(q) - 1) . \qquad (15)$$

From the D_c values the lengths L_D of the aggregates can be calculated from Broersma's formula for rigid rods [19]:

$$D_c = (kT/3\pi \eta L) \cdot (S - 0.5 \cdot (\gamma_\parallel - \gamma_\perp)) \qquad (16)$$

with $S = \ln(2L/d)$,

$$\gamma_\parallel = 1.27 - 7.4 \cdot ((1/\delta) - 0.34)^2$$

and

$$\gamma_\perp = 0.19 - 4.2 \cdot ((1/\delta) - 0.39)^2 .$$

The size of the micellar rods which were calculated both from the effective molecular weights and the hydrodynamic radii showed good agreement up to the overlap region. For higher concentrations intermicellar interactions had to be taken into account. For the overlap region the lengths of the rods were calculated from the radius of gyration R_G which is given by the equation

$$R_G = \sqrt{(R_{0°}/R_{180°} - 1) \cdot 3\lambda^2/16\pi^2 n^2} . \qquad (17)$$

One obtains for rigid rods [20]:

$$L_{rod}^2 = 12\, R_G^2 . \tag{18}$$

The relation between L_{eff} and L_{rod} is then determined by

$$L_{rod} = L_{eff}/S(q) \tag{19}$$

which is a possibility for calculating values for the structure factor $S(q)$.

Values for the lengths L_D and L_{eff} were calculated only for the dilute region. Above the overlap region the evaluation of the data led to completely wrong values.

Structural relaxation times (see Table 1) were obtained from transient and dynamic electric birefringence, from flow birefringence and, to some degree, from rheological measurements. From the slope of the $\dot{\gamma} - \chi$ plot of the flow birefringence the rotational diffusion coefficient D_{rot} and also the structural relaxation time τ can be determined by the following equations:

$$\chi = \pi/4 - \dot{\gamma}/(12\, D_{rot}) \text{ with } \dot{\gamma} \to 0 \tag{20}$$

and

$$\tau = 1/(6\, D_{rot}) . \tag{21}$$

As previously observed in other systems, it was found that the structural relaxation time is independent of whether the structures in the solution are perturbed by mechanical strain or by electric field pulse. For solutions for which the structural relaxation time could be measured with different techniques, the agreement was good.

The time constants for the chemical equilibration process were determined by temperature-jump measurements in which the change of the light scattering intensity was monitored after the T-jump.

Conclusions

On the basis of its properties, the system tetradecyldimethylaminoxide has to be arranged between the cationic and the purely nonionic surfactants. Its critical micelle concentration is lower than that of a cationic system but higher than that of a typical nonionic one where all have the same chain lengths. This abivalent nature is also expressed in the temperature dependence of the size of the micelles. With increasing temperature the system undergoes a sphere/rod transition. The rodlike micelles at first grow in size with increasing temperature, reach a plateau value at around 40 °C and finally decrease in size for even higher temperatures. The characteristic temperature for the sphere/rod transition depends on the concentration and shifts with increasing concentration to lower values. Around room temperature the system resembles nonionic sur-

factants, while at higher temperatures it behaves more like an ionic system for which the micellar rods decrease in size with temperature. Consequently the system has no cloud point. The micellar rods are shorter than for ionic micelles of surfactants with the same chain lengths. In the dilute concentration range where the rotational volumes do not overlap, the micelles grow in size with increasing concentration. At room temperature the system undergoes a sphere/rod transition at about 10 mM solutions. The overlap range begins at around 100 mM solutions. For higher concentrations the interaction between the micelles is repulsive in nature, even up to the phase boundary for liquid crystalline phases. In the semidilute range the viscosity rises continuously with the concentration up to the phase boundary.

As a first liquid crystalline phase the system forms a nematic phase which is then followed by a hexagonal, a cubic and a lamellar phase.

In the total isotropic region, the structural relaxation times are always shorter than the chemical relaxation times for the reequilibration process, so that the structural relaxation times are not kinetically controlled. The kinetic measurements indicate that the equilibration between micelles and monomers is collision controlled.

References

1. Kurzendörfer CP, Lange H, Schwuger MJ (1978) Ber Bunsenges Phys Chem 82:962
2. Zhu BJ, Rosen MJ (1984) J Coll Interf Sci 99:435
3. Ikeda S, Isunoda M, Maeda H (1979) J Coll Interf Sci 70:448
4. Angel M, Hoffmann H, Löbl M, Reizlein K, Thurn H, Wunderlich I (1984) Progr Coll & Polym Sci 69:12
5. Herrmann KW (1962) J Phys Chem 66:295
6. Chang DL, Rosano HL, Woodward AE (1985) Langmuir 1:669
7. Pospischil KH (1986) Langmuir 2:170
8. a) Yu LJ, Saupe A (1980) Phys Rev Lett 45:1000; b) Reizlein K (1983) thesis, Bayreuth
9. Hoffmann H, Rehage H, Reizlein K, Thurn H (1983) In: Shah DO (ed) Proceedings of the ACS-Symposium on Macro- and Microemulsion: Theory and Practice, ACS National Meeting, Vol 272, American Chemical Society, Washington DC, p 41
10. Thurston GB, Bowling J (1969) J Coll Interf Sci 30:34
11. Löbl M, Thurn H, Hoffmann H (1984) Ber Bunsenges Phys Chem 88:1102
12. a) Doi M, Edwards SF (1978) J Chem Soc Faraday Trans II 74:918; b) 74:560
13. Hess W (1986) private communication
14. Zero PM, Pecora R (1982) Macromol 15:87
15. Guinier A, Fournet G (eds) (1955) Small Angle Scattering of X-Rays, J Wiley & Sons, New York; Ottewill RH (1980) Progr Coll & Polym Sci 67:71

16. Berne BJ, Pecora R (eds) (1976) Dynamic Light Scattering, J Wiley & Sons, New York
17. Pusey PN, Tough RJH (1981) In: Pecora R (ed) Dynamic Light Scattering and Velosimetrie: Applications of Photon Correlation Spectroscopy, Plenum Press, New York
18. Ackerson BJ (1976) J Chem Phys 64:242
19. Broersma S (1960) J Chem Phys 32:1626, 1632
20. Kerker M (ed) (1969) The Scattering of Light and Other Electromagnetic Radiation, Academic Press, New York

Received February 12, 1987;
accepted February 12, 1987

Authors' address:

Prof. Dr. Heinz Hoffmann
Universität Bayreuth
Lehrstuhl für Physikalische Chemie
Postfach 10 12 51
D-8580 Bayreuth, F.R.G.

Progress in Colloid & Polymer Science   Progr Colloid & Polymer Sci 73:107–112 (1987)

Nonionic surfactant systems above the PIT
Part I. Phase behaviour

J. C. Ravey

Laboratoire de Physico-Chimie des Colloides UA 406 CNRS LESOC, Université de Nancy I,
Centre de 2éme cycle Vandoeuvre-lès-Nancy Cedex, France

Abstract: Just above the PIT for long nonionic amphiphiles C_mEO_n, the surfactant phase region (microemulsion area) and the water-poor domain tend to constitute a single one-isotropic phase region. According to the chemical nature of the oil (the third component), this phase domain coalescence takes place either directly toward the oil corner or for much larger surfactant/oil ratios, and give rise to typical phase diagram outlooks. Therefore, for intermediate water/surfactant ratio, a gap of solubility exists respectively for low or large oil contents, i. e. the molecular structures can be stable only in dilute or in concentrated solutions. The results are discussed in terms of the length/bulkiness and free volume of the oil molecules. We have also noted the absence of direct correlation between water solubilization and binary apolar phase behaviour.

Key words: Nonionic surfactant, phase behaviour, free volume.

Introduction

The ternary oil-water-nonionic surfactant systems may be very temperature dependent; this is particularly true for most of the one isotropic phase regions. Hence, the range of stability of the corresponding molecular structures appears rather limited, as a result of a delicate balance between hydrophobic and hydrophilic "forces", in both their enthalpy and entropy contributions. But in spite of their complexity and apparent diversity, a systematization can be proposed [1, 2, 3]. At the so-called Phase Inversion Temperature (PIT), where the affinities of the surfactant for water and the oil are equivalent, one can obtain a cosolubilisation of comparable amounts of oil and water with a minimum amount of surfactant. This PIT depends on the chemical nature of the amphiphile (i. e. its hydrophile-liphophile balance, HLB) and on the eventual presence of additives (inorganic salts). But it also depends on the chemical nature of the oil (i. e. its equivalent alcane carbon number, ACN) and on the eventual presence of hydrophobic additives (long alcohols), a phenomenon which is not yet quite understood.

In recent papers [4, 5, 6] we have proposed an empirical linear relation between PIT, HLB, ACN, salinity of the brine, etc. . . . .

$$PIT = -A + a \cdot HLB + b \cdot ACN - c \cdot S + \ldots \quad (1)$$

which is roughly valid for some range of the parameter values, at least when applied to ethoxylated nonionic surfactants. At this point, let us note that such a relation is also valid for perfluorocarbons-water cosolubilized by fluorinated surfactants, as recently recognized [7, 8]. The value of this b coefficient for linear alkanes is about $b = 2$; hence, for each additional CH_2 group in the oil the PIT is roughly shifted by 2 °C towards higher temperatures. For example [3], with the $C_{12}(EO)_5$ surfactant, the PIT is about 38 °, 46 ° and 51 °C, respectively for decane, tetradecane and hexadecane. In other words, all these systems appear roughly equivalent, but at different temperatures.

But on the other hand, our previous phase behaviour and structure studies of binary oil-surfactant mixtures have revealed quite different properties according to the chemical nature of the oil [9]. Hence, although the equivalence rule is certainly valid when the oil content is not large, i. e. in water-rich or surfactant-rich systems, we have to wonder whether it still holds for oil-rich ternary mixtures.

In the present series of paper we want to debate this matter, from both the points of view of the phase behaviour and the molecular aggregate structures, which are certainly strongly correlated.

Therefore, in a first part we shall discuss the different outlooks of the experimental phase diagrams (mainly in the oil corner), for several oil-surfactant systems. Structural results will be presented in a second part, in an attempt of correlation [10].

Experimental

Materials

A series of highly pure (98–99 %) polyoxyethylated alcohols were purchased from Nikko Chemicals (Japan), and were used as received. The oils were Merck products (purity 99 %), and used without further distillation.

A part of the data were taken from the literature. The other phase diagrams were determined by turbidimetric observation as follows. The rough limits of existence of the isotropic phases were observed at the onset of opacity of the solutions during a temperature scanning at the rate of 0.1 °C/min. For that purpose we used our experimental set up [11], where a micro-ordinator monitors the temperature cycles and stores all the data before processing. But some "critical" samples also had to be stored in baths at constant temperatures for more than a week, in order to attain a true equilibrium.

Phase behaviour

As already noted, near the PIT, comparable amounts of oil and water are cosolubilized with a minimum surfactant concentration [1–3]. The domain corresponding to this "surfactant phase" for (balanced) $C_{12}EO_n$ system often consists in a (quasi) isolated region, which is connected to a three isotropic phase triangle (Fig. 1).

Then, when the temperature is rising, there is a progressive move and coalescence of the surfactant phase domain into the isotropic water-poor oil + surfactant system regions, which pre-exist at noticeably lower temperatures [1, 9–18]. The most important point we want to emphasize in the present paper is the following one: the way this coalescence takes place strongly depends on the *chemical nature* of the oil. Therefore, all the oils should not be fully equivalent in the sense previously noted [3, 4] and as far as "inverse" structures are concerned. Since differential behaviours have already been recognized for the case of the (eventual)

apolar (binary) micellar aggregation phenomenon [9], we can then wonder whether the *apolar micellisation* ability may be correlated to the outlooks of the *ternary phase diagrams* in their oil- and water-poor regions, especially just above the PIT.

Two extreme behaviours can be described as follows (Fig. 2). In the first case (class I), the one-isotropic phase regions coalesce directly toward the oil corner; as a result, for a certain temperature range there is a relatively large domain of compositions located in this corner and adjacent to the oil-surfactant basis which allows water solubilization whatever the large oil content. In addition there is a salient microemulsion region which, when insufficient oil is present, turns into a lamellar liquid crystal phase. In any case, when temperature increases the maximum water/surfactant ratio steadily decreases. One typical example of such a behaviour is yielded by heptane/$C_{12}EO_4$ [15,18], the PIT of which is about 10°–11°C. At 16°–17°C, the maximum water/surfactant ratio is as large as 2.5 w/w, i. e. about 50 water molecules per surfactant. And for any water/surfactant ratio (below this limiting value), water may be solubilized whatever the oil content above 65–70 % w/w.

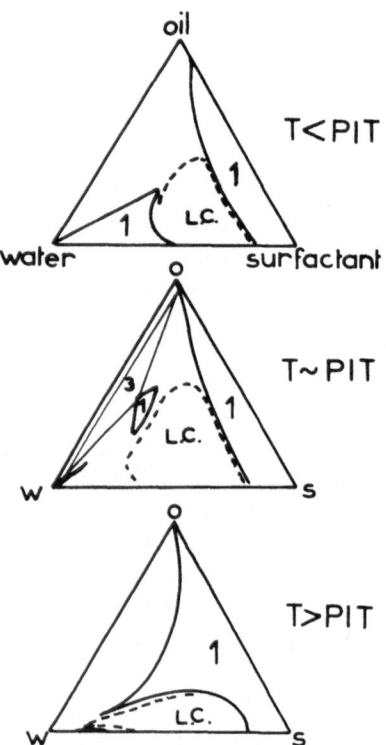

Fig. 1. Schematic evolution of the ternary phase diagram of nonionic surfactants in the vicinity of the PIT. 1 one-isotropic phase; 3 three-isotropic phases; LC lamellar liquid crystal

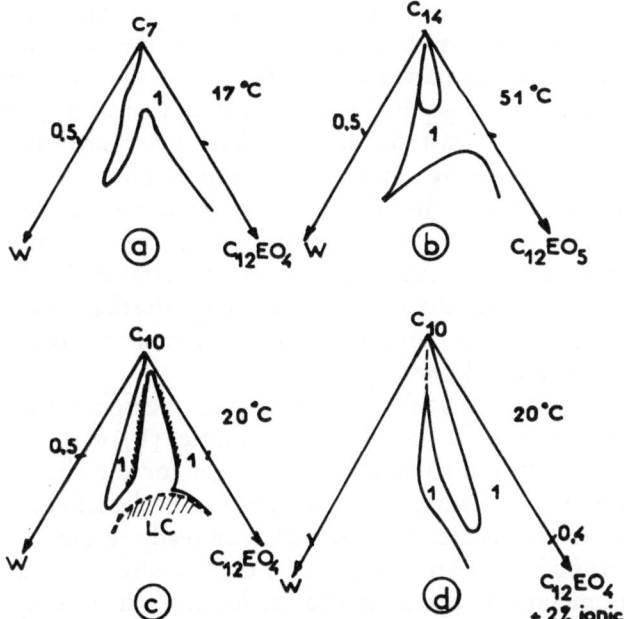

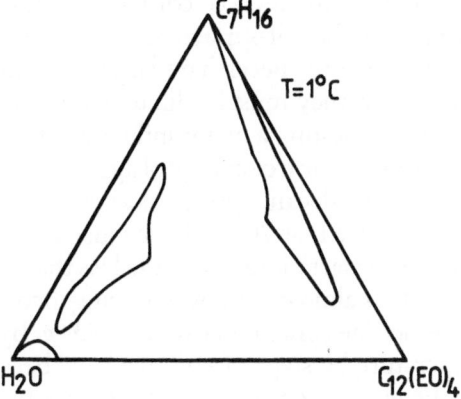

Fig. 3. A drawing (with permission from Ref. [18]) of the diagram for $C_{12}EO_4$/heptane at low temperature, showing the localization of the water-poor one-isotropic phase region.

Fig. 2. Representation of the oil-rich part of the phase diagrams. a) class I systems, see text; b) class II systems; c) $C_{12}EO_4$/decane at 20 °C; d) $C_{12}EO_4$ + 2 % $C_{12}EO_4$COONa/decane at 20 °C

For the other extreme type of diagram (class II), the coalescence between the surfactant phase region and the water-poor oil-surfactant phase area takes place for an oil/surfactant ratio as low as 1–2 w/w. But in this case, the water solubilization in the presence of larger amounts of oil, when possible, is mainly limited to very small values of this ratio, i. e. for those few compositions which already allowed this solubilization at noticeably lower temperatures. One example of such an extreme behaviour is constituted by the well known system Triton X 100 / benzene [19]. An analogous system is cyclohexane / C_6EO_6 [20]. But for most other oils, *in addition* there remains a *very narrow range* of water/surfactant ratios (practically equal to the maximum ratio) for which water can nevertheless be solubilized in the presence of large amounts of oil. An example of such a system is the couple $C_{12}EO_5$ / tetradecane [21]. Many other examples can be found in the former review by Ekwall [22] and papers by Friberg [23].

To summarize, for the longer oil molecules (or the more bulky ones), there is a gap of solubility for the larger oil contents and *intermediate water / surfactant ratios*, leading to the formation of a "nipper like" one-isotropic phase region: the one-phase region is salient *toward* the oil corner. A further temperature increase causes this gap to be filled in, and then the one phase

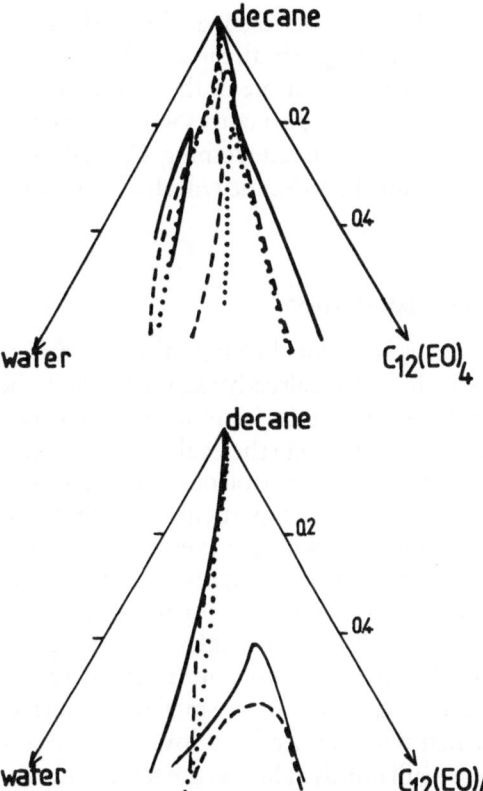

Fig. 4. Evolution of the gaps of solubility with temperature of $C_{12}EO_4$/decane. Upper figure: (———) 17 °C; (— — —) 20 °C; (······) 21 °C. Lower figure: (———) 22 °C; (— — —) 23 °C; (······) 25 °C

domain becomes extremely large, as exemplified by the system $C_{12}EO_4$ / hexadecane at 35 °C [1]. On the other hand, for the smaller oil molecules (or the less bulky ones), just above the PIT there is a gap of solubility for the smaller oil contents, and the one-phase region is prominent *from* the oil corner. In other words, according to the oil-surfactant couple, the stable molecular structures necessitate large or small amounts of the oil, i. e. they must be dilute or concentrated dispersions, suggesting a major influence of the entropic contribution to the free energy of aggregation.

The system $C_{12}EO_4$/decane appears as a very limiting case [1, 6, 14]. Indeed, at 20 °C (hence just above the PIT), for water/surfactant ratios a little less than 1, the gap of solubility is almost total, water being incorporated only for decane content in excess of 93 %. By rising the temperature by steps of 1 °C, we have been able to follow the way the coalescence takes place between the two "wings" of the one isotropic phase region, Figure 4 being self-explanatory. We have found this system particularly attractive to gain structural implications for the evolution of the phase behaviour of interest here. We have performed a lot of structure determinations according to both the sample composition and the temperature. Some of the results will be summarized in the second part of this work. For comparison, other new results concerning $C_{12}EO_n$ with hexane, cyclohexane, hexadecane will also be presented [10].

Discussion and conclusion

At this point, it is essential to note that the water-poor oil-surfactant regions already exist well below the PIT, and that their limiting (maximum) water/surfactant ratio increases little while the coalescence process takes place. Hence, this last isotropic phase, whose molecular structures appear particularly stable, has practically nothing to do with the eventual proximity of a lower consolute curve of the apolar binary mixture surfactant + oil. This can be seen for many systems, and that is particularly clear for the couple $C_{12}EO_4$/heptane [18]. In this case, this water-poor region already exists for a narrow range of small water/surfactant ratios when temperature is so low that heptane and $C_{12}EO_4$ are still immiscible, i. e. the temperature is below the lower critical point of the strictly apolar mixture (Fig. 3).

Anyway, we can conclude that the main driving phenomenon for the general evolution of the phase behaviour is the existence of very stable structures with low (but not zero) water/surfactant ratio, which could correspond to about one to two water molecules per EO group of the surfactant. Such values of the hydration degree are especially consistent with the phase behaviour of $C_{12}EO_4$/heptane.

Returning to the oil classification, the experimental data suggest that the length/bulkiness of the oil should be the driving parameter for the formation of a water solubilization gap in the presence of a large amount of this oil, at least for intermediate water/surfactant ratios. Therefore, one possibly important factor could be the ability of the oil to *penetrate without marked disturbance* into the surfactant palisade in the molecular structures. For example, an easier penetration by the small heptane molecule would result in the first type of diagram, while the more bulky hexadecane would lead to the second one.

Hence, one could wonder whether the right related parameter is the so-called "free volume" (V_f) of the oil: this fluctuation volume is the larger for the shorter alkanes [24] and rapidly decreases with the length of the oil molecule. Besides, we have already recognized the possible relevance of V_f for the micellization phenomenon in apolar media, at least as far as linear alkanes are concerned. Since it is a well known fact that V_f increases with temperature, the equivalence Equation (1) would indicate that the decremental of V_f due to one supplementary CH_2 group in the oil molecule should be balanced by a 2 °C increase of the temperature if V_f was the *only* relevant parameter. Such a simple view has to be investigated more thoroughly on the basis of thermodynamics. But at any rate, from a succinct discussion of some V_f data given in Appendix I, we can infer that the above equivalence rule *cannot* be derived from the temperature dependance of V_f alone. As a corollary, this effectively confirms the lack of evident correlation between (large) water incorporation and binary (apolar) phase behaviour. Hence, other possible parameters of importance like the dielectric constant, the Hildebrand's solubility parameter, etc, ought to be also investigated.

Nevertheless, another interesting indication of the possible relevance of this "penetration ability" parameter has been obtained in keeping *constant* the *nature of the oil*, but in bringing some very slight modification to the "surfactant component". Indeed, to reduce the penetration ability of the decane in the system decane/ $C_{12}EO_4$ (20 °C), we may try to slightly raise the radius of curvature of the surfactant interfacial film, keeping constant the area per polar head at the level of the hydrophile-hydrophobe separation in the surfactant molecule. For that purpose we have mixed a few percent of the highly pure $C_{12}EO_4COO^- Na^+$ surfactant with the original $C_{12}EO_4$. (The synthesis and purification of that ionic derivative were carried out in our laboratory according to an international patent (Claustahl, FRG)). In that way, we can gently monitor the repulsion between the hydrated polar EO groups of the surfactant chains by introducing on the internal surface of the film a few, but definite, number of electric charges. One can see that the corresponding phase diagram effectively exhibits the features tentatively ascribed to a low oil penetration [25], with a nipper-like one-phase region (Fig. 2).

To conclude this first part of the work, we have to emphasize that the above described phase behaviour evolution is not restricted to purely nonionic and hydrogenated surfactant systems. For example, let us look at the AOT amphiphile which allows the formation of microemulsions without the need of any cosurfactant: if a very bulky oil like iso-octane is used as the third component, we can verify that the corresponding phase diagram (in the oil corner) effectively exhibits this nipper-like outlook [26]. And substituting iso-octane by n-heptane (20 °C) leads to class I diagrams [27].

We have recently recognized that the general evolution of the PIT (Eq. (1)) is not limited to purely hydrogenated compounds [28, 29]. As a matter of fact we have obtained quite parallel behaviours with systems made of water and *perfluoroalkanes* cosolubilized by the parent surfactants *polyoxyethylene perfluoroalcohols* $C_mF_{2m+1}CH_2EO_n$. For example, just above the PIT, we have found that for $m = 6$, $n = 4$, the use of bulky oils like perfluorodecalin and perfluoromethylcyclohexane leads to nipper-like one-isotropic phase regions. On the contrary, if the oil molecule is linear $(C_8F_{17}CH=CH_2)$, (with the same number of C atoms as in the perfluorodecalin), we can check that the corresponding inverse system diagram effectively belongs to the right class of diagram (less bulky oils), and looks like the diagram with (hydrogeneted) decane/$C_{12}EO_4$. These findings on fluorinated compounds will be presented and discussed elsewhere [7].

But at any rate, structural investigations are necessary to assess the actual relevance of the above interpretations, and will be given in part II of this work.

Appendix I: Free volume of alkanes

The term "free volume" has been loosely used for different concepts. One is the excess of the volume occupied per mole of liquid over the sum of the volumes of all the individual molecules. Another kind of free volume is derived from the *entropy* of *expansion* of a van der Waals fluid (V_f); its value is extremely small (less than a few tenths cm³/mole), and it may represent a fluctuation volume, circumscribed by the molecules oscillating in the "cage-like" available room.

One way to calculate this last V_f is to consider the expansion from liquid to vapour states at a given temperature, characterized by the entropy change (24)

$$\Delta S_v = \Delta H_v/T = R \ln (V^g/V_f)$$

V^g being virtually the molar volume in the gazeous state.

From data in the literature for ΔH_v (25 °C), C_p(liquid)- C_p(gas) and the equilibrium vapour pressure, we can calculate V_f for various alkanes at different temperatures. A few results are shown in Figure 5, for two temperatures.

The values are the largest for hexane *and* cyclohexane, but rapidly decrease with the length of the alkane. Let us emphasize that the calculation is extremely sensitive to the value ascribed to $\Delta H_v(T)$; this is especially true for longer alkanes for which an uncertainty of a few per cent on ΔH_v may lead to a 100 % change in V_f. Hence, calculated values may be only indicative.

In another way to estimate this free volume, one assumes that the centre of each molecule moves in an equivalent spherical cage whose radius depends on the

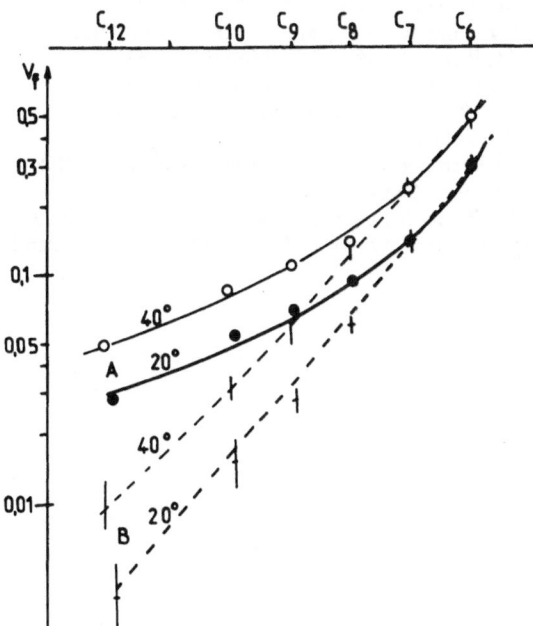

Fig. 5. Free volume (V_f) of linear alkanes calculated for two temperatures from the liquid-vapour entropy change (+) and the spherical cage model (O) (see text)

Progress in Colloid & Polymer Science, Vol. 73 (1987)

average distance between nearest neighbours and the actual volume of the molecules regarded as rigid spheres. In that case the following formula can be derived [24]:

$$V_f = (k/V^2) (R\beta/\alpha)^3.$$

$V(T)$ is the molar volume at temperature T, α is the volume coefficient of thermal expansion and β the coefficient of compressibility. A reasonable choice for k is $k = 7$, which would correspond to a coordination number of about 12. The results shown in Figure 5 are quite in agreement with the previous ones, but only as far as shorter alkanes are considered. Indeed, it is obviously conceivable that the spherical cage model is no longer suitable at all for the longer linear alkanes.

But the most important point we want to stress here is the following one: whatever the type of calculation the temperature dependence of V_f is such that *one additional* CH_2 is roughly equivalent to a *decrease of 10°* to 20 °C (and not of 2 °C) in order to keep constant the free volume of the liquid alkane.

References

1. Friberg S, Buraczewska I, Ravey JC (1979) In: Mittal KL (ed) Micellization, Solubilization and Microemulsions, vol 2, Plenum Press, New York, p 901
2. Kahlweit M, Lessner E, Strey R (1984) J Phys Chem 88:1937
3. Kunieda H, Shinoda K (1985) J Coll Interf Sci 107:107
4. Buzier M, Ravey JC (1983) J Coll Interf Sci 91:20
5. Buzier M, Ravey JC (1985) J Coll Interf Sci 103:594
6. Buzier M (1984) These Université Nancy I, France
7. Ravey JC, Stebe MJ, preceding paper
8. Mathis G, Leempoel P, Ravey JC, Selve C, Delpuech JJ (1984) J Am Chem Soc, 106:6162
9. Ravey JC, Buzier M, Picot C (1984) J Coll Interf Sci, 97:9
10. Ravey JC, Buzier M, Oberthur R (1987) following paper, Part II of this work
11. Ravey JC, Buzier M, Dupont G (1987) In: Rosano H (ed) Microemulsions, Surfactant Science Series No 24, Marcel Dekker, New York, in press
12. Ravey JC, Buzier M (1986) In: Mittal KL (ed) Surfactant in Solution, New-Dehli, submitted for publication
13. Lichterfeld F, Schmeling T, Strey R (1986) J Phys Chem 90:5762
14. Ravey JC, Buzier M (1984) In: Mittal KL, Lindman B (eds) Surfactants in Solution, Vol 3, Plenum Press, New York, 1759
15. Bostock TA, McDonald MP, Tiddy GT, Waring L (1979) SCI Chemical Society Symposium on Surface Agents, Notthingham
16. Nilsson PG, Lindman B (1982) J Phys Chem 86:271
17. Friberg S, Flaim T (1982) ACS Symposium Series no 177:1
18. Boyle MH, McDonald MP, Ross P, Wood RM (1982) In: Robb ID (ed) Microemulsions, Plenum Press, New York, p 103
19. Marsden SS, McBain JW (1948) J Phys Coll Chem, 52:110
20. Mulley BA, Metcalf AD (1964) J Coll Sci, 19:501
21. Kunieda H, Shinoda K (1982) J Dispersion Sci Techn, 3:233
22. Ekwall P (1975) In: Brown GH (ed) Advances in Liquid Crystal, Vol 1, Academic Press, New York, p 1
23. Flaim T, Friberg S (1982) In: Brown GH (ed) Advances in Liquid Crystals, Vol 5, Academic Press, New York, p 137
24. Hildebrand JH (ed) (1964) Solubility of non-electrolytes, Dover, New York
25. Ravey JC, Alibrahim M, to be published
26. Eicke HF, Kubik R, Hasse R, Zschokke I (1984) In: Mittal KL, Lindman B (eds), Plenum Press, New York, p 1533
27. Rouviere J, Couret J, Lindheimer M, Dejardin J, Marrony R (1979) J Chim Phys 76:289
28. Robert A, Tondre C (1984) J Coll Interf Sci 98:515
29. Mathis G, Ravey JC, Buzier M (1982) In: Robb ID (ed) Microemulsions, Plenum Press, New York, p 85

Received March 7, 1987;
accepted March 9, 1987

Author's address:

Dr. J. C. Ravey
Universite de Nancy I
Faculte des Sciences
Laboratories de Physico Chimie des Colloides
UA CNRS No 406
Boite Postale No. 239
Vandoeuvre-les-Nancy Cedex, France

Progress in Colloid & Polymer Science

Progr Colloid & Polymer Sci 73:113–126 (1987)

Nonionic surfactant systems above the PIT
Part II. Molecular aggregate structures

J. C. Ravey[1]), M. Buzier, and R. Oberthur[2])

[1]) Laboratoire de Physico-Chimie des Colloides, UA 406 CNRS LESOC, Université de Nancy I, Centre de 2éme cycle, Vandoeuvre-lès-Nancy Cedex, France
[2]) Institut Laue Langevin 156 X, Grenoble Cedex, France

Abstract: For systems just above the PIT, attempts of correlation between phase behaviour in some ternary oil/nonionic surfactant/water systems and molecular structures are presented. The structural investigations have been performed mainly by small angle neutron scattering, making use of the contrast variation method. Emphasis is put on the morphology of the interfacial surfactant film: thickness of the hydrophobic and hydrophilic parts, oil and water penetration into the film, area per polar head, mean curvature. Results in various oil/surfactant systems show that apolar micellization and water solubilization are not markedly correlated. A whole picture of the evolution of the inverse structures is more specifically reported for decane/$C_{12}EO_4$ concerned especially with the influence of temperature.

Key words: Nonionic surfactant, neutron scattering, aggregate structure, microemulsions, water solubilization.

Introduction

In the part I [1] of this work, we have analyzed the ternary phase diagram outlook just above the PIT for (long) nonionic surfactants as a function of a geometric characteristic of the oil molecule. It was shown that this phase behaviour is clearly oil dependent and requires a structure-based elucidation, an attempt at which we propose now.

In a first part we shall try to find whether apolar micellization and subsequent water solubilization may be correlated. Various oil-surfactant systems will be considered for this purpose.

Then, results on the system $C_{12}EO_4$/decane/water will be thoroughly discussed, bringing to light the composition of the continuous phase and the morphological properties of the molecular aggregates. The evolution of the structures with temperature will be of special concern.

As a matter of fact the study presented in this paper was essentially performed using the small angle neutron technique, but other scattering methods were also used as complementary tools.

Experimental

The scattering techniques

Small angle X-ray scattering measurements were mainly performed by photographic recording for the structural determination of multilayered aggregates. A few spectra were also obtained using a linear proportional counter, when studying fully hydrogenated systems, with typical measuring times of 1 h. Although not obtained on an absolute scale, they were used to check definitely the validity of the contrast variation method, showing that structures determined from neutron data are quite consistent with the experimental X-ray spectra.

Elastic *light scattering* experiments were carried out in order to determine the depolarization ratio for the solutions of optically anisotropic aggregates. The classical FICA apparatus fitted to precision polarizing prisms was used for that purpose.

Most of the results presented in this paper were derived from *small angle neutron* scattering measurements. They were performed at the Institute Laue Langevin (Grenoble, France) and at the Laboratory Léon Brillouin (Saclay, France). The transfer momentum (q) range was generally 0.01 to 0.2 Å^{-1}, although q values as low as 0.005 Å^{-1} were considered for the extrapolation of the scattered intensities $I(q \rightarrow 0)$.

The presentation and discussion of the method used to analyze the data can be found elsewhere [2–5]. As a summary, we have to find the particle models whose theoretical scattering spectra lead to the best fits to the experimental data. A model is satisfactory if, for a given overall oil/water/surfactant composition, the whole set of data are coherent, irrespective of the various isotopic mixtures which were actually used (generally four different "contrasts" for each sample were considered). Interparticle effects are taken into account by using the analytical structure factor derived from the Percus-Yevick solution for equivalent hard sticky spheres [6, 7].

Materials

For the structure determinations both the oil and water components were mixtures of the hydrogenated and deuterated isotopes. The oil were purchased from Merck, Sharp and Dohm (Canada), and D_2O from the CEA (France). The isotopic compositions of these solvents were chosen so that the variation contrast method was illustrative and powerful at most. On the other hand, we have checked that the exact delineation of the one isotropic phase domains was not very sensitive to the H/D substitution. Therefore we can infer only a slight structural dependence on this isotopic composition, as proved by the self-consistency of all our results.

Results and discussion

Apolar micellization and water solubilization

Results concerning the eventual micellization of $C_{12}EO_n$ surfactants in various organic solvents have already been reported [3, 8]. We now want to show whether they can be correlated to water solubilization. We had found that, according the experimental conditions, the binary system could belong to one of the three following classes.

1. In a first class, no aggregate could be detected at all. This was the case for methanol, benzene, chloroform at 20 °C, for any surfactant. Interestingly, no water incorporation is possible with benzene, unless the oil content is very high. This is due to the preferential binding of benzene molecules on the EO groups [9], preventing the autoassociation of the surfactants unless the ratio benzene/surfactant is sufficiently low. Indeed, in the surfactant rich region, the benzene rings may be *numerically* unable to cover more than typically from one half to one fourth of the EO groups; in that case, the hydration of the oxyethylene chains may

become sufficient to favour the formation of lamellar aggregates and lamellar liquid crystals [10].

Therefore, although this corresponding ternary diagram could be considered as peculiar to a "bulky-oil system" (no water solubilization at higher oil content, class II in the paper I), clearly the phase behaviour does not seem to result from these geometrical properties, but from a polarity effect. Exactly the same things happen for systems with chloroform.

The case of methanol is quite different, since huge water amounts may be incorporated into the non micellar mixture in the form of practically monomolecularly dispersed species, i. e. without formation of very micellar aggregates. In this case, we should rather consider it as a surfactant-water cosolubilization due to the highly polar methanol.

2. In a second class (B), we can detect the presence of very small aggregates, without critical concentration, whose size gently increases with the surfactant content. Typically, the aggregation number N is about 3 to 6. Examples of such systems are heptane/$C_{12}EO_4$ and cyclohexane/$C_{12}EO_4$ at 20 °C. The analogous behaviour of these two binary systems could be roughly ascribed to the similar values of their V_f (free volume) when considered as the "fluctuation volume" of the oils [11] (see part I). It could be ascribed neither to V_f considered as the simple excess volume over the sum of individual molecular volumes, neither to ε (dielectric constant) (see Table 1), since the cyclohexane should behave like decane, and this in contradistinction to the experiments. Obviously, the parameter of importance is no longer the solubility parameter. Hence, as far as the similarity in the micellization of the two systems are concerned, the entropy contribution to the free energy of the system appears quite important.

We have to note that at 20 °C the water solubilization proceeds very differently according to the chemical nature of these two hydrocarbons. As seen in Figure 1, at large oil contents (more than 80 % w/w),

Table 1

Solvent	δ Hildebrand solubility parameter	Dielectric constant ε	Free volume (cm³/mol, 30 °C) Excess vol.	Fluctuation vol.
Cyclohexane	8.2	2.01	0.48	0.3
Heptane	7.4	1.93	0.53	0.2
Decane	7.7	1.99	0.46	0.03–0.06
Hexadecane	8.0	2.40	0.36	0.01

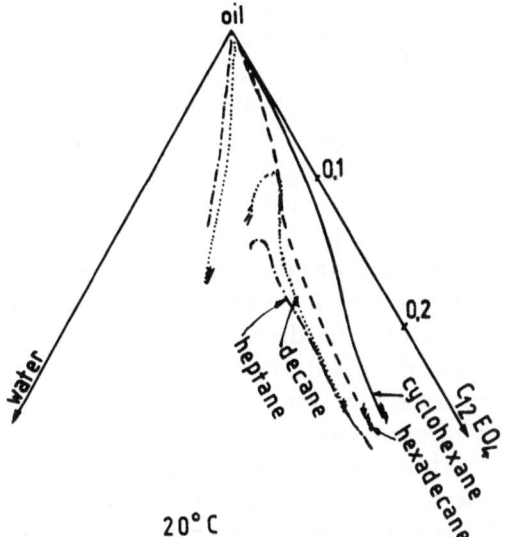

Fig. 1. The one-isotropic phase regions in the oil corner of the phase diagrams of various $C_{12}EO_4$/water/oil systems at 20°C

the order in solubilizing power is: cyclohexane < hexadecane < decane < heptane. And as previously discussed, the ternary mixtures with the first two oils should be considered as leading to "bulky oil systems" (class II in the previous part of the work), while with the last one, we would get "short-oil" ones (class I) [1].

The data in Table 1 clearly suggest that the parameter of importance for the water solubilization should be the solubility parameter δ, since its corresponding

values may be placed in exactly the same order as the reciprocal of the water solubilities. Moreover, such a classification would also be in line with the case of the benzene, where no solubilization occurs and the value of its δ parameter is about 9. Exactly the same things could be said for the chloroform. Let us note that a probable value of the solubility parameter for $C_{12}EO_n$ may be in the range 8.5–9 [12,13].

Although for these two systems (heptane and cyclohexane) there is no true CMC and the aggregates are very small in virtual absence of water, nevertheless true water-swollen micelles can be formed. A few values of the aggregation number N are shown in Table 2 (see also Fig. 2). Whatever the nature of the oil and for low water contents, there is only one way these nonionic surfactants may aggregate, i. e. as bundles of parallel molecules. More specifically, they are ordered in what we call a "hank-like" structure [3, 4, 14]: the EO chains are disposed head to tail, side by side. In other words, they form small bilayers with interdigited oxyethylene chains whose conformation is more or less extended. According to the hydration degree these few water molecules are probably located on the "central" EO groups.

As a result, each C_{12} chain has about 42 to 45 Å^2 to his command, and may either adopt a coiled conformation, or be penetrated by the (linear) hydrocarbons. Such structures have been determined by our best fitting procedure on small angle neutron data. A few examples will be discussed below.

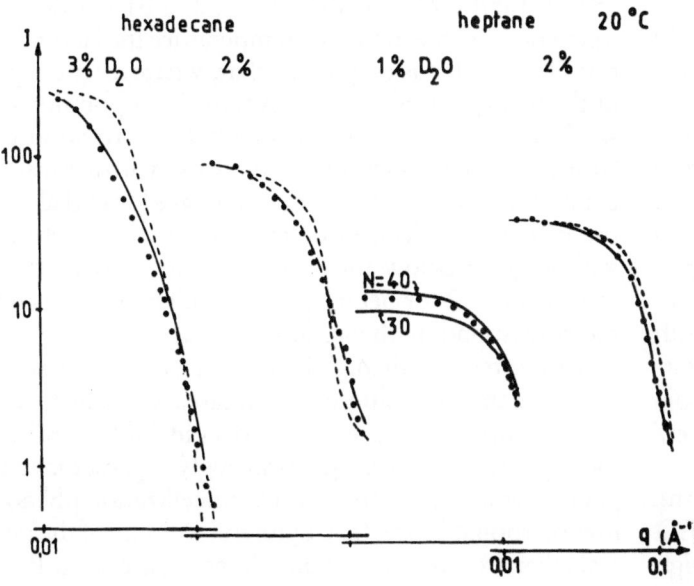

Fig. 2. A few theoretical small angle neutron scattering spectra of water swollen hank-like (———) or separate (– – –) bilayer ("lamellae") micelles compared to experimental data (●) (see text). The surfactant/(surfactant + oil) ratio is 0.1 w/w

Table 2. An evaluation of the aggregation number of water swollen (hank-like) micelles with $C_{12}EO_4$ at 20 °C. The interparticle effects are roughly taken into account by regarding the hydrophile volume plus half the hydrophobe one as a spherical hard core [3]

Solvent	$\dfrac{\text{Surfactant}}{\text{Surfactant} + \text{oil}}$ (w/w)	% Water (w/w)	N
Heptane	0.10	1	40
	0.10	2	120
	0.20	1	40
Hexadecane	0.10	1	100
	0.10	2	250
	0.10	3	900
	0.15	1	100
	0.15	2	200
	0.15	3	600
Decane	0.10	1	80
	0.10	2	200
	0.10	3	500
	0.15	1	50
	0.15	2	150
	0.15	3	300

Considering the data in Table 2, it is interesting to note that a constant water concentration the aggregation numbers follow the reverse order of the maximum water solubilities. As discussed in another paper [14], we can propose a model concerning the size-limiting effect in these bilayers: *when hydrated* any given EO group must be "surrounded" by two water molecules [15]. The conformation of such a group will be extended, while it should be coiled otherwise (i. e. in the meander form). Of course the number of EO groups *actually* "hydrated" per surfactant is not necessarily equal to the total number of EO groups in the surfactant molecule: proceeding from the core to the hydrophobic interface, it depends on the temperature (the hydration is favoured by lower temperatures) and on the length of the hydrophilic chain. But it also depends on the chemical nature of the oil. For example, the oil may induce an important coiling of the surfactant chains or may largely penetrate this film: in both these extreme cases the result will be a hydration limited to *only* the "central" EO groups, leading to the formation of larger aggregates at constant overall water/surfactant ratio.

Hence we can suggest that the "bulky" cyclohexane molecule probably *tends* to penetrate toward the EO core (as does the benzene molecule), given its high value of δ (i. e. comparable to that of the surfactant), leading to actual small hydration degree and larger aggregates. On the other hand the presence of heptane molecules would not disturb the hydrophobic tails, and would allow a full hydration of the EO chains and then lead to smaller aggregates (for the same given overall water/surfactant ratio).

At this point, we have to note that, in any case, the presence of water tends to dramatically decrease the number of free surfactant in the ternary solution as this has been clearly shown for the decane/$C_{12}EO_4$ system [3, 4, 14]. Besides there is absolutely *no central free water layer* into these low water content aggregates, as can be shown from neutron data (see Fig. 2).

3. In the third class (C), small aggregates are formed for surfactant concentrations above a certain value which could be termed as a quasi-critical concentration and whose typical values may be as large as a few percent w/w. The aggregation numbers are about 10 to 50: the micelles adopt the previously described "hank" structure for lower values of N, but seem to be true (separate) bilayers for larger N ("lamellas"). The oils of this class are longer hydrocarbons (decane, hexadecane, etc...). For given temperature and surfactant, the longer the hydrocarbon the larger the aggregation number. This same trend occurs when the temperature is lower or the hydrophilic chain is longer. Hence, a given hydrocarbon belonging to the previous class (B) at a given temperature may also lead to the formation of a (quasi) critical concentration and to larger aggregate sizes (class C) but for another temperature and/or with another surfactant. For example, this is the case for cyclohexane respectively at 10 °C with $C_{12}EO_6$ (class C) and at 20 °C with $C_{12}EO_4$ (class B). Besides, as a general rule, the lower the temperature the larger the micelles. But let us emphasize that, whatever the type of the binary micellar behaviour, the incorporation of small amounts of water promotes the formation of lamellar aggregates whose size increases with the water content, and all the water molecules are located along the EO chains: detailed description of the structures will be given below for $C_{12}EO_4$/decane above 20 °C (see also Fig. 2 for neutron data of other systems and the corresponding theoretical spectra).

To summarize, then, as far as the oils are concerned and since the free volume V_f decreases with a decrease of the temperature (see part I), data in Table 1 and all the experimental results [3, 4] strongly suggest that the parameter of importance for the micellization phenomenon should be the (entropic) free volume of the oil: a smaller free volume favours larger aggregates. But, concerning the water solubilization, we would rather consider the solubility parameter as being much more essential.

Structures in the $C_{12}EO_4$/decane/water system (above the PIT)

This system has been investigated in a particularly intensive way [4], and a few results have already been published [2, 3, 8]. They will be very briefly summarized in line with the purpose of the present paper [16]. But many new results on the more dilute system will also be presented in order to understand the water solubilization at high decane content, i. e. just at the coalescence region of the microemulsion and water-poor one-isotropic phase domains.

The water-poor region ($T \leq 20°C$) (L2 phase)

All the three scattering techniques (light, X-ray, neutron) have been used together to investigate this part of the diagram (see Fig. 12). The consistent picture which emerges from the whole set of these numerous investigations is the following (Fig. 3): as long as the oil content is less than about 93%, all the structures are "lamellar" whether they are bilayers (separate or interdigited EO chains) or multilayered grains.

Schematically, starting from the oil-rich binary oil-surfactant solution (and oil content less than 93%), the aggregation number of the water swollen micelles dramatically increases upon water incorporation, typi-

cally from $N \approx 10$ to $N \approx 1500$, till every EO group is hydrated by two water molecules [3, 14]. Except for very low water content (1–2%) where hank-like structures are more likely, the particle shapes are bilayers without any free water core. The area per polar head is nearly constant, that is about 42 Å^2.

Data suggest that the proceeding "hydration" is progressive along the EO chain and is concomitant with a change of this chain conformation, evolving from the meander form to the extended one [14]. It is more difficult to bring to light the exact conformation of the hydrophobic tails, which could be defined as semi-coiled chains allowing some (unknown) decane penetration into their palisade. At any rate, it seems to definitely coil up upon addition of surfactant, that is, due to the increase of the volume fraction of the disperse phase [3]. Once again let us emphasize that in the presence of water, the concentration of free surfactant dispersed in the oil is drastically reduced, probably less than 1% w/w, while the quasi CMC in the binary mixture is about 4% w/w. This is certainly true for samples which contain less than about 95% decane. For more dilute systems, the question has yet to be debated.

A further addition of water promotes a demixing between the same type of lamellar water-swollen micelles and a lamellar liquid crystal phase (L_α, Fig. 12) which contains the excess water as a free water central layer [7]. And at 20°C, a still further water incorporation leads to a second solubilization zone, forming inverse microemulsions made of definite water in oil globules, whatever the oil content [2] (see below).

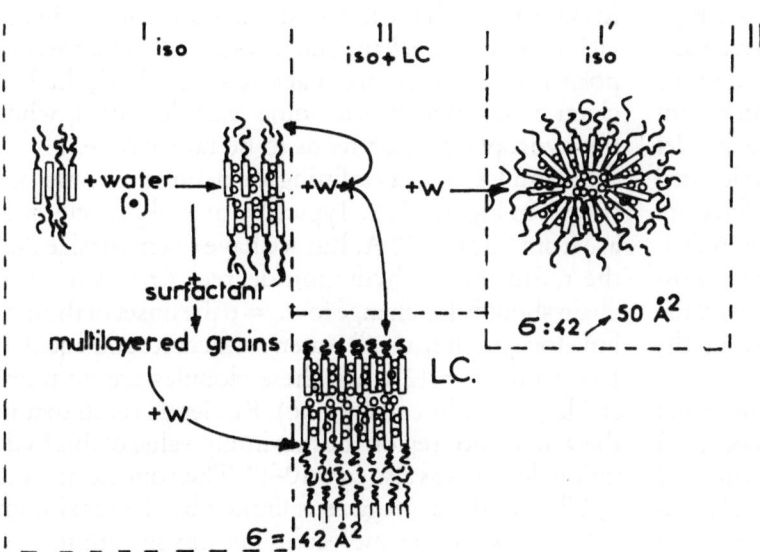

Fig. 3. Schematic representation of structure evolution when water and/or surfactant are progressively added to dilute inverse micelle systems

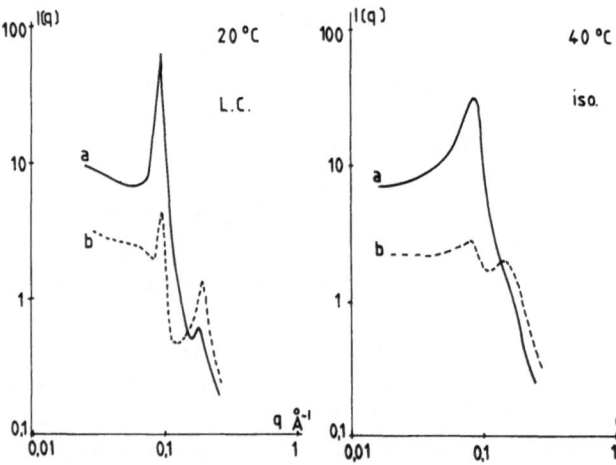

Fig. 4. Examples of spectra for a surfactant-rich system (sample X, Fig. 12) which is isotropic at 40 °C, but is a liquid crystal at 20 °C. Isotopic nature of the solvents: $C_{10}D_{22}$ and H_2O for curves (a); $C_{10}D_{22}$ and D_2O for curves (b)

For the surfactant-rich systems (more than 50 % w/w), the one isotropic phase become contiguous to a lamellar liquid crystal. Most interestingly, there is absolutely no discontinuity in the Bragg spacings when passing from the liquid crystal to the isotropic phase (obtained from X-ray and neutron spectra). For example, in Figure 4 are drawn a few neutron spectra for samples along the phase separation line (sample X in Fig. 12): at 40 °C, these samples are optically isotropic while at 20 °C they belong to the lamellar phase region. According to the temperature and the isotopic composition of the solvents (H_2O/D_2O, $C_{10}H_{22}/C_{10}D_{22}$), the first two reflection peaks may have various relative amplitudes. This has to be ascribed to the hydration sites and conformations of the surfactant chains, and will be discussed elsewhere. Here, just let us note that for these samples all the water is located along the EO chains (2 H_2O per EO group), which are in a rather extended conformation, while the hydrophobic part is probably highly coiled, leaving almost no room for any oil penetration into the film. And the area per polar head is constant everywhere in the lamellar phase (42 $Å^2$).

At this point it is particularly interesting to note that our neutron results are quite in line with recent ^2H NMR data interpretation [18]. Indeed investigations of the dynamic structure of hexadecane/$C_{12}EO_4$ in lamellar liquid crystal show that the NMR data support a model with only a small amount of penetration of the hydrocarbon between the amphiphiles molecules and a rapid exchange on the time scale between the penetrated segments and the nonpenetrating molecules, the latter of which are essentially isotropic.

Therefore, when the previous bilayers dispersed in the isotropic phase are more and more concentrated they have a tendency to form more or less dense packings of disordered multilayered grains whose typical size is 150 Å (typically made of three bilayers). And at this point, the addition of some more water promotes the formation of the lamellar liquid crystal, at constant structure.

Consequently all the systems located on all the demixing lines facing the lamellar phase domains (L_2 and microemulsion phases) are characterized by the same area per polar head and also probably by roughly the same coiling of the $C_{12}H_{23}$ chain, but not necessarily by the same hydration degree of the EO chain, as will be discussed below.

The inverse microemulsion region

From neutron scattering data, we have already shown that for any decane concentration less than 85 % w/w, the microemulsion structures are definite, although labile, water-in-oil globules [2, 4]. According to the water/surfactant ratio between 1 and 2.6, the mean surfactant aggregation number varies from about 1500 to 3000, the shapes are, on the mean, oblate globules with an average ellipticity roughly equal to 0.5 when evaluated at the hydrophobe-hydrophile interface into the surfactant film; probably, a certain polydispersity in diameter cannot be discarded, at least for samples with the lowest decane content, i. e. in the vicinity of the lamellar liquid crystal [19]. The area per polar head increases from about 43 to 58 $Å^2$; the EO chain conformation is in some extended form, while the hydrophobic part tends to be rather coiled.

Examples of curve fitting of neutron spectra are shown in Figures 7, 8. Typically, overall particle sizes are from 200 to 300 Å. But we have to emphasize that the mean molar "hydration degree" (α_f) of the EO chain should also change for $\alpha_f = 6$ the onset of the globule formation (overall water/surfactant ratio equal to 1 w/w) to $\alpha_f = 12$ when these globules are no more stable (this ratio equal to 2.6). But let us recall that in the water-poor region the maximum value of this hydration degree was also $\alpha_f = 10$–12. Then once again, the stability of the aggregates is limited by the maximum value of two water molecules per oxygen atom.

On the other hand, the formation of globules in this second solubilization region from the largest bilayers in the water-poor phase is performed at almost constant area per polar head and practically constant aggregation number [2,3]. As a result, from a pure geometry point of view, the creation of a non-zero local curvature has to be concomitant to some release of water molecules from a few EO sites into the free water core. Hence, since α_f changes from 10 to 6, even at constant temperature, a local dehydration occurs which is imposed by the presence of larger water amounts which have to be shielded from the oil contact: at that temperature (20 °C) and for this noticeably higher water content, this shielding is made possible since the remanent (strong) hydration and the mean radius of the core may be consistent with an area per polar head of about 45 Å^2: this area can be safely covered by the coiled hydrophobic chain, leaving very little room for an eventual penetration of the oil into the surfactant film.

For oil contents less than 85% w/w, it was found that the oil dilution could be performed a nearly constant structure. This can be also seen on the X-ray spectra of Figure 5, where $I(q)/C$ is represented in a logarithmic scale (C is the volume fraction of the disperse phase): at large q, I/C is independent of C. Of course larger and larger differences arise at smaller q

values as a result of interparticle interactions. Maxima occur which can be taken into account by the equivalent hard sticky sphere formalism: supposing a well width of 2 Å, an attractive potential of 1 to 2 kT is found quite consistent with the scattering data [6,7]. In any case, for such large volume fractions of the disperse phase, calculations show that $I(q \to 0)$ are not sensitive to the exact value of this potential as long as it is smaller than about 3 kT, i. e. when the system is not in the very vicinity of a critical point [20].

Although not calculated on an absolute scale, nevertheless these X-ray spectra are quite consistent with structural results obtained from neutron scattering, the scattering length density of the different nuclei being merely replaced by the electronic density of the various parts of the aggregates [2]. Hence, the present finding perfectly confirms that the H/D isotopic substitution does not modify the structures to a sensible extent [2].

A few $I(q)/C$ spectra obtained from neutron scattering for the *same water/surfactant ratio* and two oil concentrations are shown in Figure 6. For each overall oil/water/surfactant composition, four different isotopic mixtures of the oil component were used. As expected, the curves seem to coincide everywhere except for smaller q values. However, a closer inspection of the spectra corresponding to the decane containing 80%

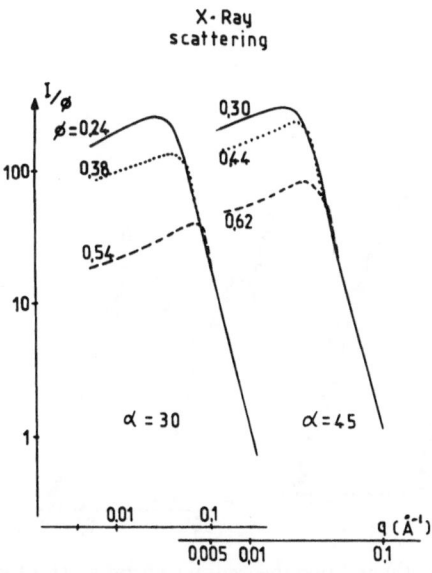

Fig. 5. X-ray $I(q)/\phi$ spectra (corrected for the transmission of the samples) for different volume fractions ϕ of the disperse phase (water + surfactant) at 20 °C. α is the molar water/surfactant ratio. Here the extrema clearly result from interparticle effects

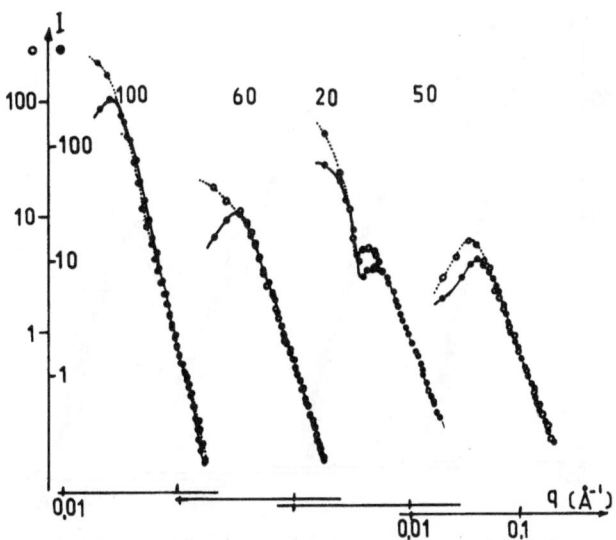

Fig. 6. Experimental neutron spectra for samples H$_2$ (O) and G$_2$ (●) ($T = 20$ °C) (see Table 3). The scales are shifted by a factor identical to the ratio of the respective volume fractions of the disperse phases. Four H/D isotopic compositions of the decane component are used: 100, 60, 50, and 20% w/w hydrogenated. Water contains 37% H$_2$O

of the perdeuterated isotope reveals the presence of secondary extrema which are not quite superposable (Fig. 6). Theoretical calculations on model globules show that such secondary extrema are very sensitive to the structure and mean curvature of the interfacial film [5]. That difference cannot be ascribed to an eventual polydispersity increase with the concentration since both these extrema exhibit the same quality of relief. Hence, we have to conclude that along a line at constant water/surfactant ratio the structures should not be strictly dilutable at constant structure: moreover, such a conclusion was also suggested by the delineations of the microemulsion domain themselves (Fig. 12).

Examples of curve fitting are shown in Figures 7, 8. There are also indicated the values for the "best" morphological parameters. As a result, the more dilute sample contains a little larger aggregates, all other parameters being constant within the precision of the determination, except for α_f which is also a little larger. The influence on the spectra of the thickness of the hydrophobic shell (eventually penetrated by some solvent) can be seen only at larger q values (Fig. 7). A comparison of the data with model calculations strongly suggest that the probable thickness should be

about 7–10 Å. Therefore, the hydrophobic tail should be considered in a rather coiled conformation without important oil penetration into the film.

Hence, each of these samples should actually belong to two different dilution lines, if any: the actual water/surfactant ratio *in the globule* is the largest for the more dilute sample. Consequently, the dilution lines cannot rigorously abut to the oil corner. In other words the "oil continuous phase" must in fact be constituted by a solution of a few percent of *monomeric* surfactant molecules.

In order to determine the surfactant content in these continuous phases as a function of the overall composition we have measured the neutron intensities scattered at $q = 0$ for two series of samples (series P and D in Fig. 9). Moreover, for each sample two different isotopic compositions of the solvent have been used: water was a mixture 50/50 of H_2O and D_2O; solvent A was the decane 100% hydrogenated and the solvent B was a mixture 50/50 of $C_{10}H_{22}/C_{10}D_{22}$. Neglecting the contribution of the monomeric surfactants, the intensities scattered at $q = 0$ are:

$$I(0) = C \cdot K(C) \cdot N \left[\frac{b_s + \alpha_g \cdot b_w}{V_s + \alpha_g \cdot V_w} - \frac{\beta b_o + b_s}{\beta V_o + V_s} \right]^2.$$

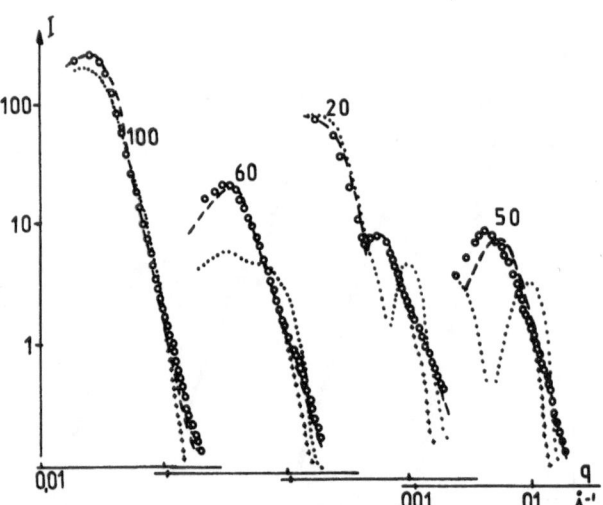

Fig. 7. Neutron spectra for sample G_2 for different isotopic compositions of the decane component, at 20 °C: experimental data (O) and three theoretical spectra for $N = 1500$, hydrophile thickness $L_2 = 14$ Å (extended EO chain), molar hydration degree of the surfactant film $\alpha_f = 6.5$. (– – –) spheroid, $p = 0.5$, hydrophobe thickness $L_1 = 7$ Å (coiled C_{12} chain), best fit; (+ + +) spheroid, p 0.5, $L_1 = 14$ Å (extended C_{12} chain); (·····) bilayer with water core (see text)

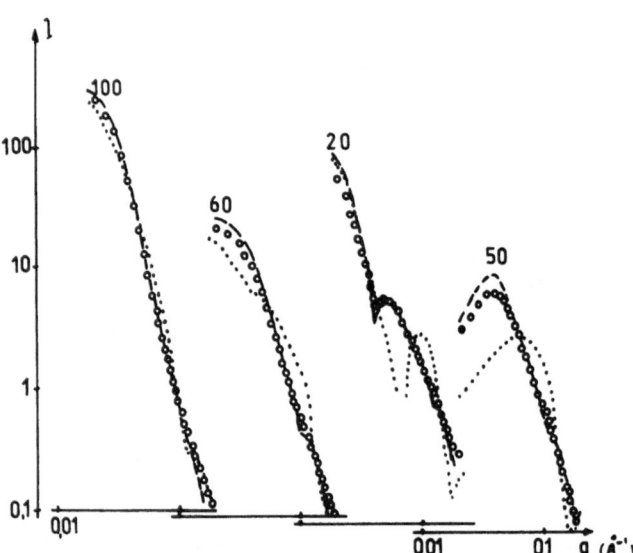

Fig. 8. Experimental (O) and two theoretical neutron spectra for sample H_2 (20 °C): $N = 2000$, $\delta \approx 46$ Å², $\alpha_f = 8$, $L_1 = 10$ Å, $L_2 = 14$ Å; (– – –) spheroid, $p = 0.6$, best fit; (·····) bilayer with water core, "deformed bilayer". The free surfactant concentration in the continuous phase is 2% w/w

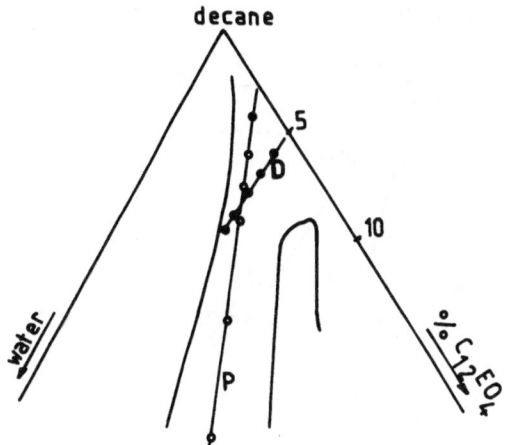

Fig. 9. Phase diagram of decane/$C_{12}EO_4$/water at 20 °C in the oil corner, and samples investigated along lines (D) and (P)

Such a relation supposes all the water molecules to be in the globules.

N is the aggregation number of the surfactant in the globules, β is the oil/surfactant molar ratio in the continuous phase, α_g is the water/surfactant molar ratio in the globules, b_s, b_o, b_w are the scattering lengths of the surfactant, oil and water components, and V_s, V_o, V_w are their molar volumes, C is the volume fraction of the disperse phase, which depends on α_g, $K(C)$ also depends on the interparticle effects.

In the present case, the unknown parameter is β. For a given overall oil/water/surfactant composition and from the experimental values of $I(0)$, one apparent value of N can be calculated for each value ascribed to β. In Figure 10 are reported the values of N calculated for samples along the line (P), supposing that the surfactant concentration in the continuous phase (C_1) is 0, 2 and 3 % w/w. Since the true aggregation number N should be practically independent of the scattering length of the solvents, we find that the actual composition of the continuous phase for all the samples along line (P) is probably not far form 2 % w/w.

Similarly, from experiments and calculations performed on samples (D), one can propose a drawing for the lines at constant continuous phase composition (Fig. 11). Although the values of N are more than merely indicative, they cannot be very precise (let us say within 30 % for N, i. e. about 15 % for the size) since we are not aware of the exact evolution of the attractive interparticle potential with the overall composition. (The $I(q)$ spectra were not determined at larger q for these samples and cannot confirm these values of N.)

Nevertheless that determination of C_1 does not depend on the interparticle interactions, and these values of C_1 appear quite consistent with both the delineations of the microemulsion domain and the value of the quasi-CMC in the binary apolar system (≈ 4% w/w).

Besides, these lines are not really dilution lines: indeed, we can see in Figure 10 that there is a drop in the aggregate sizes when the volume fraction of the

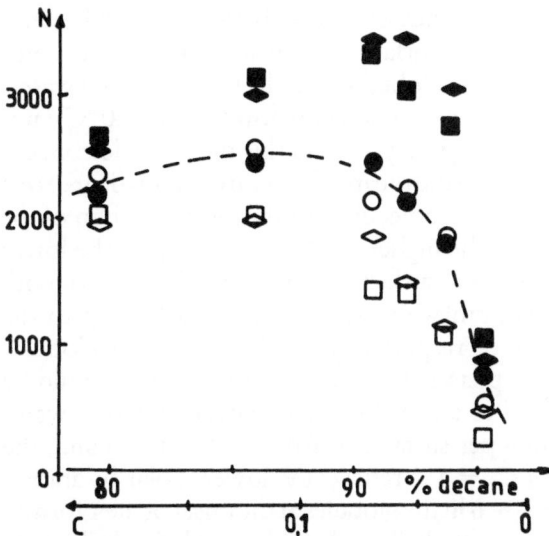

Fig. 10. Apparent aggregation numbers calculated for samples along line (P) (Fig. 9), according to the value ascribed to C_1 (surfactant composition of the continuous phase): open symbols: contrast B; full symbols: contrast A (see text); ($\square$) $C_1 = 0$; ($\bigcirc$) $C_1 = 2$% w/w; ($\diamondsuit$) $C_1 = 3$% w/w

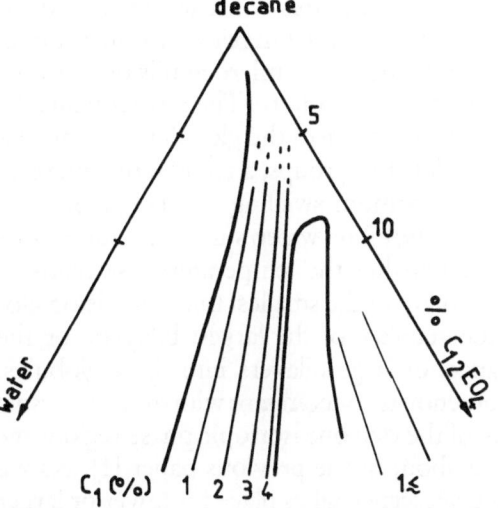

Fig. 11. A drawing of likely iso-C_1 lines (see legend of Figs. 9 and 10)

disperse phase (C) is less than about 2%: when C changes from 4 to 2%, N drops from 2000 to about 500; however, these aggregate sizes at 2% are not very small, since the diameters of the globules are still as large as 100 Å. Then, as shown by neutron scattering data, the structures in the oil-rich region are relatively large water-in-oil globules which clearly deserve the name of microemulsion (however, this may not be true for very small water/surfactant ratios).

The fact that these globules are dispersed in a oil continuous phase which contains several percent w/w of monomeric surfactant clearly proves the difference in nature of the surfactant films whether they are committed to these globules or to the bilayers of the water-poor region for oil content below 93% (at 20 °C): it is very tempting to try to bind this monomeric concentration (C_1) to the mean effective hydration degree (α_f) of a surfactant molecule into the interfacial film of the aggregates: the higher this hydration degree the lower the monomeric surfactant content C_1. Indeed, for both the maximum water/surfactant ratios leading to the demixing of respectively the lamellar- and globular aggregate phases, the very low values of C_1 could be correlated to a α_f value a little larger than two water molecules per surfactant. And on the other hand, the value of $C_1 = 4\%$ reflects the lower molar hydration degree $\alpha_f \approx 6$ in the globules at the onset of their formation. Hence, we believe that this correlation is the only one which could really be worked out between water solubilization and properties of the binary apolar mixtures.

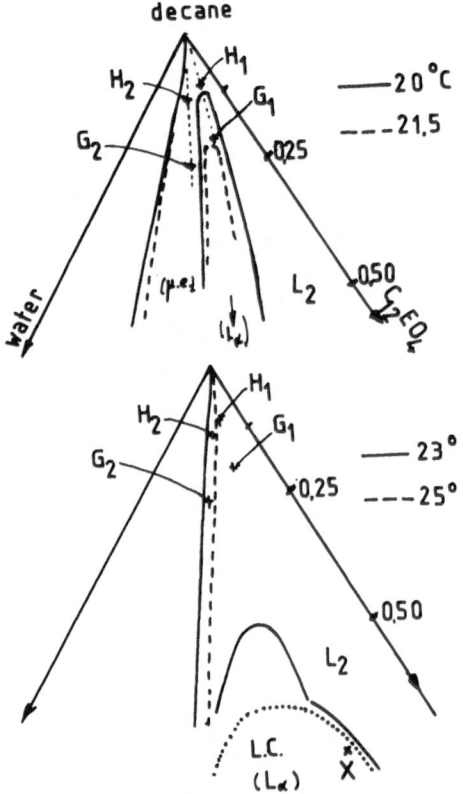

Fig. 12 A part of the phase diagram of $C_{12}EO_4$/decane/water at different temperatures. H_1, H_2, G_1, G_2 and X represent the samples investigated by neutron scattering. L_2 is the isotropic phase, L_α is the lamellar liquid crystal

Evolution of the structures with temperature

From the first part of this work we already know that, just above the PIT, the maximum water/surfactant ratio confining the "microemulsion region" decreases with the temperature. Then by reducing the mean radius of curvature of the globules by some dehydration of a few EO groups, a temperature increase will limit the maximum swelling of these globules.

But for *intermediate* water/surfactant ratios one may wonder whether the temperature rise tends to favour the stability of the smallest microemulsion globules, or the stability of the largest bilayers, or the transformation of large bilayers into these globules. This phenomenon is concomitant with the progressive coalescence of the two one isotropic phase regions we were talking about in the previous paper [1]. As we know, this coalescence takes place for lower or larger surfactant/oil ratios accordingly as the oil is "short" or

"long" (or bulkier), compared to the hydrophobic chain of the surfactant.

Decane may be considered as the limiting "short" oil for $C_{12}EO_4$. Phase diagrams at different temperatures are shown in Figure 12. In an attempt to correlate this phase behaviour to structural changes, we have studied more specifically the neutron scattering spectra of samples noted H_1, H_2, G_1, G_2, whose compositions are given in Table 3. A few preliminary results have already been published elsewhere [17]. For each sample four different isotopic compositions of the solvent have been used.

Samples H_1 and G_1 are one-phase systems at 21.5°, 23° and 25 °C. Sample G_1 is not yet stable at 20 °C. Samples H_2 and G_2 were studied at 20°, 21.5° and 23 °C; they are no more stable at 25 °C; their spectra at 20 °C were given in Figures 7, 8 and already discussed.

As for H_2 and G_2, the spectra $I(q)/C$ for H_1 and G_1 are not exactly superposable *everywhere* (at larger q

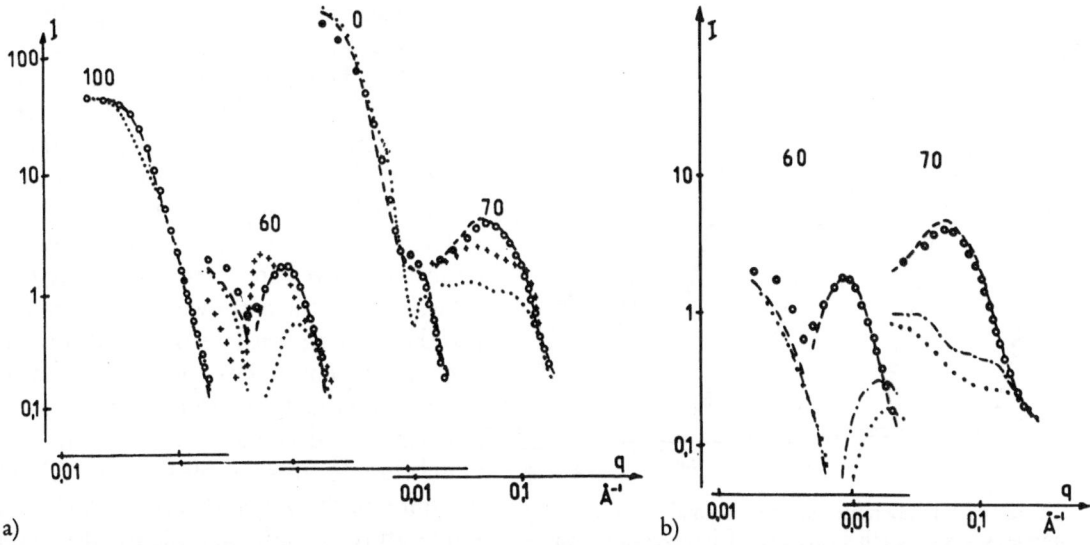

Fig. 13. Experimental (O) and a few theoretical neutron spectra for sample G_1 at 21.5 °C, for several isotopic compositions of decane. The water component contains 80% D_2O ($N = 1000$, $L_1 = 7$ Å). a) (– – –) bilayer with 10 Å water core ("deformed bilayer"), $L_2 = 14$ Å, $\alpha_f = 6$, best fit; (·····) bilayer without water core, $\alpha_f = 12$, $L_2 = 14$ Å; (+ + +) spheroid, $\alpha_f = 4.5$, $L_2 = 14$ Å. b) (– · – ·) bilayer with water core, $\alpha_f = 10$, $L_2 = 7$ Å; (·····) bilayer with water core, $\alpha_f = 7$, $L_2 = 7$ Å; (– – –) bilayer with 10 Å water core ("deformed bilayer"), $L_2 = 14$ Å, $\alpha_f = 6$, best fit

Table 3. Compsition (w/w) of the samples H and G investigated by small angle neutron scattering

	H_1	G_1	H_2	G_2
$\dfrac{\text{Water}}{\text{Surfactant}}$ (w/w)	0.5	0.5	1.1	1.1
% water (w/w)	3.4	7.0	7.2	14
$\dfrac{\text{Surfactant}}{\text{Decane}}$ (w/w)	0.076	0.178	0.076	0.178

account for *all* the four very different spectra obtained from the four isotopic mixtures of the solvent.

For the sample H_2 and G_2 at 20 °C, we have already seen that the aggregates are oblate globules of *mean* ellipticity $p = 0.4-0.6$, area per polar head $\sigma \approx 46$ Å^2, extended EO chain and likely coiled hydrophobic tail; the aggregation numbers are respectively *about* 1500

values) for certain "sensitive" isotopic compositions of the solvent, whatever the temperature (Figs. 13, 14). Hence, the structures must be somewhat different in these samples.

Relatively minor morphological modifications also arise from the temperature changes as indicated by the variations in both the position and level of the extrema at relatively large q (Fig. 15). Let us recall that these extrema are purely *structural*, i. e. do not result from interparticle effects. Indeed, for one of the solvent (100% $C_{10}H_{22}$), there is obviously no such extremum at all. The difficult problem is to find *the* good model which, for a given sample and temperature, can

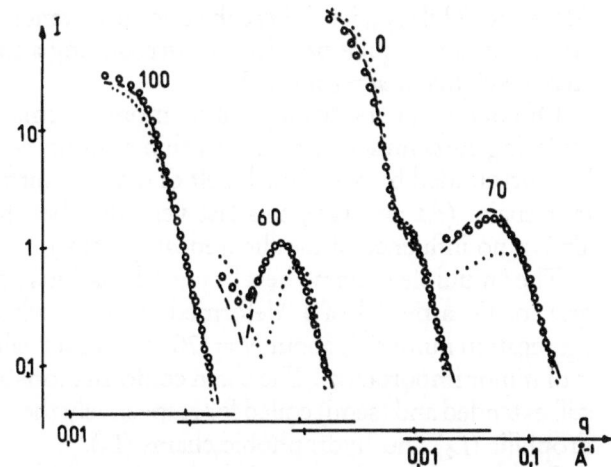

Fig. 14. Experimental (O) and a few theoretical neutron spectra for sample H_1 at 21.5 °C. The water component contains 80% D_2O. ($N = 800$, $\alpha_f = 5$, $L_1 = 10$ Å, $L_2 = 14$ Å); (– – –) spheroid; (·····) deformed bilayer

                                                                               Progress in Colloid & Polymer Science, Vol. 73 (1987)

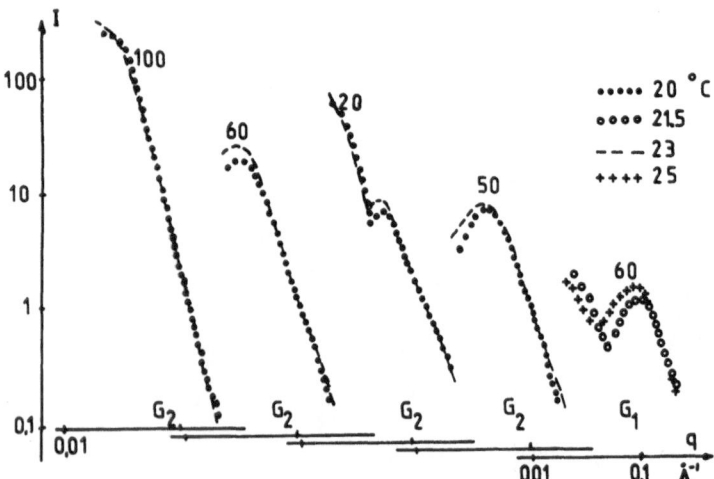

Fig. 15. Change with temperature of experimental spectra for G_1 and G_2 samples, suggesting a slight increase of the aggregation number with temperature (compare to Fig. 6)

and 2000 (the composition of the continuous phase being not identical, containing about 3–4 % w/w surfactant), and the hydration degrees α_f are respectively 6 and 8: the larger α_g in the globule the larger N and the larger α_f.

Now, sample G_1 at 21.5 °C (which is *biphasic* at 20 °C) is particularly interesting. Clearly, neither a typical globular shape nor the pure bilayer without water layer can account for the experimental data. However, the globule model appears to be not far from the real one. So we have introduced the concept of a "deformed" bilayer, i. e. where the two surfactant layers are no more quite parallel but surrounding a thin "lens-like" *free water core* [17].

Of course our modelling will be more schematic, replacing the central water lens by a thin disk-like core, but surrounded by a 2 Å thick belt of dry EO surfactant chain. (At any rate, this last very thin belt has almost no influence of the theoretical spectra.)

The "multiple spectra" best fitting calculations suggest for G_1 a model of a "deformed" bilayer, whose aggregation number is about $N = 1200$ (the exact value is of minor importance). The chain conformations are still extended and (semi) coiled for respectively the hydrophilic (L_2) and hydrophobic chains (L_1).

But the most striking result is the presence, in the bilayer, of a free water film of about $e = 10$ Å, and a somewhat *reduced* hydration degree of the EO groups $\alpha_f \approx 6$.

The experimental spectra are particularly sensitive to the values ascribed to e and α_f when the oil solvent is 40 or 30 % deuterated and the water component contains 63 % D_2O (see Figs. 13, 14). (The curve fitting also takes systematically into account the interparticle effects, which come here from a hard core potential plus a 1–2 kT attractive energy on a 2 Å distance.)

Then, this very small temperature jump promts some hydration water of the bilayers to be released and to be *added* to the extra water we want to incorporate into the system: this loss of (hyration) energy should be overcompensated by a gain from the entropic term arising from the dispersion into small particles of large lamellar crystal elements (one of the two phases in equilibrium for this sample G_1 at 20 °C). Therefore the vanishing of the water solubility gap by increasing the temperature is made possible thanks to the formation of "deformed" bilayers, which are clear intermediates between bilayers (without free water film) and globules (with globular free water core). Incidentally, one can note that to accommodate for this non zero curvature of the film there must be some decrease of the aggregation number as compared to the *neighbouring* bilayers at 20 °C which yet contain a little less water on the whole.

Of course all these findings are in accordance with the geometrical constraints induced by the type of model and the molar volumes of the different molecules since the modelling just consists of building a possible aggregate from individual molecules, taking into account their volumes and the chain lengths.

The way the coalescence takes place is quite in line with the above interpretation. Indeed, higher and higher temperatures are necessary to form the isotropic dispersion of these particular bilayers. This just means that a very little more "hydration energy" has to be spent since the volume fraction of this disperse phase will be larger and larger, i. e. the counterbalancing entropic gain is less and less.

The sturctures obtained for sample H_1 at 21.5 °C give full support to the above scheme. (Let us recall that H_1 is already in the one isotropic phase just below 20 °C.) From the best fit calculations we can infer that this sample already contains globular "microemulsion" particles, with $\alpha_f \approx 4.5$–5 and $N \approx 800$ (see Fig. 14). It is possible that the hydrophobic tail is comparatively less coiled, since we can no longer distinctly determine its conformation.

Returning to samples H_2, G_2, the fits in Figures 7, 8 indicate that the structures are actually quite globular ($\alpha_f = 6$–8), i. e. they definitely cannot accommodate for the "deformed" bilayer morphology, only valid for smaller water/surfactant ratios (what we previously designated "*intermediate*" ratios).

When temperature is still rising, the spectra of these four samples also change a little (Fig. 15). The small modifications of "sensitive" curves could be interpreted in terms of the onset of the transition between deformed bilayers and spheroids, together with a certain increase in the aggregate size. But more intensive calculations are necessary to eventually achieve firmer conclusions.

Conclusion

The present structural investigations show how small angle scattering can supply important details on the most probable morphology of the molecular aggregates in microemulsions or water swollen micelles, managing to approximately take into account the interparticle effects (use of an analytical structure factor we have derived from the solution of the Percus-Yevick approximation for equivalent hard sticky spheres.) Hence we acknowledge that the actual particle sizes cannot be obtained to a very high degree of precision. But let us emphasize that the fitting procedure concerns the logarithm of the intensities on an absolute scale: in other words, roughly the *same* quality of fit can be achieved (about 5 %) for every q in the whole q range investigated and for the several orders of magnitude of the scattered intensities.

In particular, by using judicious isotopic mixtures of the solvents (water and oil), we can determine the central water cores, hydration degree and conformation (or rather the shell thickness) of the hydrophobic and hydrophilic parts of the surfactant film, taking into account its penetration by the solvent molecules. All these parameters describe the mean curvature of the aggregates, which for a given overall composition of the sample can be consistent only with a narrow range of aggregation numbers, whose a typical value is also obtained from the fits at the smaller q values.

The question of the polydispersity remains open, but we believe it was not essential for the relatively dilute systems considered here. Besides, calculations show that a different probability distribution ought to be introduced for each of such so different isotopic compositions of the solvent which are used for a same given sample. On the other hand, experimental data tend to strongly suggest that to any representative point in the phase diagram would probably correspond one relatively well defined structure of the interfacial film. And for a given overall composition, this cannot be geometrically achieved for arbitrary size/shape distributions. Hence when we are able to find one good representative mean particle model, we choose to disregard this polydipersity aspect. Of course this cannot be generalized to every system: but in a convincing case of polydispersity the contrast variation method can still be used, by investigating essentially the intensities scattered at $q = 0$, as it was shown for oil-in-water fluorinated nonionic microemulsions [19].

To conclude, a correlation between phase behaviour and molecular structures of inverse systems just above the PIT actually exists, but it can be gained only from the investigation of many systems, with various compositions and different oil and surfactant molecules. Clearly this correlation rests on the evolution of the geometric properties of the surfactant interfacial film, which can only be described thanks to the sophisticated contrast variation method in small angle neutron scattering. However, even as far as simple true ternary systems are concerned where distinct particles actually exist, much remains to be done given the difficulties arising from interparticle effects and size/shape polydispersity.

But what is still much more unknown is the thermodynamic quantifying of each parameter of importance. For example, what is the influence of the conformation of the *hydrophobic* chain and how to evaluate it as a function of the temperature, composition, nature of the oil solvent etc.? Of course the entropic components are certainly essential, but it is difficult to propose simple rules: the mere intoduction of simple notions like the free volume or bulkiness of the oil are far from being satisfactory. The dielectric constant or enthalpic terms vaguely represented by the solubility parameter give only some indication for a classification of the systems: is the eventual oil penetration into the film of a primary importance or not, etc.? Of course these entropy terms mut be considered together with the entropy contribution of the dispersion phenomenon itself.

The potential energy in the interparticle interactions also certainly delimits a part of the stability area of the particles, and may be responsible for the existence of the critical point.

Nevertheless, in the present papers we have not merely described what the structures could be, but we have attempted to correlate film curvature, hydration degree, entropy of the dispersion, to a molecular property of the solubilizing oil.

Then, data and discussions would indicate that the micellar aggregation phenomenon in binary apolar medium is largely disconnected from the question of the water solubilization. One possible simple driving parameter for this micellization could be the free volume of the oil, which is essentially temperature dependent. This would suggest some intimate mixing of oil and hydrophobic chains in the micelles.

As far as the water solubilization is concerned, the hydration of the EO groups (probably by a quantum of two water molecules per oxygen atom) is fundamental for the whole picture of the evolution of the phase diagram with temperature.

But apparently one important characteristic of the oil could be its solubility parameter compared to that of the surfactant. Hence the notion of oil penetration seems much less indispensable to be brought in, at least for these water-in-oil structures: probably the entropic contribution due to an important penetration into the hydrocarbon palisade is unfavourable as compared to the dissolution of this oil into its own liquid, and will be avoided when the hydrophobic tail of the surfactant molecule can be sufficiently coiled to cover all the area per polar head, and then to efficiently shield water and EO groups from the oil medium. This assumption is quite in line with the structural results obtained from our neutron data, but have to be confirmed in another way. (Let us recall they are consistent with recent NMR results in liquid crystals [18].)

Hence, most promising are our further investigations on other types of "hydrophoby", and we place much hope in our comparative studies with fluorinated compounds.

References

1. Ravey JC (1987) preceding paper
2. Ravey JC, Buzier M (1984) In: Mittal KL, Lindman B (eds) Surfactant in Solution, vol 3, Plenum Press, New York, p 1759
3. Ravey JC, Buzier M, Picot C (1984) J Coll Interf Sci 97:9
4. Buzier M (1984) Thèse Université Nancy I, France
5. Ravey JC, Buzier M, Dupont G (1987) In: Rosano H (ed) Microemulsions, Surfactant Science Series 24, Marcel Dekker, New York 24:163–182
6. Baxter J (1968) J Chem Phys 49:2770
7. Ravey JC, to be published
8. Ravey JC, Buzier M (1984) In: Luisi PL, Straub BE (eds) Reverse Micelles, Plenum Press, New York, p 195
9. Christenson H, Friberg S, Larsen DW (1980) J Phys Chem 84:3633
10. Marsden SS, McBain JW (1948) J Phys Coll Chem 52:110
11. Hildebrand JH (ed) (1964) Solubility of non-electrolytes, Dover, New York
12. Little RC (1978) J Coll Interf Sci 65:587
13. Muller N (1978) J Coll Interf Sci 63:383
14. Ravey JC, Buzier M (1987) J Coll Interf Sci 116:30
15. Kizling J, Olofsson G, Stenius P (1986) J Coll Interf Sci 111:213
16. Ravey JC, Buzier M (1986) In: Mittal KL (ed) Surfactant in Solutions, New-Dehli, submitted for publication
17. Ravey JC, Buzier M (1985) ACS Symposium Series 272:254
18. Ward AJ, Friberg S, Larsen DW, Rananavare SB (1985) Langmuir 1:24
19. Ravey JC, Stébé MJ, Oberthur R (19??) In: Mittal KL, Bothorel P (eds) Surfactants in Solutions, Plenum Press, New York 6:1421
20. Regnault C, Ravey JC (19??) submitted for publication

Received March 7, 1987;
accepted March 9, 1987

Authors' address:

Dr. J. C. Ravey
Universite de Nancy I
Faculte des Sciences
Laboratoire de Physico Chimie des Colloides
U.A.C.N.R.S. No 406
Boite Postale No 239
Vandoeuvre-les-Nancy Cedex, France

Progress in Colloid & Polymer Science

Progr Colloid & Polymer Sci 73:127–133 (1987)

Phase and structure behaviour of fluorinated nonionic surfactant systems

J. C. Ravey and M.-J. Stébé

Laboratoire de Physico-Chimie des Colloides UA 406 CNRS LESOC, Université de Nancy I, Vandoeuvre-lès-Nancy Cedex, France

Abstract: The phase behaviour of many (apolar and aqueous) binary and ternary systems with series of polyoxyethylene perfluoroalcohols are presented. Emphasis is put on their relation to the structures of the micellar and microemulsion aggregates. The influence of the fluorocarbon tail on the hydrophily properties of the EO chain is discussed, as compared to hydrogenated compounds, whether aqueous binary- or ternary systems are concerned. In particular, the evolution of the L_2 and L_3 phases is studied as a function of the chemical composition of the surfactant, bringing out the enhanced stability of lamellar liquid crystals.

Key words: Fluorinated surfactants, nonionic surfactants, microemulsions, micelles, phase diagram.

Introduction

As it is apparent from our previous structure and phase properties studies [1–4], the nonionic fluorinated surfactant systems are most interesting from a basic point of view, in order to get a better understanding of the hydrophobic forces in nonionic systems. Indeed, a *full comparative* study of these fluorinated mixtures with those obtained from homologous series of the parent hydrogenated compounds is certainly essential to clarify the influence of the hydrophobic tail of the amphiphiles on the formation of nonionic micelles and microemulsions, and on their mutual interaction [5]. For example one can wonder whether one given $(EO)_m$ chain has the same hydrophilic properties when committed to a hydrogenated or a fluorinated surfactant molecule. But, whether we are interested in the basic or an applied point of view, any serious investigation must be grounded on the phase behaviour of these systems. Therefore, this part of our study will concern their binary and ternary phase diagrams, obtained on a sufficiently detailed level of description. Since the evolution of the phase behaviour with the experimental conditions (temperature, chemical nature of the oil and emulsifier . . .) is obviously strongly correlated to the molecular structures of the surfactant aggregates [3, 6–9], such a phase investigation is a pre-requisite for the structural determinations we are currently developing (mainly by scattering techniques: neutron, X-ray, visible light).

We propose to present new and essential phase behaviour results of some fluorinated compounds; for comparison, a few references to results on hydrogenated systems will also be made. More specifically, we shall be almost exclusively dealing with polyoxyethylene alcohols which have been the far most investigated compounds. The structural aspect will be only occasionally quoted.

Experimental

Materials

The surfactants of interest here are the nonionic semifluorinated amphiphiles $C_mF_{2m+1}CH_2(EO)_n$, and were synthesized according to an original method [10, 11]. Typical compounds are such that $m = 6, 7$, and $n = 3, 4, 5, 6$.

The fluorinated oils are mainly the perfluorodecalin, the perfluoromethylcyclohexane purchased from Ventron GmbH, $C_4F_9CH=CHC_4F_9$ (noted C4) donated by Dr. Riess J. G. (Université de Nice, France) and $C_8F_{17}CH=CH_2$ (noted C8) obtained from Atochem, France. They were used as received. Let us note that all these oils contain the same number of carbon atoms: they differ essentially by their bulkiness and their free volume in the liquid state.

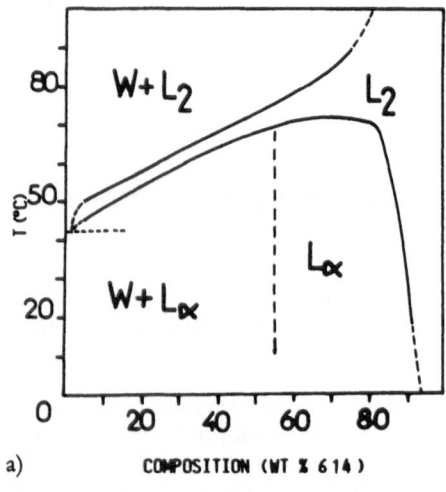

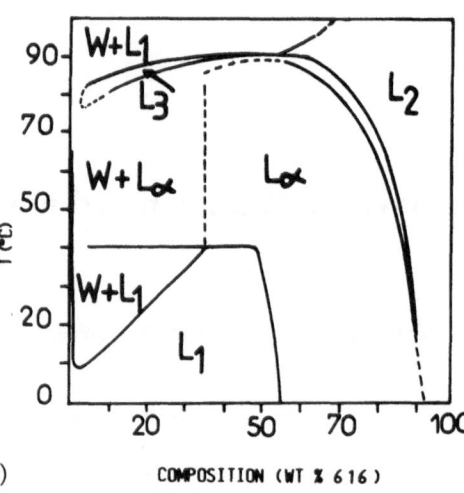

Fig. 1. Phase diagrams of surfactant/water systems (composition in wt % surfactant),
a) $C_6F_{13}CH_2(EO)_4$,
b) $C_6F_{13}CH_2(EO)_6$

Experimental techniques

1. Neutron scattering

Most of the small angle neutron scattering measurements have been performed at the Laue-Langevin Institute in Grenoble (France), using the D11, D16 and D17 instruments. A few others have also been carried out on the PACE spectrometer at the Laboratory Léon Brillouin (Saclay, France).

The experimental spectra were interpreted by using a best fit procedure, making use of the variation contrast method [7,12]. For that purpose, theoretical spectra for model aggregates were calculated, taking into account the oil/water penetration into the surfactant palisade, the molecular characteristics of the molecules (molar volumes, chain conformation), the interparticle effects (based on the Percus-Yevick solution for the equivalent hard sticky spheres [13]), and all the orientations of the actual anisometric scatterers.

2. X-ray scattering

The X-ray beam was "point collimated" by a collimator made of two pinholes, and the intensity was photographically recorded. The q-range available was $q > 0.02$ Å^{-1}. Such an apparatus was mainly used for the determination of the Bragg spacings in lamellar aggregates and liquid crystals.

3. Light scattering

Elastic light scattering was mainly used to determine the depolarization ratios of the light scattered by the small optically anisotropic aggregates. The measurements were essentially performed at the scattering angle $\theta = 90°$, since the "dust free" samples exhibited a scattering with a very low angular dissymetry.

4. Surface tension

Surface tension measurements were carried out by using the Wilhelmy plate method, in order to determine the CMC of the aqueous surfactant solutions. Temperature was monitored between 10° and 60°C. Platinum plates were used.

Results

1. Binary aqueous systems

The phase equilibria have been determined between 0 and 100 °C for a number of $C_mF_{2m+1}CH_2(EO)_n$ compounds, in the whole range of water/surfactant compositions. Examples of such diagrams are shown in Figure 1a and 1b, which correspond to amphiphiles with $m = 6$, and $n = 4, 6$. They exhibit a relatively simple behaviour when the number of oxyethylene groups is less than 6. The most striking result is the existence of very large domains with a lamellar liquid crystal (noted L_a), roughly for surfactant contents in the range 50–90 % w/w. No other type of liquid crystal seems to exist. Typically the L_α phase melts only at about 70°–80 °C, that is at a noticeably higher temperature than for the parent hydrogenated compounds [14].

More specifically, the system $C_6F_{13}CH_2(EO)_4/H_2O$ does not form any classical L_1 isotropic phase, i. e. does not form the usual aqueous micellar phase: as a summary, this surfactant appears always "water insoluble", except for a very narrow salient region, which is prominent from the stripe which exists whatever the temperature for very high surfactant content; this domain is classically noted L_2. The temperature of emergence of this phase in the dilute region is about 40 °C.

Therefore, this diagram roughly appears as a mere translation towards a lower temperature of the one corresponding to a more hydrophilic surfactant. This can be seen in Figure 1b, which is the diagram for $C_6F_{13}CH_2(EO)_6$. But at high temperatures the salient region becomes disconnected from the high content

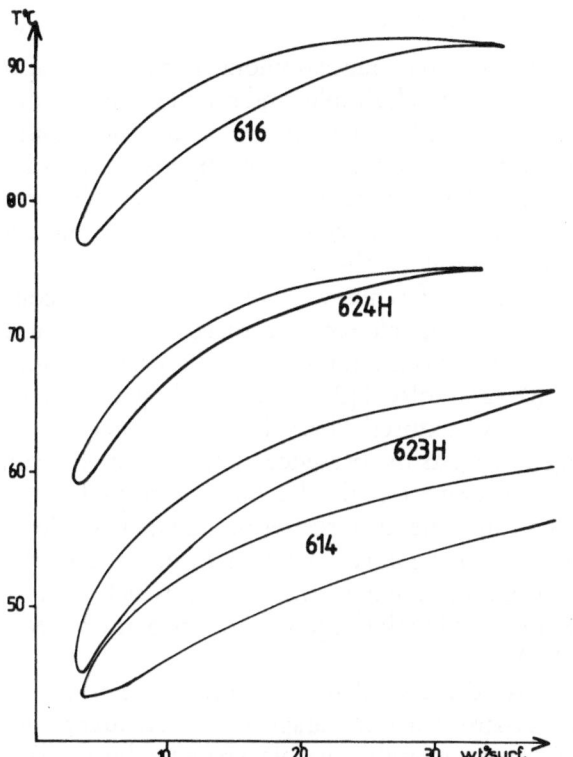

Fig. 2. L_2 (or L_3) phases for different surfactant/water systems. The surfactants are $C_6F_{13}CH_2(EO)_n$ with n = 4,6 (noted 614 and 616) and $C_6F_{13}CH_2CH_2CONH(EO)_m$ with m = 3,4 (noted 623H and 624H)

surfactant stripe, then forming the new isotropic phase realm classically noted L_3. The L_1 phase may exist till about 40 °C; the corresponding liquid-liquid consolute curve [3] presents the well known lower critical point for temperatures between 15° and 5 °C according the purity of the sample. Indeed, this lower part of that consolute curve is highly sensitive towards excessively tiny amounts of (probably hydrophobic) contaminants, the lower the (apparent) critical point the larger this amount.

As far as these phase diagrams are concerned, the most similar hydrogenated systems seem to be respectively $C_{12}(EO)_3$ and $C_{12}(EO)_5$ [14], although the corresponding melting/emergence temperatures of the various phases are not quite identical. From this comparison, two important points can be noted.

Firstly, the "equivalent" hydrophobic chain is not the $C_{10}H_{21}$-chain, as it should be if the following "rule" (derived for *ionic* surfactants) were strictly valid, that is: "2CF$_2$ groups are equivalent to 3CH$_2$ groups". Instead, and as far as the *general outlook of the whole* bi-

nary phase diagram is concerned, the equivalence ratio in chain length should be 1 : 1.8 (since $6 \times 1.8 + 1 \approx 12$) [15, 16].

Secondly, one EO group in a surfactant chain has not the same hydrophilic properties whether the hydrophobic part is fluorinated or is hydrogenated. Quantitatively this can be seen by comparing the emergence temperatures of the L_2 or L_3 phase in the dilute (5–10 %) systems: for hydrogenated surfactants, data from the literature [14] show that (when the L_2/L_3 phases exist), each additional EO group shifts this temperature of about 10° to 15 °C, while this change seems to be about 20 °C for the fluorinated ones.

Besides, in Figure 2 are shown the L_2/L_3 phases for two other fluorinated surfactants from other series, which are $C_6F_{13}(CH_2)_2$-CONH-(EO)$_n$, with n = 3 and 4. The results clearly indicate that the –CH$_2$-CONH– group is roughly equivalent to one EO group, and that the above value of the temperature shift is still valid, thus, confirming the influence of the *nature* of the hydrophobic chain on the "hydrophily" of the EO chain. In other words, one EO group in a fluorinated surfactant appears more "effective" (may be, more "hydrogen bound") than when it is in a hydrogenated one, since we need a larger temperature rise to obtain a L_2/L_3 phase or to melt the L_a phase.

As far as the CMC values are concerned, a direct comparison with hydrogenated nonionics $C_m(EO)_n$ with the same number of EO groups [17, 18] shows that the equivalence in the hydrophobic chain length seems to be about 1 : 1.7. For example, the CMC for $C_6F_{13}CH_2(EO)_6$ is respectively 4 and 3.10^{-4} mol/l at 15 °C and 30 °C, which should be very similar to that of $C_{11}(EO)_6$ (since then $6 \times 1.7 + 1 \approx 11$).

Concerning the aqueous micellar structures, the only "water soluble" surfactant system we have investigated was $C_6F_{13}CH_2(EO)_6$ [4]. The structural determinations were rather ticklish, given the proximity of a critical point at very low temperature. Clearly, the critical behaviour is the main phenomenom in the very close vicinity of this consolute point [30]. However, we are convinced that a so low temperature and a so low critical concentration have to be correlated to the presence of relatively large and flexible elongated micelles [4, 19].

Indeed, we have shown that the experimental neutron scattering data could be perfectly fitted to a model of semi rigid "worm like" particle, with a possible persistence length of about 150 Å. The hydrophobic chain is in an extended conformation, while the polyoxyethylene one should be in the meander form: conse-

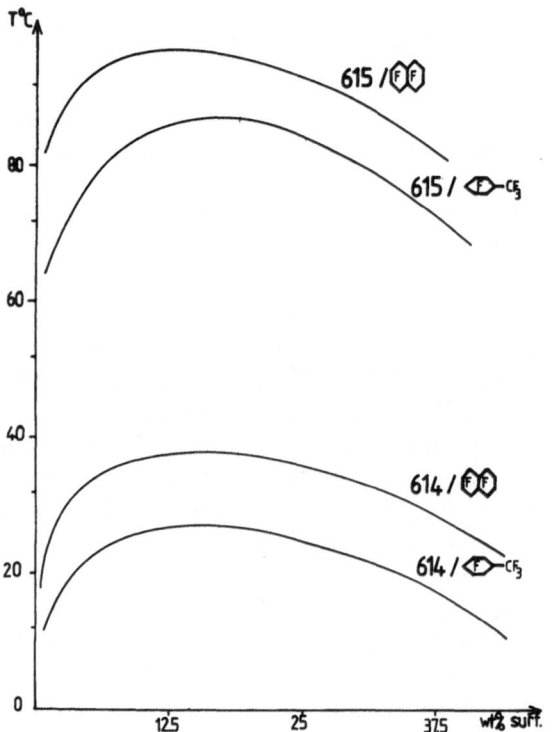

Fig. 3. Lower consolute curves for different systems surfactant/oil. The surfactants are those noted 613, 614, 615 and 616. The oils are perfluorodecalin and perfluoromethylcyclohexane

quently, the area per polar head (measured at the hydrophilic/hydrophobic interface) is about 60 Å², and there is room for 18–20 water molecules inside the surfactant film for such an interface curvature.

On the basis of the sticky interparticle potential it is quite impossible to represent this micellar system as interacting small spherical particles. But on the other hand, our interpretation of the data are perfectly coherent with the presence of anisometric micelles, which, as far as the *structure factor* of the solution is only concerned, may be equivalent to (larger) spheres interactive via an attractive potential of depth less than about 4 kT and a thickness of 2 Å. Typically, at a temperature 5 °C below the critical point, the aggregation number should be about 1500–2000.

Other complementary techniques clearly tend to support such a model. They are the Kerr effect and the changes in the neutron scattering by micelles submitted to a hydrodynamic flow, the results of which will be reported in separate papers [20].

2. Binary apolar mixtures

The phase behaviour of nonionic surfactants in apolar solvents is much simpler and mimics that of ordinary organic liquids: generally, we obtain lower liquid-liquid consolute curves with an upper critical point, these phases being isotropic. Some results are shown in Figure 3, for two surfactants in perfluorodecalin and perfluoromethylcyclohexane.

The main difference with ordinary organic compounds lies in the value of the actual critical concentration [17]. For surfactant molecules this critical point is about 10–12 % vol/vol [2], while it is about 0.5 vol/vol for ordinary mixtures. This value of this concentration strongly suggests the presence of quasi-isometric surfactant aggregates [13] that has been confirmed by recent small angle neutron scattering. Incidentally, let us recall that, for the binary aqueous L_1 phase, this critical concentration may be only 1–2 %, which then has to be correlated to the large anisometry of the aqueous (flexible) micelles.

Clearly, the critical temperature depends on the chemical nature of both the species. As a matter of fact, some systems are fully miscible whatever the temperature above 0 °C: that is for example the case for $C_6F_{13}CH_2(EO)_n$, with $n = 4,5$, in the *linear* oils previously noted C8 and C4.

A more hydrophilic surfactant leads to higher consolute temperatures: as seen in Figure 3, an additional EO group corresponds to a shift of 60 °C. On the other hand, at constant molecular weight, (i. e. 10 carbon atoms in the present case), the critical point depends on the stereochemistry of the oil molecules: the more bulky the solvent molecule, the higher that temperature. For example, there is a shift of about-10 °C when the perfluorodecalin is replaced by the less bulky perfluoromethylcyclohexane.

3. Ternary phase diagrams

Aqueous systems

The "microemulsion" domains are more or less deformed sector-like regions which about to the water corner (see e.g. Fig. 4). Examples of such phase diagrams have been shown in other papers [3, 21].

As a general rule, these "aqueous" domains progressively rotate towards higher oil/surfactant ratios when one rises the temperature, till they become disconnected from the water corner and form quasi-isolated regions [3]. At this point, the respective affinities of the

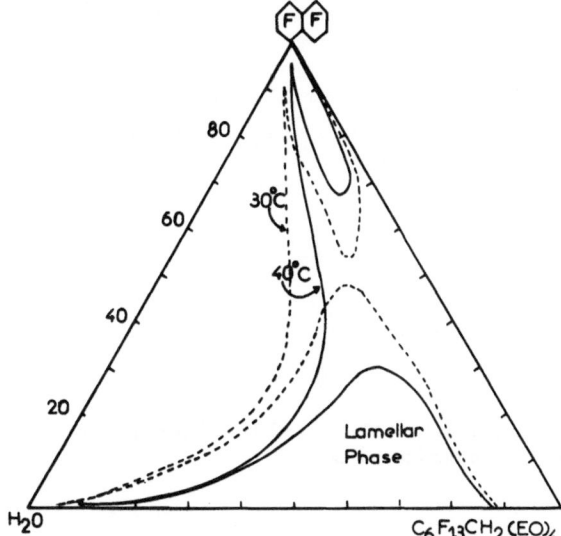

Fig. 4a) Phase diagram at 30°C and 40°C for the system 614/perfluorodecalin/water

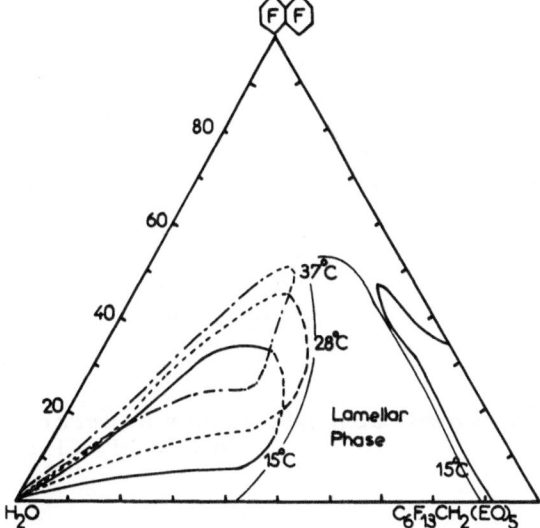

Fig. 4b) Phase diagram at 15° 28° and 37°C for the system 615/perfluorodecalin/water

surfactant for the oil and water tend to be exactly balanced, and the corresponding temperature may be called the Phase Inversion Temperature (PIT) [22]. Therefore, these trends quite parallel those of hydrogenated systems, a result we have already noted some time ago [5, 23].

As far as the structures are concerned, we have shown that the most concentrated systems behave like a dispersion of polydisperse disk-like particles [1]. Assuming that both the thickness (T) and the diameter (D) of these globules follow a log-normal distribution, typically their variances σ^2_T and σ^2_D are respectively (0.05–0.3) and (0.1–2) when the volume fraction increases from 0.4 to 0.5. Correlatively, the main thickness slightly decreases and tends to a value in accordance with the elementary cell dimension of the lamellar liquid crystal for which the area per polar head is about 50 Å^2, the conformations of the EO and CF$_2$ chain being respectively extended and semi-coiled.

For a moderately concentrated system [24] ($\Phi \simeq 0.05 - 0.02$) the experimental neutron data can be fitted to the theoretical spectra of mean oblate globules. The aggregation number slightly increases with temperature, but is more dependent on the oil/surfactant ratio: for example, it changes from 900 to 1700 when this ratio increases from 0.8 to 1.4, while the area per polar head keeps an almost constant value of about 50 Å^2 (there is room for two water molecules per EO group into this surfactant film).

When more and more water is added (for oil/surfactant ratios which make this dilution possible) the globules become ellipsoids which are more and more prolate, at roughly constant N. Correlatively the local curvature increases, the area per polar head tends to about 60 Å^2, while the conformation of the EO chain is progressively coiling, the hydrophilic thickness changing from about 18 to 10 Å in order to keep a roughly constant "hydration degree". These structural results have already been discussed elsewhere [1].

Surfactant-rich systems

For all the systems studied so far (C$_m$F$_{2m+1}$-CH$_2$(EO)$_n$, with $m = 6,7$, and $n = 4,5,6$, and the above mentioned fluorocarbons), a one isotropic phase L_2 exists for the higher surfactant contents. Moreover a large domain of a lamellar liquid crystal (L_a) extents widely throughout the diagram. At the PIT, it lies inbetween the L_2 region and the "surfactant phase" realm, i.e. the preceding isolated isotropic phase region.

As for the hydrogenated compounds, the extent of this lamellar phase decreases with temperature [25]. Examples of such diagrams at different temperatures are shown in Figure 4 for the system perfluorodecalin/C$_6$F$_{13}$CH$_2$(EO)$_4$ [2, 26].

From the values of the depolarization ratio, the light scattering results strongly suggest that this part of the

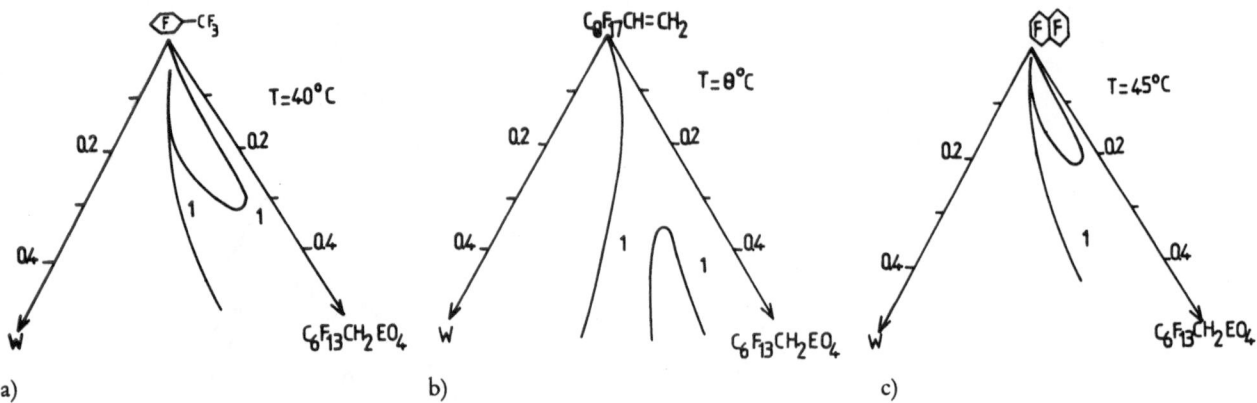

Fig. 5. Phase diagrams (oil corner) for systems with different oils and the same surfactant 614; a) at 40 °C with perfluorodecalin; b) at 8 °C with $C_8F_{17}CH = CH_2$; c) at 45 °C with perfluorodecalin

isotropic phase L_2 must be viewed as a random dispersion of small anisotropic grains, made of several surfactant bilayers.

Considering the rings and peaks measured in X-ray and neutron spectra as originating from Bragg reflections, we can calculate the lamellar thickness (d) of the samples likened to stacks of bilayers [14, 23, 27].

In the *lamellar phase* L_α the area per polar head (σ) of the surfactant molecule (supposed oriented perpendicularly to the plane of the lamellas) is *practically constant* everywhere throughout the whole liquid crystal domain. Typically, this area is $47 \pm 1\text{Å}^2$ for $C_6F_{13}CH_2(EO)_4$. It has to be compared to $42\,\text{Å}^2$ for the hydrogenated $C_{12}(EO)_4$ surfactant [28]. Therefore, σ is not entirely determined by the polar head, even in planar structures, but also *depends* on the nature of the hydrophobic tail.

In the *"pretransitory"* isotropic phase L_2, the structures can be determined only by taking advantage of the contrast variation method (use of H_2O/D_2O mixtures). To explain our neutron scattering data, we had to devise a structural model where the aggregates consist in multilayered and disordered grains, whose dimension is typically 200 Å. These aggregates are stacks of surfactant films and of "oil" layers, i.e., *without water layer*. A fulll picture of these aggregates is given in another paper [20].

Phase behaviour above the PIT

In the oil-rich regions, and for temperatures just above the PIT, there is a progressive coalescence of the (isolated) surfactant phase domain with the oil + surfactant regions [26]. As already noted for hydrogenated compounds [25, 28, 29], the way this coalescence takes place mainly depends on the penetration ability of the oil into the surfactant film in the micellar aggregates. This can be seen in Figure 5, which represents the oil corners of the ternary diagrams for three fluorocarbons of different bulkyness. For the oil molecules which allow some penetration into the film (e.g. the C8 oil), the coalescence of the surfactant phase takes place directly at the oil corner. And for the more bulky oils (e.g. for perfluorodecalin and perfluoromethylcyclohexane), the phase domains merge together in regions corresponding to much higher surfactant/oil ratios.

Whatever the case, there is a one isotropic phase area which extents towards the oil corner, and which is a true "inverse microemulsion" region, i. e. containing water-in-oil globules. However, for the bulky oils, this salient domain exists only for an extremely narrow range of the water/surfactant ratio.

Conclusion

The evolution of the phase behaviour of the fluorinated polyoxyethylene alcohol/perfluorocarbon systems reflects the progressive change of the relative affinities of the surfactant respectively for the water and the oil. Although the binary oil-surfactant systems do not exhibit special features as compared to ordinary organic mixtures, the binary aqueous phase diagrams clearly express a very peculiar association between water and the fluorinated surfactants. The presence of the fluorine atoms in the surfactant chains essentially promote the formation of lamellar structures, whether they belong to isotropic- or liquid crystal phases. The stability of these lamellar structures depend on the temperature, as for any other nonionic system, but to a much less extent.

And this is also true whether binary aqueous or ternary system are concerned, whatever the composition the system unless the water content is quite large. The important parameter is the area per polar head which seems rather independent of this composition. Since, at constant EO chain length, this area is larger for fluorinated surfactants than for hydrogenated ones, this enhanced stability for the lamellar packings has to be related to stronger hydration forces among the more effectively hydrated EO chains. Therefore, this "lamellar structure" stability could merely result from the geometrical constraint imposed by the sole *bulkiness* of the fluorinated chain, although the cohesive energy itself of fluorocarbons is noticeably less than that of hydrocarbons. As a conclusion, the apparent enhanced hydrophobic property of fluorinated surfactant could be a quite indirect consequence of purely geometrical property of CF_2 group: their important size.

Acknowledgements

We thank Dr. R. Oberthur, ILL, Grenoble, and Drs. J. Teixeira and J. P. Cotton, LLB, Saclay, for their help in the SANS experiments, and L. Mansuy for the synthesis of the surfactants.

References

1. Ravey JC, Stébé M-J, Oberthur R (1987) In: Mittal KL, Bothorell P (eds) Surfactants in Solution, Plenum Press, New York 6:1421-1430
2. Ravey JC, Stébé MJ (1986) In: Mittal KL (ed) Proceedings of 6th International Symposium on Surfactants in Solution, New-Dehli, Inde, submitted for publication
3. Mathis G, Leempoel P, Ravey JC, Selve C, Delpuech JJ (1984) J Am Chem Soc 106:6162-6171
4. Mathis G (1982) Thesis University of Nancy I, France
5. Mathis G, Ravey JC, Buzier M (1982) In: Robb ID (ed) Microemulsions, Plenum Press, New York 85-102
6. Ravey JC, Buzier M, Picot C (1984) J Coll Interf Sci 97:9-25
7. Ravey JC, Buzier M, Dupont G (1987) In: Rosano H, Dekker M (ed) Microemulsions, Surfactant Science Series, New York 24:163-182
8. Ravey JC, Buzier M (1984) In: Shah O (ed) Emulsion and Microemulsion, ACS Symposium Series 272, American Chemical Society, Washington 253-263
9. Nilsson PG, Lindman BJ (1982) Phys Chem 86:271-279
10. Selve C, Castro B, Leempoel P, Mathis G, Gartiser T, Delpuech JJ (1983) Tetrahedron 39:1313-1316
11. Gartiser T, Selve C, Mansuy L, Robert A, Delpuech JJ (1984) Chem Res (S) 292-293
12. Ravey JC, Buzier M (1984) In: Mittal KL, Lindman B (eds) Surfactants in Solutions, Plenum Press, New York 3:1759-1779
13. Baxter RJ (1968) J Chem Phys 49:2770-2774
14. Mitchell DJ, Tiddy GJT, Waring L, Bostock T, McDonald MP (1983) J Chem Soc, Faraday Trans 1, 79:975-1000
15. Shinoda K, Masakatsu H, Hayaski T (1972) J Phys Chem 76:909-914
16. Kunieda H, Shinoda K (1976) J Phys Chem 80:2468-2470
17. Shinoda K (1974) Principles of Solution and Solubility, Dekker M, Inc, New York
18. Becher P (1967) In: Schick MJ (ed) Nonionic Surfactants, M Dekker Inc, New York 478-515
19. Ravey J-C (1983) Coll Interf Sci 94:289-291
20. Ravey J-C, Stébé M-J, Oberthur R, to be published
21. Stébé M-J, Serratrice G, Delpuech J-J (1985) J Phys Chem 89:2837-2843
22. Shinoda K, Kunieda H (1977) In: Prince LM (ed) Microemulsions, Theory and Practice, Academic Press 57-89
23. Friberg S, Buraczewska I, Ravey J-C (1977) In: Mittal KL (ed) Micellization Solubilization and Microemulsions, Plenum Press, New York 2:901-911
24. Stébé M-J, Serratrice G, Ravey J-C, Delpuech J-J (19??) In: Mittal KL, Bothorel P (eds) Surfactants in Solution, Plenum Press, New York, in press
25. Kunieda H, Shinoda K (1982) J Dispersion Sci and Techn 3:233-244
26. Robert A, Tondre C (1984) J Coll Interf Sci 98:515-522
27. Buzier M (1984) Thesis University of Nancy I, France
28. Ravey J-C, Buzier M (1986) In: Mittal KL (ed) Proceedings of 6th International Symposium on Surfactants in Solution, New-Dehli, Inde, submitted for publication
29. Bostock TA, McDonald MP, Tiddy GJT (1984) In: Mittal KL, Lindman B (eds) Surfactants in Solution, Plenum Press, New York 3:1805-1820
30. Zulauf M, Weckstrom K, Hayter JB, Degiorgio V, Corti M (1985) J Phys Chem 89:3411-3417

Received March 7, 1987;
accepted March 9, 1987

Authors' address:

Dr. J. C. Ravey
Universite de Nancy I
Faculte des Sciences
Laboratoire de Physico Chimie des Colloides
U.A. C.N.R.S. No. 406
Boite Postale No. 239
Vandoeuvre-les-Nancy Cedex, France

Progress in Colloid & Polymer Science   Progr Colloid & Polymer Sci 73:134–141 (1987)

Electrical properties of polymerized, planar, bimolecular membranes

R. Rolandi[1]), S. R. Flom, I. Dillon, and J. H. Fendler

Department of Chemistriy, Syracuse University, Syracuse, New York, U.S.A., and
[1]) Dipartimento di Fisica, Universitá di Genova, Genova, Italy

Abstract: Electrical measurements have been made on bilayer lipid membranes (BLMs) made from four different polymerizable surfactants. Polymerization leads to a stabilization of the membrane. Membranes polymerized near the polar head of the surfactant are stabilized less than those polymerized in the tail. The duration and mechanism of rupture are dependent on the spreading solvent, as well as on the surfactant itself.

Key words: Membranes, bilayer lipid membranes, BLMs, polymerized BLMs, surfactant membranes, polymerizable surfactants.

Introduction

During the past decade, membrane-mediated processes have become an area of increasingly active research [1]. Thin, stable, and permselective polymeric membranes have been used in industry to facilitate microfiltration, ultrafiltration, dialysis, electrodialysis, and reverse osmosis [2, 3]. The structure of naturally occurring cell membranes has been studied and seems to be best described in terms of the "fluid mosaic" model [4, 5] which attributes separation processes to the lipid bilayer of the membrane, while proteins incorporated into the membrane mediate the membrane's transport properties. Further information as to the detailed nature of the lipid bilayer in naturally occurring membranes has been gleaned from studies of artificial biomimetic systems, such as vesicles and planar bilayer (or black) lipid membranes (BLMs) [1, 6].

The study of artificial biomimetic systems has the fundamental advantage of chemical homogeneity. For example, the BLM is most commonly used to model the electrical properties of cell membranes. At equilibrium, the planar lipid membrane consists of an oriented bimolecular film, the electrical properties of which – surface potential, capacitance and conductance – yield information as to its thickness and long-term dynamics. Subsequent to understanding these fundamental properties, the BLM has been advantageously exploited for the investigation of biological transport mechanisms [7, 8].

Much of this work is hampered by the lack of long term stability of the BLM. BLMs rarely last longer than a few hours. Polymer coating and polymerization are two promising avenues that can lead to an enhancement of the longer term stability of BLMs. Indeed, coating the BLM with polysaccharide derivatives leads to long-term stabilization [9]. Unilamellar vesicles made from amphiphiles containing polymerizable moieties such as diacetylene [10–12], methacryloyl [13–15], dienoyl [16–18], vinyl [19], and styryl [20, 21] groups have been stabilized by polymerization. Planar BLMs, prepared from similar polymerizable surfactants, may also gain long-term stability upon polymerization.

An investigation of the effect of polymerization on the BLM electrical properties has been reported by Benz et al. [22]. They described capacitance and conductance changes in a polymethacrylamide BLM. The observed changes of the conductance and capacitance of the membrane upon polymerization were attributed to penetration of water into "clefts" formed by the clustering of the polymerized head groups. Cleft formation was also observed upon two dimensional surface polymerization of vesicles [23].

In order to examine more fully the possible stabilization of BLMs through polymerization, we have measured the electrical properties of BLMs made from four surfactants containing polymerizable, styrene-like groups. Three of the surfactants have the polymerizable moiety located in the polar head of the amphiphile,

$$[C_{15}H_{31}CO_2(CH_2)_2]_2N^+[CH_3]-$$
$$[CH_2C_6H_4CH = CH_2], Cl^- \qquad (1)$$

$$[C_{18}H_{37}]_2N^+[CH_3][CH_2C_6H_4CH = CH_2], Cl^- \qquad (2)$$

$$[C_{18}H_{37}]_2N^+[CH_3]-$$
$$[(CH_2)_2CO_2C_6H_4CH = CH_2], Br^- \qquad (3)$$

while the fourth has a vinyl benzamide group at the end of one of its hydrocarbon chains:

$$[CH_2 = CHC_6H_4CONH(CH_2)_{11}O]-$$
$$[C_{16}H_{33}O]\, PO_2^-. \qquad (4)$$

Experimental

The synthesis of compounds (1)–(4) have been previously described [21, 24]. Each of the surfactants was checked for the presence of polymers by HPLC or TLC and recrystallized from acetonitrile or ethyl acetate, if necessary. The solvents (Aldrich, Gold Label) were used without further purification.

The membranes were formed at room temperature (22 °C) by hydrophobic apposition of two surfactant monolayers using the method of Montal and Mueller [25]. The BLM cell was a two-compartment version of that described by Schindler and Feher [26]. The membrane supports were made from 6 μm thick teflon. The holes were made by focusing the 514.5 nm line from an Argon Ion laser (Spectra-Physics) on the teflon film darkened by a marker. The diameters of the holes ranged from 100 μm to 500 μm.

The procedure for formation of the BLM was as follows. The surfactants were dissolved in n-decane or n-hexane (2–5 mg/ml). 5–25 μl of the surfactant solution was spread on the surfaces (1.3 cm²) of salt solutions (1–50 mM KCl) contained in the two compartments of the cell. The levels of the two solutions were alternately raised by two syringes. Usually, the membrane was formed within a few trials. Upon membrane formation, the two levels were equalized until a minimum in the membrane capacitance was observed. This was considered to be the beginning of the membrane lifetime.

A Supracil quartz window allowed the illumination of the membrane with ultraviolet light. Two UV sources were used. The first was a 150 W Xenon lamp (Oriel) delivered to the membrane via a fused silica fiber optic cable mounted approximately 1 cm from the membrane. The measured UV intensity at the membrane was 0.7 mW cm⁻². The second source was the doubled output of a synchronously mode-locked, argon-ion-pumped, R6G dye laser (Spectra-Physics), which was cavity dumped at 800 kHz. The resulting 298 nm (8–10 ps FWHM, 300 pJ/pulse) light was focused on the membrane with a 50 cm focal length quartz lens.

Capacitance and conductance were measured simultaneously using a Genrad 1689M RLC Digibridge. The instrument was connected to the membrane via two Ag/AgCl electrodes. The electrode capacitance and resistance were measured before and after each membrane measurement. An 11.7 Hz sinusoidal signal, typically 20 mV peak to peak, was applied to the membrane. For the experiments on membranes made from compound (4), the data were automatically acquired using the GPIB option of Digibridge to transfer the data to a Zenith Z-100 microcomputer. The capacitance (conductance) values reported are the average of five values. Some of the conductance measurements were made by a Dagan Patch Clamp using a 10 mV square wave with an inversion period of 100 s. The output was recorded either on an x-t recorder or a memory oscilloscope.

Results and discussion

Non-polymerized membranes

Each of the surfactants compounds (1)–(4) is sparingly soluble in pure n-alkanes. Therefore, either a small amount (1–4 %) of ethanol was added to the solu-

Table 1. Specific capacitances, conductances and dielectric breakdown voltages for BLMs formed from polymerizable surfactants and two kinds of lipids

Lipid and solvent	Specific capacitance ($\mu F \times cm^{-2}$)	Specific conductance ($\Omega^{-1} \times cm^{-2}$)	Dielectric breakdown (mV)
(1), n-hexane, ethanol	0.55 ± 0.10	$\sim 10^{-5}$	70
(1), n-hexane	0.59 ± 0.05	$\sim 10^{-5}$	70
(2), n-hexane, ethanol	0.68 ± 0.06	$\sim 10^{-5}$	70
(2), n-hexane	0.49 ± 0.06	$\sim 10^{-5}$	70
(3), n-hexane, ethanol	0.80 ± 0.10	$\sim 10^{-5}$	70
(3), n-hexane	0.49 ± 0.06	$\sim 10^{-5}$	70
(4), n-decane, ethanol	0.32 ± 0.05	$\sim 10^{-5}$	90
glyceryl-mono-oleate n-decane[c]) [a])	0.383 ± 0.008	$\sim 10^{-7}$	> 100
glyceryl-mono-oleate n-hexane[b])	0.745 ± 0.001	$\sim 10^{-7}$	> 100
lecithin, n-decane[c]) [a])	0.385 ± 0.013	$\sim 10^{-8}$	> 100
lecithin (18), n-hexane[b])	0.721 ± 0.001	$\sim 10^{-8}$	> 100

Capacitance values of compounds (1), (2) and (3) membranes are mean values of different membranes collected between 15 and 25 mins after formation. [a]) Capacitance from Reference [27]; [b]) capacitance from Reference [28]; [c]) prepared by the brush technique

tion, or the lipid suspension was heated to 50 °C to yield a clear solution. As will be described below, a solvent dependence can be observed in the electrical properties and lifetime of the BLM.

The observations of formation of planar BLMs made from long-chain, quaternary ammonium salts or phosphoesters are not appreciably different from observations of BLMs formed from phospholipids, though the final electrical properties are somewhat different (see Table 1). Membrane formation is characterized by large changes in the cell conductance and capacitance, both of which decrease dramatically when the membrane is initially formed. Typically, the change in capacitance observed is from several nF before formation to hundreds of pF upon initial formation, depending on the cross section of the hole in the membrane support and the thickness and curvature of the membrane. Changes in the observed cell resistance are from $k\Omega$ before formation, to $M\Omega$ afterwards.

The initial capacitance is insensitive to the level difference of the supporting electrolytes, i. e., the pressure

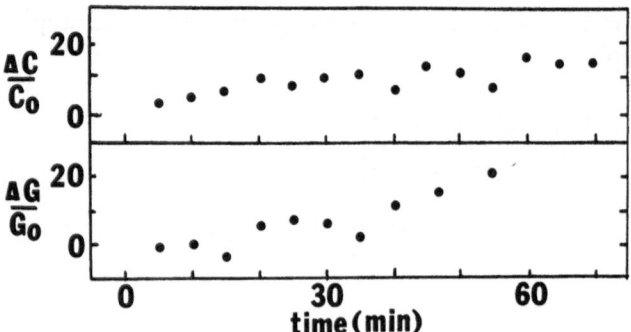

Fig. 2. Percent variation of the capacitance and conductance of compound (1) spread with pure hexane. Values are the mean of six different membranes

differential across the membrane. After a short period of time, which depends on the spreading solvent, the membrane capacitance begins increasing and becomes dependent on the pressure differential across the membrane. The membrane capacitance reaches a minimum upon equilibration of the solution levels. Thereafter, the capacitance slowly increases with time at a rate which seems to depend on both the spreading solvent and the surfactant used.

Figures 1 and 2 quantitatively show the changes of the electrical properties of membranes prepared from compounds (1)–(3) in hexane with and without added ethanol, respectively. Figure 1a shows the percent variation in the capacitance of membrane of compounds (1) and (3) after equilibration of the solvent levels (the percent of variation of the capacitance of membrane of compound (2) is almost identical to that of compound (1) and has been omitted for the sake of clarity). Figure 2 shows the same plot for membranes made from compound (1) using pure hexane as a spreading solvent. Membranes of compounds (2) and (3) have very similar characteristics, though compound (1) membranes last about twice as long. Comparison of Figures 1 and 2 indicates that the presence of ethanol in the spreading solvent augments the rate of increase of both the capacitance and conductance. It also seems to prolong the membrane lifetime. One further effect of the spreading solvent should be noted — membranes made from compound (4) using hexane/ethanol last a few minutes at most, while those made using decane/ethanol form long-lasting membranes whose capacitance and resistance are quite stable, as shown in Figure 3.

BLMs are generally modeled as analogs to parallel plate capacitors. It is known that the presence of sol-

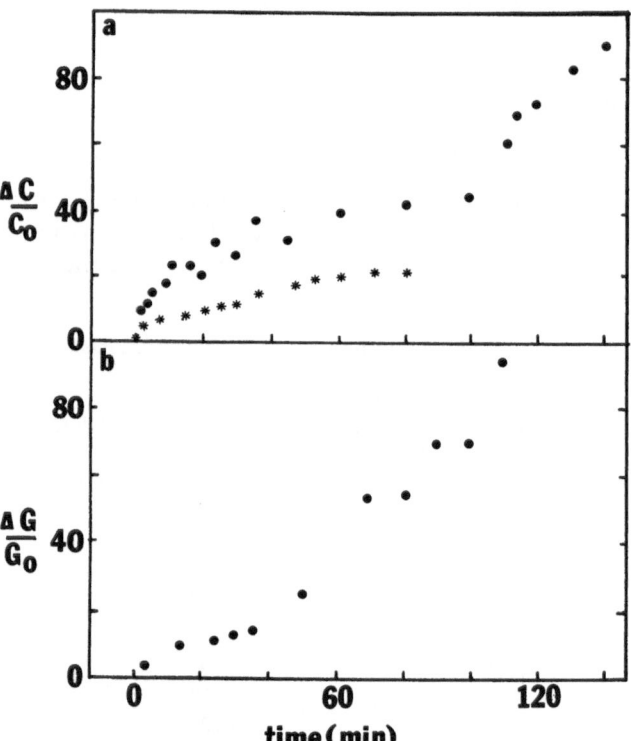

Fig. 1. Percent variation of the capacitance of compounds (1) and (3) spread with 100 : 1 (v/v) hexane/ethanol (1 mM KCl electrolyte. (a) Capacitance of coumpound (1) (●) and (3) (∗); (b) conductance of compound (3)

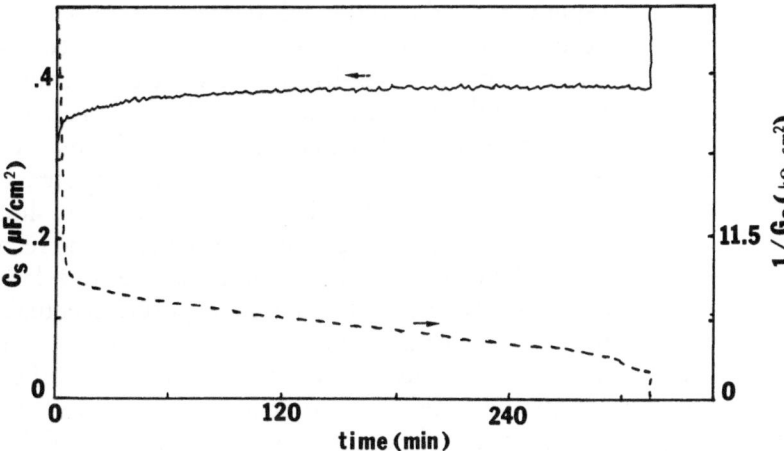

Fig. 3. Specific capacitance (solid line) and resistance (dashed line) of compound (4) spread with decane/ethanol (100 : 4 v/v)

vent thickens the membrane [27]. These statements allow one to interpret the physical processes associated with each of the changes of the capacitance. The initially observed low capacitance, which is insensitive to the pressure differential, corresponds to a solvent loaded membrane. The rapid increase in capacitance is associated with bilayer formation. The pressure dependent capacitance changes are a result of deformation of the membrane, which increases the surface area of the membrane and, thus, its capacitance. The slow increase of the capacitance shown in Figures 1–3 is probably due to slow leaching of solvent out of the membrane.

It is our experience that this last process happens more quickly than we have observed above, in membranes made from phospholipids under the same conditions. Moreover, membranes made from phospholipids usually attain a steady value before rupturing in contrast to those made from compounds (1)–(3). Further comparison of the properties of some phospholipid membranes and those studied above can be found in Table 1. Direct comparison of the specific capacitance of membranes of compounds (1)–(3) with glycerol-monooleate, which has the same hydrocarbon chain length, indicates that these membranes probably contain more solvent. The higher specific capacitances observed for surfactants spread with ethanol may well be due to the alcohol increasing the membrane's effective dielectric constant [29]. As a final comparison, membranes made from compound (4) have a capacitance that is similar to phospholipid membranes containing n-decane formed by the Mueller and Rudin brush technique [29].

The conclusions that may be drawn from membranes made from these four surfactants are firstly that each ruptures before attaining a thickness that corresponds to a "solvent free" membrane; secondly, since BLM usually exist only in a liquid, crystalline-like state [22], one may infer a mechanism for breakage of these membranes. As solvent leaches out of the membrane, it becomes a "super-saturated" solution and undergoes an abrupt phase transition which ruptures the membrane. Finally, the slow rate of the capacitance increase and longer lifetime of membranes made from compound (4) using the less volatile n-decane tend to support this explanation. The driving force for the solvent leaching out of the membrane phase must be the equalization of the chemical potential of the components in the different phases. In our open system, the vapor pressure of the solvent should determine the rate of increase of the capacitance.

Polymerized membranes

Membranes made under identical conditions to those described above were exposed to ultra-violet light to initiate polymerization. Generally, irradiation had begun 15–40 min after membrane formation and was of 15–20 min in duration. The data presented in Figures 4a–d show that the electrical properties of the BLM change rather dramatically upon irradiation. Control experiments using the cell filled with electrolyte alone and on membranes made from dioctadecyldimethylammonium chloride (DODAC) or phosphatidyl serine (PS) ensured that the observed changes in the BLM's electrical properties were not due to elec-

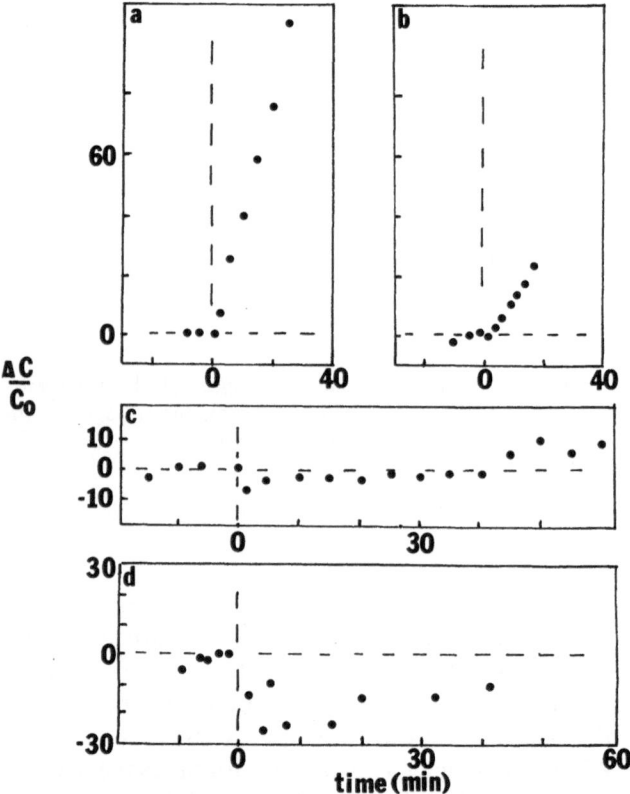

Fig. 4. Percent variation of the capacitance of compounds (1), (2) and (3). Each of the membanes were illuminated at $t = 0, 15$–40 min after formation: (a) compound (1) spread with hexane/ethanol; mean of three membranes; (b) compound (2) spread with hexane/ethanol; mean of two membranes; (c) compound (1) spread with pure hexane; mean of four membranes; (d) compound (3) spread with hexane/ethanol; mean of two membranes

trode photo-corrosion artifacts. The electrical properties of the DODAC and PS BLM remained constant when exposed to UV light.

Figures 4a, b, and d show UV-induced changes in capacitance observed for compounds (1), (2), and (3) when hexane/ethanol is used as a spreading solvent. The increases in capacitance during irradiation exhibit by compounds (1) and (2) (Figs. 4a and b) are consistent with results described by Benz et al. [22]. The decrease in capacitance exhibited by compound (3) (also spread with hexane/ethanol) is anomalous. Compound (1) exhibits the same (though smaller) anomalous trend when spread with pure hexane (Fig. 4c).

Further experiments using membranes made from compound (1) spread with pure hexane and irradiated with the laser (298 nm) yield similar results. In these experiments, the observed decrease is larger and occurs more quickly. The size and rate of the decrease is related to the power density of the excitation source. Further, if the excitation duration is limited to ca. 30 s, an immediate decrease in the capacitance is observed. After the excitation source is blocked, the capacitance increases to its original value. Subsequent short periods of irradiation again induce decreases and recoveries of the capacitance. The decreases are not as large as the initial decrease. Repetitious exposure eventually results in no change in the capacitance. Presumably, at this point, the membrane is fully polymerized. If the membrane is initially exposed for a period of 4–5 min, the capacitance first decreases, then stabilizes, and then begins to slowly increase at approximately the same rate as it did before exposure. Subsequent blocking of the excitation and re-exposure induces no further changes. It is clear that the light is causing a change in the membrane capacitance and the change has some saturation point. The most likely source of this change is photopolymerization of the styryl moiety. The polymerization of the membrane results in a modest 1.5 increase in its average lifetime.

While these experiments help to identify the source of the changes in the electrical properties induced by UV, they in no way clarify the variations with spreading solvent. Taken by itself, the observed decrease in the capacitance is not difficult to rationalize. Boheim et al. [30] have described the preparation of folded bilayers. The phase transition of the lipid bilayer is accompanied by contraction of the headgroups and an increase in the thickness of the membrane which would result in a decrease in the capacitance. A decrease of membrane capacitance due to a phase transition has also been reported by Gliozzi et al. [31]. Polymerization near the head group may cause the same kind of contraction which forces the alkyl chains into a more linear configuration, thickening the membrane. UV irradiation of monolayers made from compounds (1), (2), and (3) results in a decrease in the area per molecule [32]. Nevertheless, it is unclear what the role of ethanol is unless it in some way assists in the formation of clefts in polymerized compounds (1) and (2). Clearly, more research is needed to address the questions which these solvent effects evoke.

In contrast to the membranes formed from compounds (1)–(3), membranes formed from compound (4) exhibit no change in the capacitance upon exposure to UV, as shown in Figure 5. Rather, some of the compound (4) membranes exhibit a marked decrease in their resistance. If the membrane is exposed to UV

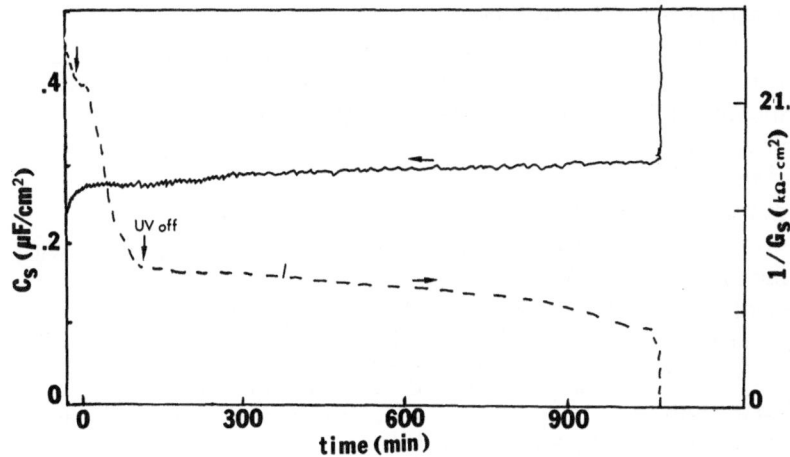

Fig. 5. Specific capacitance (solid line) and resistance (dashed line) of compound (4) spread with decane/ethanol. The membrane was illuminated with 298 nm light at $t = 0$, 30 min after formation. All conditions were identical to those in Figure 3

before it has appreciably thinned, no change in the electrical properties is observed. However, if the membrane has thinned to less than ca. 200 Å (as calculated from $C/A = \varepsilon\varepsilon_o/d$, where C/A is the specific capacitance, ε is the membrane dielectric constant, ε_o is the permittivity constant, and d is the membrane thickness) and is then exposed to UV, a precipitous drop in its resistance is observed, as illustrated in Figure 5.

There is good qualitative evidence that the membrane is polymerized. Each time a compound (4) membrane is exposed to UV, it remains visibly intact after rupture. In other words, the mechanism for rupture is the membrane pulling away from the teflon support. The adhering thin film stays partially attached to the septum and looks very much like a valve opening and closing when a small pressure gradient is applied. In contrast, when the membrane has not been exposed to UV, it breaks by bursting like a bubble.

Figure 6 quantitatively demonstrates this "valve"-like effect. Shortly after exposure to UV (denoted in the figure by arrow (a)), the membrane depicted in Figure 6 ruptured as indicated by the abrupt increase in the capacitance (arrow (b)). The value of the capacitance did not, however, return to the electrode value, nor did the resistance reflect that of an electrolyte solution with an unimpeded hole. The sudden spike in the plot, arrow (c), corresponds to visibly "opening and closing the valve" by careful manipulation of the solution levels, i. e., application of a gentle pressure gradient. The measured capacitance of the cell returned to the value of the electrode capacitance only with gross changes in the solution level and removal of the membrane by washing it away.

Evaluation of the stability of polymerized compound (4) can be conservatively estimated by comparison of Figures 3 and 5. The two membranes were

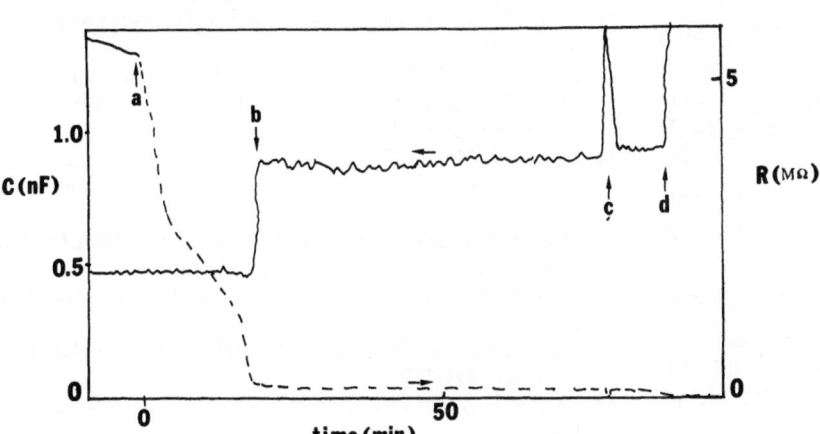

Fig. 6. Raw capacitance and resistance of compound (4) membrane: (a) = the start of UV irradiation; (b) = partial electrical rupture-detachment of the membrane from the teflon support; (c) = indicates a spike in the capacitance produced by manipulation of the batting solution levels; (d) = indicates a large change in solution levels. The membrane is pulled away from the support

produced, as nearly as possible, under identical conditions. The polymerized membrane lasted more than three times longer than the non-polymerized membrane. We have made polymerized membranes (albeit with specific capacitance that indicate a large amount of solvent contained in them) that have lasted approximately 44 h. These very long lasting membranes were polymerized immediately upon formation. The increase in the capacitance was very slow, which seems to indicate that the solvent is trapped by the polymer. The fact that large changes in the resistance are not observed unless the membrane is allowed to thin appreciably could be due to interlayer polymerization. In other words, if the two layers are separated by too much solvent, there can be no linkage between the opposing layers.

In retrospect, it is not surprising that compound (4) BLM shows a greater stability than the BLM made from surfactants which contain polymerizable groups near the polar head. The balance of forces that hold the BLM together is the opposition of the dispersive forces: hydrocarbon-hydrocarbon interaction vs. the coulombic repulsion of like charges. Replacing some of the dispersive forces with chemical bonds leads to a greater stability.

Conclusion

The electric measurements described above indicate that BLMs made from each of the surfactants of compounds (1)–(4) undergo polymerization upon ultraviolet irradiation. Polymerization near the polar head group results in a minor stabilization of the BLM. The observed changes in the capacitance indicate that the structure of the BLM is altered upon irradiation. The mechanism for rupture of the membrane probably does not change.

Polymerization in the alkyl chain of the surfactant results in a stabilization of the BLM by a factor greater than three. The lack of changes of the capacitance of the membrane may indicate that its structure is unaltered. The large changes in the resistance of the BLM can be attributed to partial detachment of the membrane from its support. The observation of the visibly intact membrane indicates that the mechanism for rupture of the polymerized membrane is different from that of the nonpolymerized BLM. The observation of detachment as a mechanism for rupture suggests a direction for future research — the use of support materials which have a greater affinity for the surfactant.

Acknowledgements

Support of this work by the National Science Foundation is gratefully acknowledged. We are grateful to our coworkers for the many helpful suggestions received during the course of this work.

References

1. Fendler JH (1982) Membrane Mimetic Chemistry, Wiley-Interscience, New York
2. Lonsdale HK (1982) J Membr Sci 10:81
3. Kesting RE (1985) Synthetic Polymeric Membranes, A Structural Perspective, 2nd Ed, John Wiley, New York
4. Singer SJ, Nicolson GL (1983) Science 173:720
5. Hoppe W, Lohman W, Markl H, Ziegler H (1983) Biophysics, Springer-Verlag, Berlin
6. Tien HT (1974) Bilayer Lipid Membranes (BLMs), Theory and Practice, Marcel Dekker, New York
7. Gliozzi A, Rolandi R (1985) In: Colombetti G, Lenci F (eds) Membranes and Sensory Transduction, Plenum Press, New York, p 1
8. Montal M, Darszon A, Schindler H (1981) Q Rev Biophys 1:1
9. Moellerfeld J, Prass W, Ringsdorf H, Hamazaki H, Sunamote J (1986) Biochim Biophys Acta 857:265
10. Johnston DS, Sanghera S, Pons M, Chapman D (1980) Biochim Biophys Acta 602:57
11. Hub H, Hupfer B, Ringsdorf H (1980) Angew Chem Int Ed Engl 19:938
12. O'Brien DF, Whitesides TH, Klingbiel RT (1981) J Polym Sci, Polym Lett Ed 19:95; O'Brien DF, Whitesides TH (1982) J Am Chem Soc 104:305
13. Regen SL, Czech B, Singh A (1980) J Am Chem Soc 102:6638
14. Akimoto A, Dorn K, Gros L, Ringsdorf H, Schupp H (1981) Angew Chem Int Ed Engl 20:90
15. Regen SL, Singh A, Oehme G, Singh M (1982) J Am Chem Soc 104:791
16. Gros L, Ringsdorf H, Schupp H (1981) Angew Chem Int Ed Engl 20:305
17. Dorn K, Klingbiel RT, Specht DP, Tyminski PN, Ringsdorf H, O'Brien DF (1984) J Am Chem Soc 106:1627
18. Tyminski PN, Latimer LH, O'Brien DF (1985) J Am Chem Soc 107:7769
19. Tundo P, Kippenberger DJ, Klahn PL, Pieto NE, Jao TC, Fendler JH (1982) J Am Chem Soc 104:456
20. Fendler JH, Tundo P (1984) Acc Chem Res 17:3
21. Reed W, Guterman L, Tundo P, Fendler JH (1984) J Am Chem Soc 106:1897
22. Benz R, Prass W, Ringsdorf H (1982) Angew Chem Suppl 869
23. Nome F, Reed W, Politi M, Tundo P, Fendler JH (1984) J Am Chem Soc 106:8086
24. Serrano J, Mucino S, Millan S, Reynoso R, Fucugauchi LA, Reed W, Nome F, Tundo P, Fendler JH (1985) Macromol 18:1999
25. Montal M, Mueller P (1972) Proc Natl Acad Sci USA 69:3561
26. Schindler H, Feher G (1976) Biophys J 16:1109
27. Fittiplace R, Andrews DM, Haydon DA (1971) J Membr Biol 5:277
28. Benz R, Froblich O, Lauger P, Montal M (1975) Biochim Biophys Acta 394:323
29. Mueller P, Rudin DO, Tien HT, Wescott WC (1962) Nature 194:979

30. Boheim G, Hanche W, Eibl HJ (1980) Proc Natl Acad Sci USA 77:3403
31. Gliozzi A, Rolandi R, De Rosa M, Gambacorte A (1983) J Membr Biol 75:45
32. Rolandi R (1986) unpublished results

Received December 26, 1986;
accepted January 20, 1987

Authors' addresses:

Dr. Ranieri Rolandi
Dipartimento di Fisica
Universitá di Genova
Via Dodecaneso 33
I-16146 Genova, Italy

Prof. Dr. J. H. Fendler
Syracuse University
Department of Chemistry
Syracuse, New York 13244-1200, U.S.A.

Progress in Colloid & Polymer Science   Progr Colloid & Polymer Sci 73:142–145 (1987)

Calculation of self-diffusion for a water-in-oil microemulsion model

U. Genz, J. K. G. Dhont, and R. Klein

Fakultät für Physik, Universität Konstanz, Konstanz, F.R.G.

Abstract: The long-time self-diffusion coefficient D_s is calculated to first order in the volume fraction ϕ for a model of a water-in-oil microemulsion. The model contains the hard-core repulsion, an attractive square-well potential due to the overlap of surfactant tails and the hydrodynamic interactions. The calculation is based on the solution of the Smoluchowski equation.

Key words: Microemulsions, self-diffusion, Brownian motion, Smoluchowski equation.

1. Introduction

There is an increasing interest in the various physical properties of microemulsion systems. The water-in-oil type of microemulsion, which is considered here, may form spherical symmetric "particles" consisting of a water core surrounded by surfactant molecules with their hydrophilic head group in the water core and the hydrophobic tail in the oil-phase. With respect to the concentration dependence of the self-diffusion coefficient, these systems are characterized by the interactions between the particles, which consist of direct potential interactions and hydrodynamic interactions. In Section 2 the model is described. Section 3 briefly sketches the theory. The detailed calculation will be published in a separate paper. In Section 4 some numerical results for the self-diffusion coefficient are considered.

2. Model

We assume the surfactant-coated water droplets to be spherically symmetric and monodisperse. As direct interactions and hydrodynamic interactions are taken into account, the two-particle potential and the two-particle diffusion tensors have to be specified. The geometrical sizes relevant for this work are the two-particle minimum distance $2b$ and the interparticle distance $2a$ at which the surfactants of different globules touch. $h = a - b$ gives the interpenetration depth.

The two-particle potential consists of a hard-core repulsive part and a short-ranged attractive interaction acting at distances, where the surfactant 'tails' interpenetrate. So we assume the two-particle potential to be of the simple form:

$$\beta V = \begin{cases} \infty & r < 2b \\ -\beta V_o & 2b < r < 2a \\ 0 & r > 2a \end{cases} \tag{1}$$

r is the interparticle distance, and $\beta = 1/k_B T$ is the Boltzmann factor. It follows from experiments [1, 2] and theoretical arguments [3], that the potential strength depends on the overlap volume of the interpenetrating surfactant tails. For $h \ll a$, this leads to a proportionality between the depth of the potential well and the particle radius,

$$\beta V_o = \text{const} \frac{\pi h^2}{6} \left(3a - \frac{h}{2} \right) \approx P a, \tag{2}$$

where the proportionality constant P is introduced.

The hydrodynamic interaction is described by diffusion tensors $\underline{D}_{ij}$, which give the linear relation between the velocity of a particle i, $\underline{V}_i$, and the force $\underline{F}_j$ which acts on the j-th particle. To get a result correct to first order in concentration, the two-particle diffusion tensors depending on the relative coordinate $\underline{r} = \underline{r}_1 - \underline{r}_2$

have to be considered. For hydrodynamically equal particles these are given by the relation:

$$\begin{pmatrix} \underline{V}_1 \\ \underline{V}_2 \end{pmatrix} = \beta \begin{pmatrix} \underline{D}_{11} & \underline{D}_{12} \\ \underline{D}_{12} & \underline{D}_{11} \end{pmatrix} \cdot \begin{pmatrix} \underline{F}_1 \\ \underline{F}_2 \end{pmatrix}. \quad (3)$$

For two spherical particles the $\underline{D}_{ij}$s can be decomposed in a part proportional to the projector on the line of centers $\underline{P}$ and the corresponding orthogonal projector $\underline{1} - \underline{P}$:

$$\underline{D}_{11} = a_{11}(r)\,\underline{P} + b_{11}(r)\,(\underline{1} - \underline{P}) \quad (4)$$

$$\underline{D} := 2(\underline{D}_{11} - \underline{D}_{12}) =: A(r)\,\underline{P} + B(r)\,(\underline{1} - \underline{P}), \quad (5)$$

where Equation (5) defines the relative diffusion tensor. The two-particle diffusion tensors for hard spheres are known as series expansions [4–6] in $1/r$ and expansions valid at small interparticle distances [5]. The hydrodynamic interaction tensors, taking into account the surfactant tails, however, are not known. This forces us to make some approximations.

For the region $r > 2a$, that is for non-overlapping particles, we assume the hydrodynamic interaction to be similar to the interaction between hard spheres of the radius b, and employ the b/r expansion given by Jones and Burfield [6] up to the order of $\left(\dfrac{b}{r}\right)^{20}$. For overlapping particles ($2b \leq r \leq 2a$) in perpendicular motion it is assumed that the arrangement of the surfactant tails prevents the particles from gliding along one another. This situation is similar to the case of touching spheres having stick boundary conditions. To have an estimate for the $(\underline{1} - \underline{P})$ part of the diffusion tensors, the values [5] for the functions B and b_{11} for the case of touching hard spheres having a diffusion coefficient D_o at infinite dilution are employed:

$$B = 0.802\, D_o\,, \quad b_{11} = 0.891\, D_o\,; \quad 2b \leq r \leq 2a. \quad (6)$$

Assuming that the surfactant tails do not influence the parallel motion as strongly as the perpendicular motion, the b/r expansion given by Jones and Burfield is employed for the function a_{11} if $2b \leq r \leq 2a$. The $1/r$ expansion does not give the correct result for the function $A(r)$ in the hard-sphere case, if the spheres are (nearly) touching. So we approximate the function $A(r)$ by a constant in the overlap region. The presence of the surfactant tails should decrease the parallel component of the relative mobility. Therefore the function

A should be smaller than the corresponding hard-sphere function A_{HS}. So we assume for interparticle distances between $2b$ and $2a$ that:

$$A =: A_- = f \cdot \frac{1}{2h} \int_{2b}^{2a} dr\, A_{HS}(r) =: f A_{HS}\,,$$

$$2b \leq r \leq 2a\,, \quad (7)$$

where f is a factor smaller than or equal to one.

3. Theory

The calculation of the self-diffusion coefficient is along the lines of Batchelor's approach (relaxation method) [7]. On the basis of the Smoluchowski equation this method gives the same long-time result to first order in concentration of the Brownian particles as the approach using projection operator techniques [6]. The line of thought is the following: a steady force $\underline{F}^{\text{ext}}$ is applied to the tagged particle 1 only. This leads to an anisotropy of the stationary pair distribution function $g(\underline{r}_1 - \underline{r}_2)$. From this "deformed" $g(\underline{r}_1 - \underline{r}_2)$ the mean velocity of particle 1, $\langle \underline{V}_1 \rangle$, is calculated. D_s then follows from:

$$\langle \underline{V}_1 \rangle = \beta\, D_s\, \underline{F}^{\text{ext}}. \quad (8)$$

The force acting on a particle j, $\underline{F}_j$, is a sum of direct interaction forces, 'Brownian' forces, and, if $j = 1$, the external force. So $\langle \underline{V}_1 \rangle$ and therefore D_s is composed of a term resulting from potential interactions (D_K), a term being related to the 'Brownian' forces (D_B) and a mean-field term (D_M). D_M determines the short-time behaviour, while $D_s = D_M + D_B + D_K$ gives the long-time self-diffusion coefficient. To first order in volume concentration $\phi_b = 4\pi N b^3 / 3V$, D_s is expressed in the form

$$D_s = D_o + \phi_b\, \frac{d_s^{(0)} + d_s^{(1)}\, e^{\beta V_o} + d_s^{(2)}\, e^{2\beta V_o}}{1 + C\, e^{\beta V_o}}$$

$$=: D_o + d_s \phi_b\,, \quad (9)$$

where the dependence of the potential depth enters only explicitly. $d_s^{(0)}, d_s^{(1)}, d_s^{(2)}$ and C are functions of the ratio of the outer and inner radius, a/b, and the constant A_- (respecting the factor f) introduced in Equation (7).

Results

The theory, which was briefly sketched in the preceeding section, predicts a linear dependence of D_s on ϕ_b with a slope depending on a/b, A_- (resp. f) and βV_o. The functions in Equation (9) are shown in Figures 1–4. From the general features of these functions it can be seen, that d_s will be negative; this means that the self-diffusion coefficient will be decreased by the presence of other particles. This decrease will become stronger with increasing potential strength. The dependence on the factor f will be weak as long as a/b and/or V_o are not too large.

As there is a relation between the potential and the geometrical sizes, it is convenient to present results depending on a different set of variables. These variables are the potential constant P, defined in Equation (2), the outer radius a and the penetration depth h. Keeping h fixed, results depending on the particle radius are obtained. Figure 5, curves (a), show d_s vs a for parameters h and P given by Huang [1] for an AOT system. Notice that d_s is given in units of D_o here, which also depends on the radius. The upper curve (b) gives the mean-field (short-time) result. For small potential the additional long-time contribution is small, as can be seen from the difference between curves (a) and curve (b). For comparison, the van den Broeck [8] result, which neglects hydrodynamic interaction, is also

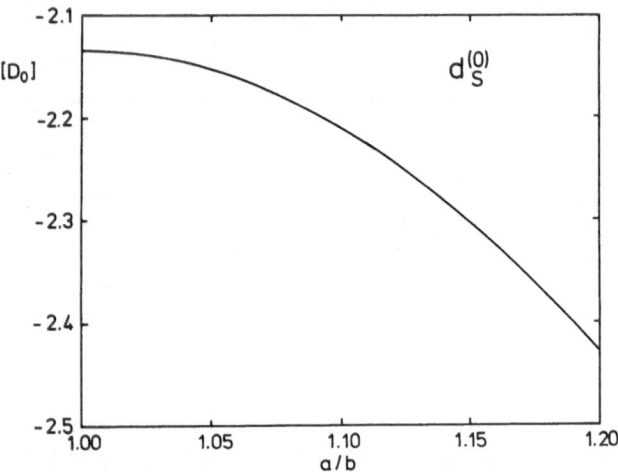

Fig. 1. $d_s^{(0)}$ as a function of a/b in units of D_o. $d_s^{(0)}$ is independent of A_-, respective f

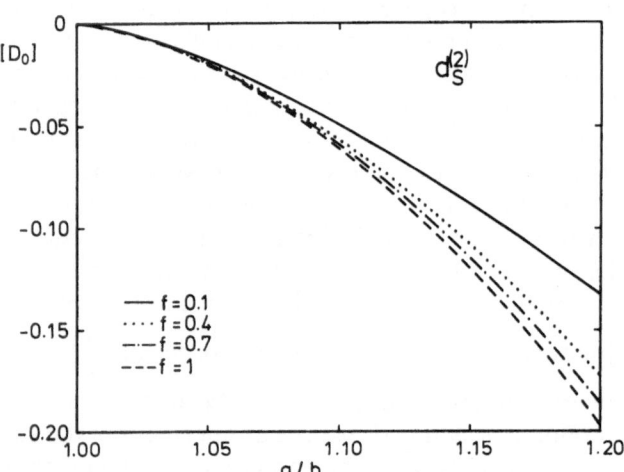

Fig. 3. $d_s^{(2)}$ vs. a/b for different factors f in units of D_o

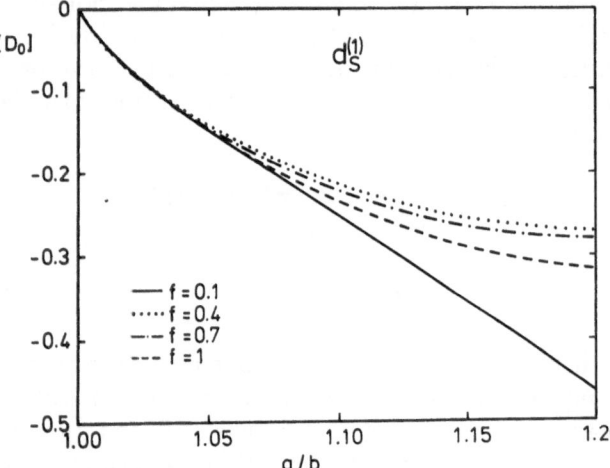

Fig. 2. $d_s^{(1)}$ vs. a/b for different factors f in units of D_o

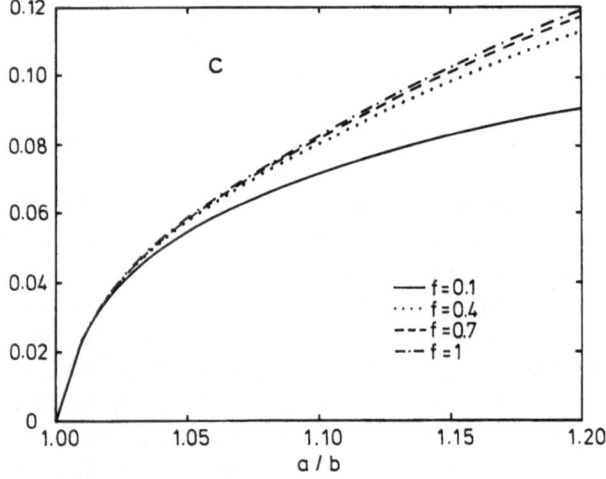

Fig. 4. C as a function of a/b for different factors f

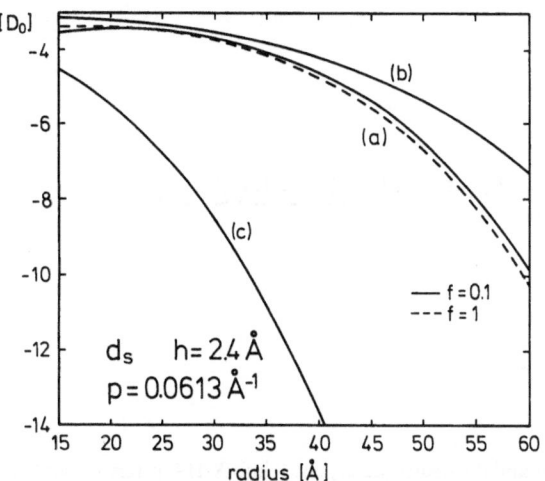

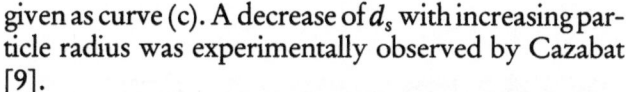

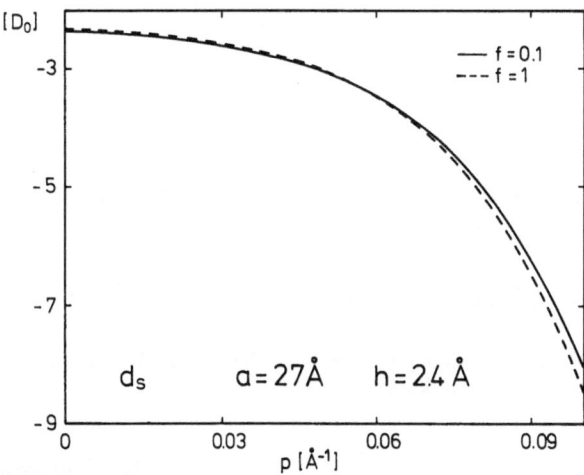

Fig. 5. The slope of D_s vs. ϕ_b, that is d_s, for the parameters $h = 2.4\,\text{Å}$, $P = 0.0613\,\text{Å}^{-1}$ as a function of the radius a in terms of D_o. Curves (a): long-time result for different fs; curves (b): short-time (mean-field) result; curves (c): d_s without hydrodynamic interaction, data obtained from Reference [8]

Fig. 6. d_s for a fixed geometry ($a = 27\,\text{Å}$, $h = 2.4\,\text{Å}$) against the potential constant P in units of D_o

given as curve (c). A decrease of d_s with increasing particle radius was experimentally observed by Cazabat [9].

In Figure 6 the potential constant P is varied, keeping the geometrical sizes constant. It can be seen, that d_s can be a measure for the potential strength, if the relevant geometrical sizes are known. The choice of a refers to an AOT system studied by Clarke et al. [10], containing particles with a hydrodynamic radius of about 27 Å.

Discussion

In this work the effect of an attractive short-ranged interaction and hydrodynamic interaction on the self-diffusion coefficient is investigated. One result is that D_s vs ϕ decreases more steeply with increasing potential. Neglecting hydrodynamic interaction, this feature is also found, but the magnitude of d_s can be much larger in the latter case. For a small potential the results are comparable in magnitude, as in the hard-core case. But the reasons for the decrease of D_s are different. Without hydrodynamic interaction, the decrease is due to the long-time contribution only, while, including hydrodynamic interaction, the decrease is mainly due to the mean-field (short-time) term; the additional long-time contribution is smaller in magnitude than the additional long-time contribution in the case where hydrodynamic interaction is neglected. This is supposedly due to the fact that the relative mobilities of the particles decrease strongly when their separations are small and direct interaction becomes important.

Our hard-core result ($a/b \to 1$) is $d_s = -2.13\,D_o$, which compares well to Batchelor's [7] results of $d_s = -2.10\,D_o$, where the behaviour of the hydrodynamic functions for very small interparticle distances is correctly taken into account.

References

1. Huang JS et al (1984) Phys Rev Lett 53:592; Huang JS (1985) J Chem Phys 82:480
2. Kim MW, Dozier WD, Klein R (1986) J Chem Phys 84:5919
3. Lemaire B, Bothorel P, Roux D (1983) J Phys Chem 87:1023
4. Felderhof BU (1977) Physica 89A:373; Schmitz R, Felderhof BU (1982) Physica 116A:163
5. Jeffrey DJ, Onishi Y (1984) J Fluid Mech 139:261
6. Jones RB, Burfield GS (1985) Physica 133A:152
7. Batchelor GK (1983) J Fluid Mech 131:155; (1976) 74:1; (1983) 137:467
8. Van Den Broeck C (1985) J Chem Phys 82:4248
9. Chatenay D, Urbach W, Cazabat AM, Langevin D (1985) Phys Rev Letters 54:2253
10. Clarke JHR, Nicholson JD, Regan KN (1985) J Chem Soc Faraday Trans I 81:1173

Received December 24, 1986;
accepted January 29, 1987

Authors' address:

Prof. Dr. Rudolf Klein
Universität Konstanz
Fakultät für Physik
Postfach 55 60
D-7750 Konstanz 1, F.R.G.

Membrane proteins in a lamellar phase of a polyglycol ether hydrate of Triton X-114®*)

F. Kopp[1]) and R. Heusch[2])

[1]) Diabetes-Forschungsinstitut an der Universität Düsseldorf, Düsseldorf, F.R.G.
[2]) Bayer AG, Leverkusen, F.R.G.

Abstract: In mixtures of water and the neutral detergent Triton X-114 (p-tert-octylphenol polyoxyethylene with 7 to 8 oxyethylene units per molecule) a pure lamellar liquid-crystalline phase is formed at a water content such that the structure formed can be attributed to a polydihydrate complex. A similar lamellar phase is separated in the lower part of the miscibility gap. Solubilized hydrophobic intrinsic membrane proteins coacervate with this lamellar phase as visualized by freeze fracture electron microscopy.

Key words: Cloud point, freeze fracturing, lamellar phase, membrane protein, miscibility gap.

When aqueous solutions of detergents of the polyglycol ether type, such as Triton X-100 and Triton X-114, are heated above their cloud point, a detergent-rich phase separates out. Intrinsic membrane proteins solubilized by Triton X-114 coacervate with the detergent-rich phase, whereas more hydrophilic plasmic proteins remain in the aqueous phase [1]. During the last few years, several authors have taken advantage of this behaviour and have successfully separated proteins and peptides originating from different biological sources on the basis of hydrophobicity for either preparative or analytical purposes [2–6].

Electron microscopic investigation by freeze fracturing of the detergent-rich phase of Triton X-114 separated in the miscibility gap revealed its lamellar ultrastructure, as shown in the micrograph Figure 1. For freeze-fracturing, an aqueous solution containing 1% (w/w) of detergent was heated above the cloud point up to 30 °C. The detergent-rich phase which separated at this temperature was collected by a short centrifugation (10 min at 1800 g). The final detergent concentration was between 15% and 17% (w/w). Microlitre size droplets of the detergent-rich phase were placed onto small gold plates (diameter 3 mm) and shock-frozen from 30 °C. For optimal freezing (avoidance of the

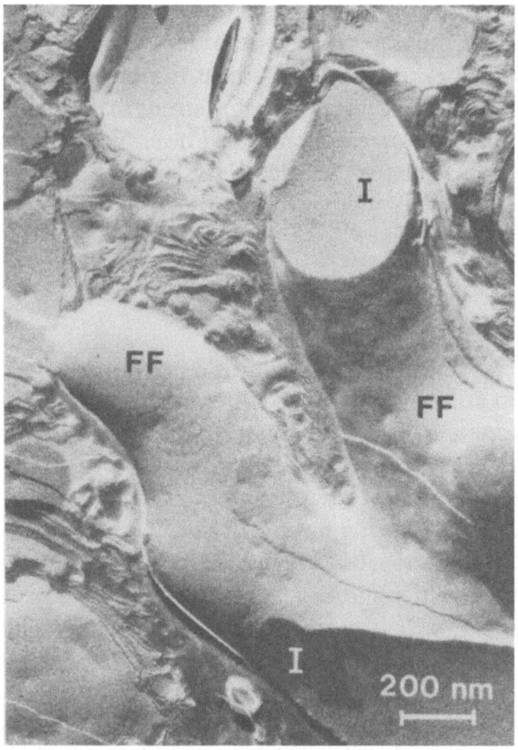

Fig. 1. Freeze fracture electron micrograph of a detergent-rich phase of Triton X-114 separated in the lower part of the miscibility gap. Cross-fractured stacks of detergent bilayers, smooth internal fracture faces of the bilayers (FF) and ice (I) originating from free water are shown (from [8]). Magnification × 50,000

*) ® Product of Rohm & Haas Deutschland GmbH.

Leidenfrost phenomenon), quenching was performed by plunging the specimens into a mixture of liquid propane and propylene which was at the temperature of liquid nitrogen. In this way, freezing velocities of about 10^3 Ks^{-1} are achieved in the interior of the specimen. A significantly higher freezing rate may be expected close to the surface of the droplet in contact with the cryogen.

Flexible bimolecular multilayers form extended dendritic tubes that are filled with ice originating from free water. Crossfractured multilayers and extended smooth internal fracture faces of the cleaved bilayers are clearly visible.

The phase diagram of the system Triton X-114/water (Fig. 2) has been discussed elsewhere [7].

At room temperature only a single mesophase is found, which is attributed to a polydihydrate, i. e. a complex in which two water molecules are hydrogen-bound to each oxygen atom of the polyether chain.

The texture of this mesophase, as it is visible in the polarizing light microscope is given in Figure 3.

The picture (magnification 180 times) has been taken from a sample in a 0.1 mm glass cuvette after

Fig. 3. Polarization light micrograph taken of the lamellar polydihydrate mesophase of Triton X-114 at 20°C. Detergent content 63% (w/w). Magnification × 180

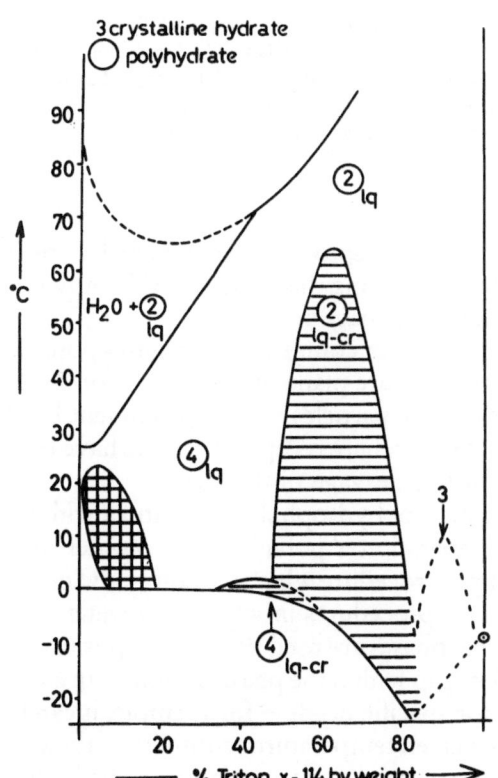

Fig. 2. Phase diagram of the system Triton X-114/water

melting and subsequently cooling down to room temperature. Stacks of lamellae are visible that had grown in different directions.

A freeze fracture electron micrograph of the mesomorphic polydihydrate phase is given in Figure 4.

It shows that the polydihydrate forms infinite bimolecular sheets which are arranged parallel to each other. Again cross fractured layers and extended smooth internal fracture faces are found, but no free water is left to form intramembranous ice as with the detergent-rich phase obtained by phase separation (Fig. 1). The amount of water in the mixture was just sufficient to form the polydihydrate complex.

Since the lamellae formed in the lower part of the miscibility gap (Fig. 2) cannot be distinguished from the polydihydrate mesophase lamellae, we concluded, that the former also consist of molecules in the polydihydrate form. In contrast to the mesophase lamellea however, the lamellae formed in the miscibility gap

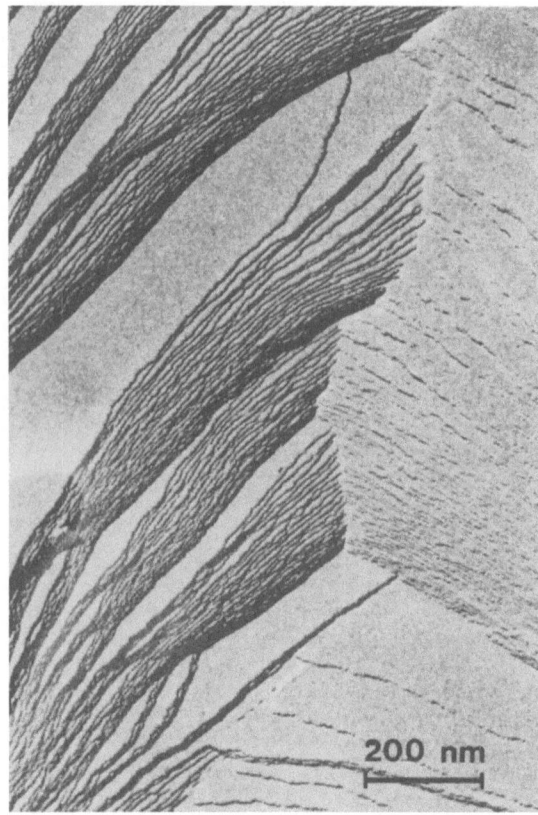

Fig. 4. Freeze fracture electron micrograph of 55% Triton X-114 in water. Shock-freezing was performed as described in the text (to produce Fig. 1) except that the starting temperature was 20° ± 0.5 °C. The polydihydrate forms a lamellar mesophase. Magnification × 75,000

Fig. 5. Freeze fracture electron micrograph of Band-3-protein inserted into lamellae of Triton X-114. The protein molecules appear as membrane particles in the smooth hydrophobic internal fracture face of detergent bilayers. Experimental conditions as those which produced Figure 1. Magnification × 75,000

coexist with a water solution of a very low detergent content close to the critical micellar concentration [1].

Hydrophobic membrane proteins solubilized by Triton X-114 and coacervated by phase separation are expected to be inserted into the hydrophobic part of the polydihydrate bilayer and should therefore be visible after freeze fracturing.

Figure 5 shows erythrocyte band-3-protein isolated with Triton X-114 and inserted into the detergent lamellae concomitant with phase separation. As expected, the protein molecules appear as intramembranous particles in the smooth hydrophobic fracture face of the freeze fractured detergent lamellae. The size of the particles is comparable with the size of protein particles found in biological membranes. The number of exposed band-3-protein particles increases concomitantly with increasing protein concentration as we have shown recently [8].

Finally, insulin receptor protein which has been made visible by applying this method [5] is shown in Figure 6 at a higher magnification.

The insulin receptor complexes are again exposed as particles in the fracture plane of the split Triton X-114 bilayer. Besides the particles visible, prominent holes can be seen at sites where receptor particles have been pulled out during cleavage of the bilayer.

The separation of hydrophobic substances with the aid of phase separation of polyglycolether/water solutions is a simple procedure which has some advantages over separation procedures in which two water-containing organic phases are used [9–11]. The position of the miscibility gap within the phase diagram can easily be adapted for specific needs — for example, it can be adjusted for a given temperature sensitivity of the substances to be separated — by a variation in the hydrophilic polyglycol ether chain length. By altering the

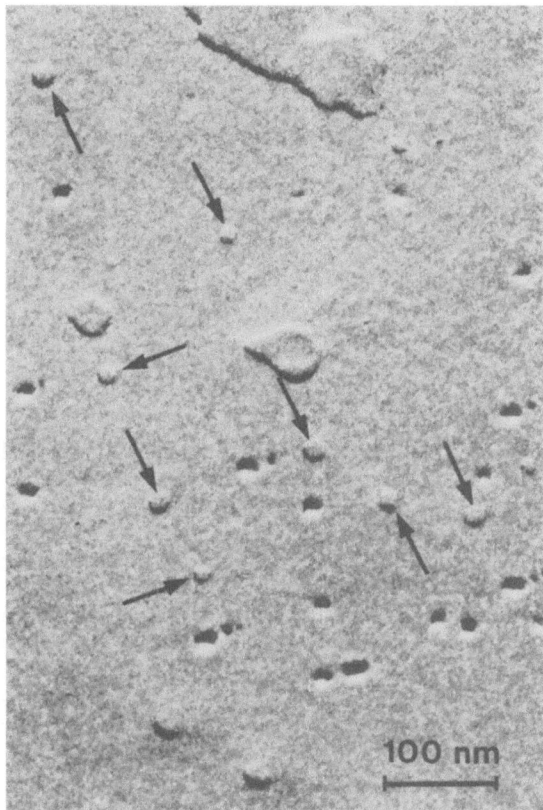

Fig. 6. Freeze fracture electron micrograph of purified insulin receptor protein, inserted into lamellae of Triton X-114, concomitant with phase separation. Experimental conditions as those which produced Figure 1. 1 µl of the starting solution of 1% Triton X-114 contains about 10 ng of protein. The insulin receptor protein molecules appear as intramembranous particles (arrows). Holes are visible at sites where the molecules have been pulled out of the detergent layer during cleavage in the frozen state. Magnification × 150,000

hydrophobic carbohydrate portion, the polyglycol ether might be adapted to the hydrophobicity demands of the substances investigated.

Acknowledgements

We thank Dr. Charlene Hohl for kindly improving the English of the manuscript. This work was supported by the Deutsche Forschungsgemeinschaft (SFB 113), the Ministerium für Wissenschaft und Forschung des Landes Nordrhein-Westfalen and the Bundesministerium für Familie, Jugend und Gesundheit.

References

1. Bordier C (1981) J Biol Chem 256:1604–1607
2. Alcaraz G, Kinet J, Kumar N, Wank SA, Metzger H (1984) J Biol Chem 259:14922–14927
3. Bricker TM, Sherman LA (1984) Arch Biochem Biophys 235:204–211
4. Nicolson DW, McMurray WC (1986) Biochim Biophys Acta 856:515–525
5. Kopp F, Meyer HE, Reinauer H (1985) Biol Chem Hoppe-Seyler 366:695–698
6. Pryde JG (1986) TIBS 11:160–163
7. Heusch R (1986) Biotechn Forum 3:2–8
8. Kopp F, Meyer HE, Reinauer H, Heusch R (1986) Tenside Detergents 23:119–124
9. Albertsson PA (ed) (1971) Partition of Cell Particles and Macromolecules, 2nd ed, Wiley, New York
10. Hustedt H, Kroner KH, Kula M-R (1986) In: Walter H, Brooks DE, Fischer D (eds) Applications of Phase Partitioning in Biotechnology, Partitioning in Aqueous Two-Phase Systems: Theory, Methods, Uses and Applications to Biotechnology, Academic Press, New York
11. Hampe MJ (1986) DPA 3 606 048.8

Received December 18, 1986;
accepted February 11, 1987

Authors' address:

Dr. Friedrich Kopp
Diabetes Forschungsinstitut
Auf'm Hennekamp 65
D-4000 Düsseldorf, F.R.G.

Progress in Colloid & Polymer Science   Progr Colloid & Polymer Sci 73:150–155 (1987)

Reactivity in microemulsion media:
Oxidation or reduction of cystine residues in keratin

P. Erra, C. Solans, N. Azemar, J. L. Parra, M. Clausse[1]) and D. Touraud[1])

Instituto de Tecnología Química y Textil (C.S.I.C.), Barcelona, España
[1]) U.A. C.N.R.S. no. 858. Département de Génie Biologique Université de Technologie de Compiègne, Compiègne Cedex, France

Abstract: The oxidation or reduction of cystine residues of keratin fibres by chemical agents incorporated in microemulsions of a model system was investigated. The microemulsion yielding system water/sodium dodecyl sulfate/1-pentanol/n-dodecane was selected for the following reasons: (a) At values of SDS/C_5OH mass ratio of 1/2 or so, the microemulsion domain in a pseudo-ternary diagram spans a wide range of compositions; (b) The extent and general configuration of the microemulsion domain are not significantly affected by the additon of chemical agents such as thioglycolic acid or hydrogen peroxide nor by temperature changes in the range 20 °C – 50 °C. It was found that the reactivity of cystine, as evaluated by the amount of cysteic acid formed (hydrogen peroxide as reagent), or of cysteine formed, (thioglycolic acid as reagent), could be related to the microstructure of the microemulsions used as the reaction media.

Key words: Microemulsion, phase diagram, keratin, cystine, reactivity.

Introduction

According to the most widely accepted definition, the term microemulsion refers to systems of water, oil and amphiphile which are single optically isotropic and thermodynamically stable liquid solutions [1]. There has been an extraordinary growth of research in the field of microemulsions since the last decade. In additon to their intrinsic theoretical interest, due to their characteristic properties, microemulsions constitute a promise for a wide range of technological applications [2–4].

The microstructured organization of microemulsions with more or less well defined hydrophobic and hydrophilic domains makes them appropriate media for specific chemical reactions. In fact, reaction studies in microemulsion media have been reported in the recent years [5–9]. However, due to the complexity of the processes involved in aggregate formation, the exact treatment of the data still poses considerable questions.

In an attempt to contribute to a better understanding of the role of microemulsions as reaction media, we have studied the modification of chemical reactivity of

keratin cystine by the action of an oxidant (hydrogen peroxide) or a reductive (thioglycolic acid) agent incorporated in microemulsion media. Classical reaction studies of keratinic materials are carried out using oxidative or reductive regular-type solutions, i. e. aqueous or water/alkanol solutions. These reactions are especially relevant for their application in a wide range of technologies such as cosmetic, textile, etc.

The well known system water/sodium dodecylsulfate/n-pentanol/n-dodecane was chosen as a model microemulsion system. The realm of existence of microemulsions in the pseudo-ternary phase diagram (surfactant/cosurfactant, mass ratio, k_m, equal to 1/2) is an all-in-one block domain extending from its water apex to the close vicinity of its hydrocarbon-surfactant edge [10]. Due to the continuity existing between water-rich and water deficient composition regions it was predicted and experimentally evidenced [10–11] that microemulsion structure is highly diverse and varies smoothly with composition. Consequently an additional goal of this research was to find out if cystine reactivity could be related to the structural state of microemulsions.

Experimental

Materials

Sodium dodecylsulfate (99 %) (SDS) and n-dodecane (99 %), were supplied by Merck, n-pentanol (98 %) was purchased from Fluka. They were used without any further purification. Water was deionized and twice distilled in glass and its conductivity was 30.0 μS cm^{-1} (0.0030 Sm^{-1}) Hydrogen peroxide (33 %), used as oxidative reagent, was supplied by Panreac. Thioglycolic acid (99 %), used as reductive reagent, was supplied by Fluka.

Merino 64's wool fibres and human hair of Caucasian type were used as keratin substrate. The former were washed with ethyl ether, ethanol and deionized water, extrated in a Soxhlet with dichloromethane and dried at room temperature. The human hair was extracted in a Soxhlet with ethyl ether, washed with ethanol and deionized water and finally dried at room temperature.

Methods

Phase diagram

The isothermal pseudo-ternary phase diagrams of the water/ SDS/n-pentanol/n-dodecane system with a weight ratio SDS to n-pentanol, k_m, equal to 1/2 and containing 3 % (w/w) of hydrogen peroxide or thioglycolic acid were determined by progressive titration. The boundaries of the microemulsion domains were determined by visual inspection of the samples and the presence of liquid crystalline phase detected with polarized light.

Microemulsion properties and cystine reactivity were determined along P_w and P_s experimental paths (P_w = mass fraction of water and P_s = mass fraction of surfactant/alcohol mixture). In the following, W = water, O = n-dodecane, S = sodium dodecylsulfate and n-pentanol mixture.

Electrical conductivities were performed with a Crison conductimeter model CDTM 525 using a dip cell containing a platinezed platium electrodes.

Viscosities were determined with a Ferranti-Shirley MK III viscometer.

In all cases, the measurements were carried out at 25° $\pm$ 0.2 °C

Keratin fibers treatments

Either wool fibres or human hair were treated under thermostated conditions with different microemulsion compositions containing 3 % (w/w) of the chemical reagent. After the treatment, the keratin fibres were washed with water (3 × 100 ml) and dried in a vacuum dryer.

Parallel treatments were carried out either in aqueous or n-pentanol media. The mass ratio keratin fibres to microemulsion compositions was 1 : 50 (w/w).

Cysteic acid determination

Keratin samples (100 mg) were hydrolyzed at 105 °C for 24 h with 20 ml of 6N-HCl in sealed glass tubes under N$_2$ atmosphere. The HCl was removed by evaporation in vacuo and the content of cysteic acid in the hydrolyzate estimated by ion-exchange chromatography with an automatic amino acid analyser (Biotronik IC 5000).

Cysteine determination

The cysteine formation on the keratin fibre was estimated according to the procedure described by Meichelbeck [12] using the Ellman's reagent [13]. All the values were corrected according to the sorption of surfactant by the hair fibre, as described below.

Surfactant sorption determination

The total amount of anionic surfactant sorbed by the hair fibre was determined according to the method described by De Boss [14]. The surfactant is quantitatively removed from the keratin fibre with n-propanol/water (1 : 1) solution at pH 11 and titrated by the modified Epton two-phase titration method [15].

Results

Phase diagrams, electrical conductivity and viscosity determinations

The general configuration of the microemulsion domain in the pseudo-ternary phase diagram of the model system water/sodium dodecyl sulfate/n-pentanol/ n-dodecane (with a weight ratio SDS/n-pentanol, k_m, equal to 1/2) was not significatively affected by addition of a 3 % (w/w) of hydrogen peroxide or thioglycolic acid (Fig. 1). It should be pointed out, how-

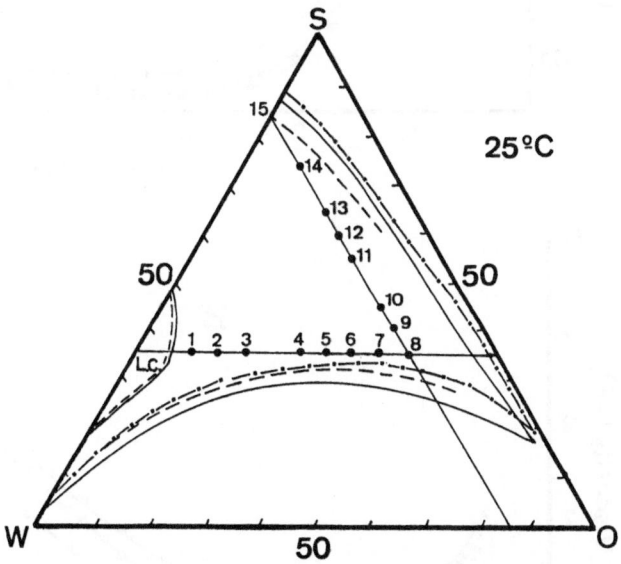

Fig. 1. Mass pseudoternary phase diagrams showing the microemulsion regions of the systems: (——) water/SDS/n-pentanol/n-dodecane (Model system); (———) water/SDS/n-pentanol/n-dodecane with 3 % (w/w) H$_2$O$_2$; (—·—·—) water/SDS/n-pentanol/ n-dodecane with 3 % (w/w) HSCH$_2$COOH. W = water, O = n-dodecane, S = sodium dodecyl sulfate and n-pentanol mixture, weight ratio $k_m = {}^1/_2$. Numbers 1 to 14 indicate microemulsions used as reaction media. L.C. denotes regions containing a liquid crystalline phase

ever, that the presence of thioglycolic acid caused destabilization of the lamellar liquid crystalline phase which emanates from the W-S edge into the microemulsion region.

Variations of electrical conductivity and viscosity, at 25 °C as a function of surface active mixture to water ratio S/W (Fig. 2) along the experimental $P_s = 0.35$ and $P_w = 0.15$ paths also revealed that microemulsions containing hydrogen peroxide or thioglycolic acid showed similar properties that those of the model system.

These results supported the early assumption made on the basis of phase diagram determinations. The additon of 3 % w/w either hydrogen peroxide or thioglycolic acid does not seem to affect the general structure of microemulsions.

On the grounds of information available regarding microemulsion structure [10,11], along the experimental $P_s = 0.35$ path, there is an evolution from a preferentially water external-type microemulsions to typically inverse structures of the W/O type, as the water content decreases. The sharp decrease of elec-

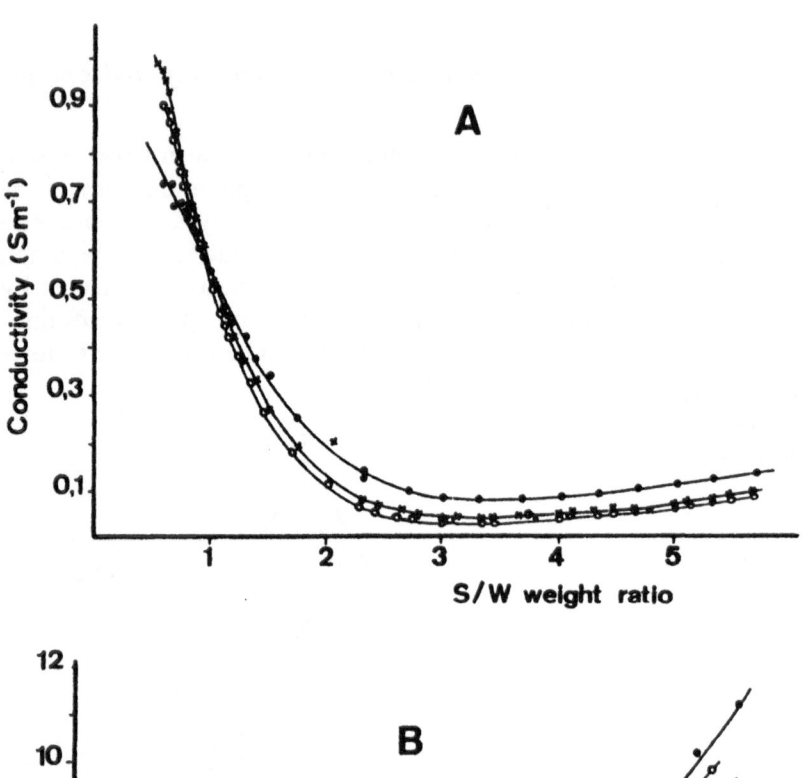

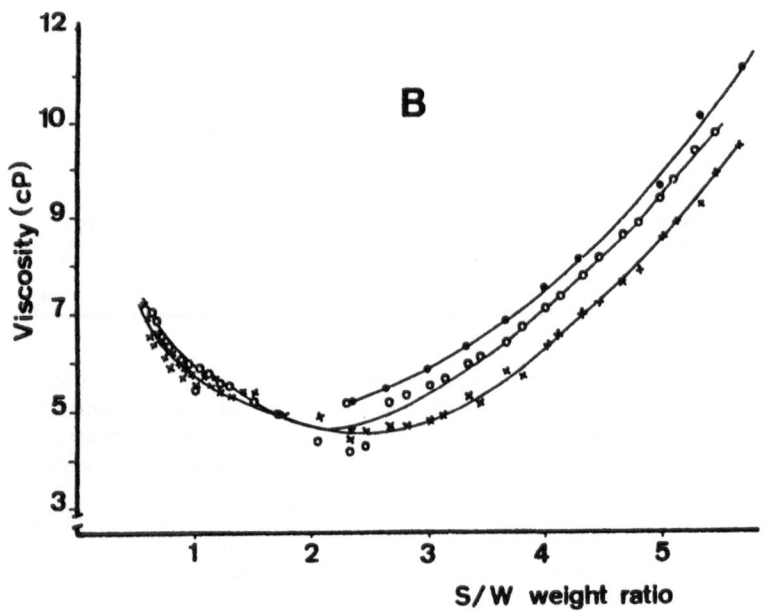

Fig. 2. Variations of microemulsion electrical conducivity (A) and viscosity (B) along $P_s = 0.35$ and $P_w = 0.15$ composition paths as a function of surface active mixture to water ratio, S/W (—○—) Model system; (—●—) Model system with 3 % (w/w) H_2O_2 (—×—) Model system with 3 % (w/w) $HSCH_2COOH$

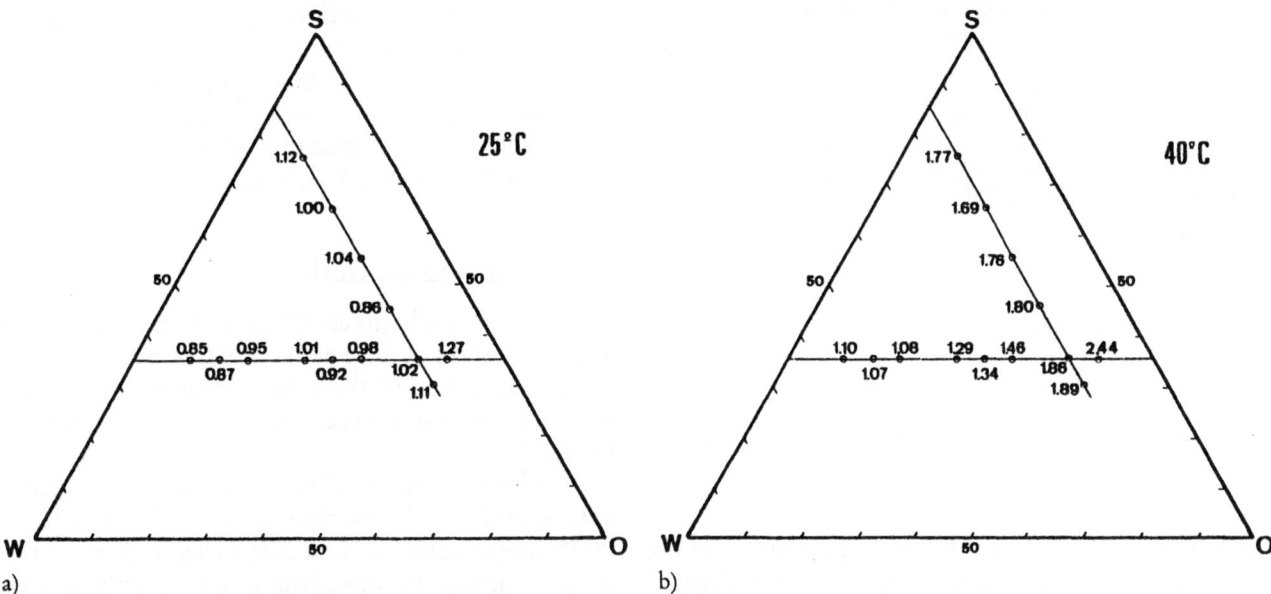

Fig. 3. Cysteic acid formation at 25 °C a) and 40 °C b) in wool fibres treated for 50 min with different microemulsion compositions containing 3 % (w/w) H_2O_2

troconductivity along this P_s path (Fig. 2 A, S/W weight ratios up to 2.3) reveals the modifications that microemulsion structure undergoes as the water concentration decreases since the content of sodium dodecylsulfate in the microemulsion remains constant. Along the $P_w = 0.15$ experimental path, the variations in microemulsion structure are not as pronounced. However, a progressive transition in the external phase occurs as the S content increases, hydrocarbon being replaced by n-pentanol. Obviously, these changes are not evidenced by microemulsion electroconductive behaviour (Fig. 2 A, S/W weight ratios higher than 2.3). The lowest values of viscosity (Fig. 2 B), were found at the region consisting of typically inverse microemulsions.

Cystine reactivity

The extent of cystine reactivity was determined by measuring the amount of cysteic acid or cysteine formed (oxidative or reductive treatments, respectively) according to the methods described in Section 2. Microemulsion compositions containing 3 % (w/w) of hydrogen peroxide or thioglycolic acid, labelled 1 to 14 (Fig. 1) were chosen as reaction media. Reactivity tests were first carried out at 25 °C and the results obtained expressed in percentage of cysteic acid or cysteine residues formed are given in Figures 3a and 4, respectively. As it can be observed, the variations in cys-

teic acid formation along experimental $P_s = 0.35$ and $P_w = 0.15$ paths were not as significant as those of cysteine formation at this temperature. In order to obtain more significant results, the oxidative treatments were carried out at a higher temperature, 40 °C (Fig. 3b).

The sorption of surfactant into the keratin fibre was low and uniform in treatments with hydrogen pero-

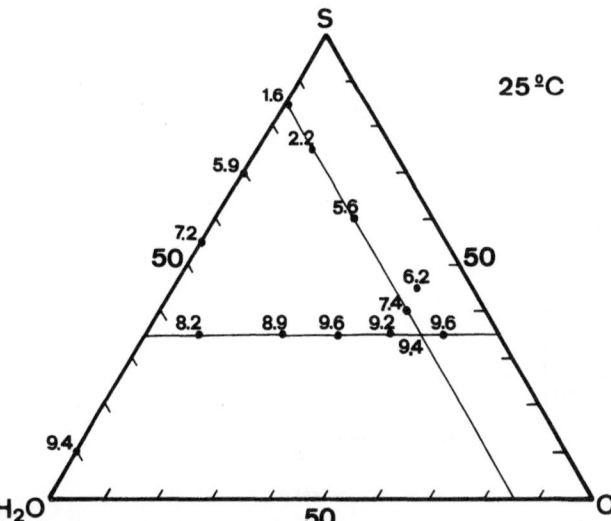

Fig. 4. Cysteine formation at 25 °C in human hair treated for 30 min with different microemulsion compositions containing 3 % (w/w) $HSCH_2COOH$

Table 1. SDS sorption on either wool fibres or human hair treated with different microemulsion compositions containing 3 % (w/w) reactive agent

Composition of the treatment solution % (w/w) W/S/O	SDS sorption in treatments with H_2O_2 at 40 °C (μ mol SDS/g wool)	SDS sorption in treatments with $HSCH_2COOH$ (μ mol SDS/g hair)
55/35/10	64	102
50/35/15	–	105
40/35/25	–	62
35/35/30	71	–
30/35/35	–	56
15/65/20	68	63
15/35/50	83	54

xide (Table 1) while it was found to be dependent on the water content of the media in treatments with thioglycolic acid, as is also shown in Table 1.

Surfactant sorption values lower than 80 μmols SDS/g fibre did not modify significantly the results of cystine reactivity. Therefore cysteic acid values given in Figure 3 are uncorrected values and cysteine results given in Figure 4 are corrected values taking into account surfactant sorption on the fibre. The low and uniform adsorption of surfactant in the oxidative treatment could be a consequence of the increase of negative charge of the fibre when cysteic acid forms, consequently the ionic interaction between sodium dodecylsulfate and the keratin fibre is not favoured.

Irrespective of the chemical reagent, keratin substrate and temperature considered cystine reactivity was found to be favoured in microemulsions belonging to the hydrocarbon-rich region.

Discussion and conclusions

When the results given in Figure 3b and 4 are plotted as a function of surface active mixture to water ratio, S/W, (Fig. 5 and 6) the influence of microemulsion structure on cystine reactivity appears quite straightforward.

The highest values of reactivity for both systems correspond to hydrocarbon-rich microemulsions (S/W weight ratios around 2.3) a clear indication that inverse microemulsions of the W/O type enhance cystine reactivity. These results are consistent with previous ones [8, 16] obtained in a similar system using heptane instead of dodecane to form the microemulsions and hydrogen peroxide, thioglycolic acid and sodium bisulfite as chemical agents. A decrease of the reaction rates with the increase of water content in reactions investigated in inversed aggregates has also been reported [17].

The reactivity behaviour along $P_w = 0.15$ (values of S/W higher than 2.3, Figures 5 and 6) is not as coincident as in the $P_s = 0.35$ path for both systems. Cysteic acid formation decreases as S/W increases reaching a minimum at S/W = 4.2 and increases again, (Fig. 5). In

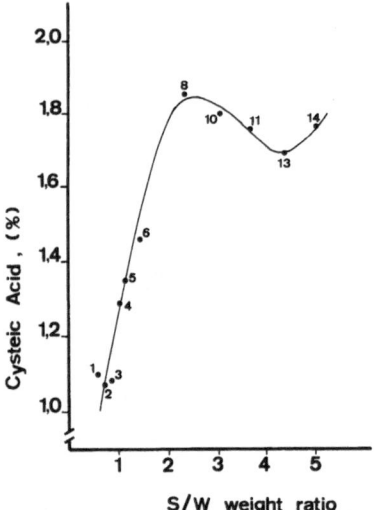

Fig. 5. Cysteic acid formation at 40 °C, as a function of surface active mixture to water ratio along experimental $P_s = 0.35$ and $P_w = 0.15$ paths. Numbers 1 to 14 correspond to microemulsions shown in Figure 1

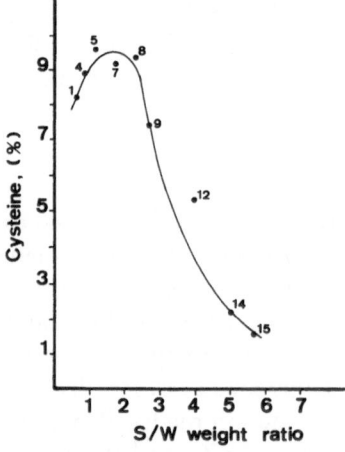

Fig. 6. Cysteine formation at 25 °C as a function of surface active mixture to water ratio along experimental $P_s = 0.35$ and $P_w = 0.15$ paths. Numbers 1 to 14 correspond to microemulsions shown in Figure 1

contrast cysteine formation decreases continuously with the increase of S/W, (Fig. 6). Cysteic acid formation was found to be favoured when reactions were carried out in pentanol saturated with a 3 % aqueous solution of H_2O_2 while the formation of cysteine was practically null in pentanol containing 3 % of thioglycolic acid. These results seem to indicate that in addition to the influence of microemulsion structure on cystine reactivity, other factors, like polarity of the continuous organic medium, may play also an important role.

Acknowledgements

This work was performed in the context of an Integrated Action (66-023/099) between the Université de Technologie de Compiègne (Compiègne, France) and The Instituto de Tecnología Química y Textil (Barcelona, Spain). Our gratitude to the Spanish and French authorities for their financial support. The authors also express their appreciation to Mrs. I. Yuste, M. Dolcet and I. Carrera for their valuable assistance.

References

1. Danielsson I, Lindman B (1981) Coll Surf 3:391
2. Robb ID (ed) (1982) Microemulsions, Plenum Press, New York
3. Mittal KL, Lindman B (eds) (1984) Surfactants in Solution, Vols 1–3, Plenum Press, New York
4. Robinson BH (1986) Nature 320:309
5. Kiwi J, Gratzel M (1978) J Am Chem Soc 100:6314
6. Jones CA, Weaner LE, Mackay RA (1980) J Phys Chem 84:1495
7. Fendler JH (1984) Chem Engng News 2:25
8. Parra JL, García Domínguez JJ, Solans C, Comelles F, Sánchez J, Pelejero C, Balaguer F (1985) Int J Cosmetic Sci 7:127
9. Solans C, Parra JL, Erra P, Azemar N, Clausse M, Touraud D, in press
10. Clausse M, Zradba A, Nicolas-Morgantini L (1985) Coll & Polym Sci 263:767
11. Zradba A, Clausse M (1984) Coll & Polym Sci 262:754
12. Meichelbeck H, Hark AG, Sentler C (1968) Z Gesamte Textilind, 70:242
13. Ellman GL (1959) Arch Biochem Biophys 82:70
14. De Boss AG, Finimore ED (1982) Tenside Detergents 19:262
15. Reid VW, Longman GF, Heinerth E (1967) Tenside 4:292-304
16. Erra P, Solans C, Azemar N, Parra JL, unpublished results
17. Bardez E, Monnier E, Valeur B (1986) J Coll Interf Sci 112:200

Received January 21, 1987;
accepted January 26, 1987

Authors' address:

P. Erra
Head of Protein Research Group
Instituto de Tecnologia
Quimica y Textil (C.S.I.C.)
Jorge Girona Salgado 18–26
080 34 Barcelona, Spain

Progress in Colloid & Polymer Science   Progr Colloid & Polymer Sci 73:156–160 (1987)

Water in oil microemulsions of the potassium oleate/hexanol/n-dodecane/ water system: A magneto-optical investigation

S. Bucci[1]), M. Carlà[1])[2]), C. M. C. Gambi[1])[2]), M. Neri [1])[2]), and D. Senatra[1])[2])

[1]) Department of Physics, University of Florence, Florence, Italy
[2]) CISM (of the M.P.I) and GNSM (of the C.N.R.) groups

Abstract: The optical effects induced by a magnetic field up to 0.8 tesla in a potassium oleate/hexanol/n-dodecane/water microemulsion system of the phase diagram monophasic region (w/o type) are investigated at constant alcohol/surfactant and active mixture/oil ratios vs. the increase of the water content. A small variation of the light polarization state with characteristic times of several minutes is observed. An analysis has been done to characterize the phenomenon. No magnetic birefringence is detectable, but a depolarization of the light occurs in crossing the sample. The role of both temperature and magnetic field is investigated. A sample-magnetic field interaction, due to the small but finite conductivity of the microemulsion, cannot be excluded; however, the temperature seems to play the main role and the resulting thermal stabilization of 0.5 °C is not accurate enough to investigate phenomena which are clearly of a hydrodynamic nature.

Key words: Ionic microemulsion, potassium oleate, magneto-optical effects, hydrodynamics, thermal diffusion.

Introduction

Microemulsions are homogeneous transparent dispersions of either water-in-oil or oil-in-water, stabilized by an amphiphilic film. For recent revision of the field, see references [1–3]. The microemulsion studied in our laboratory [4–6] is a potassium oleate/hexanol/n-dodecane/water system exhibiting a large monophasic domain of w/o type in the pseudoternary phase diagram [7] with the constant alcohol/surfactant ratio = 1.6 wt/wt. The investigation regards samples with the constant active mixture/oil ratio = 0.41 wt/wt studied as a function of increasing water concentration up to the one preceeding the occurrence of lyotropic mesophases. In this paper, we report an investigation about the optical effects induced in this microemulsion by a magnetic field up to 0.8 tesla, using a set-up like the one employed for the detection of the Cotton-Mouton effect. Magnetic and electric birefringence studies on different w/o microemulsion systems have been performed by several authors [8–12] with the aim of obtaining structural information; the induced birefringence being interpreted in terms either of deformation of microemulsion globules or orientation of globule aggregates and/or of parts of the interfacial film. In our case only a small variation of the light polarization state has been observed (with characteristic rise and decay times of the order of several minutes) which cannot be described in terms of the above mentioned magnetic birefringence. The analyses done to characterize the phenomenon and to present evidence for the parameters which play the main role are here reported.

Experimental

Materials

n-dodecane was used as supplied by Merck while the n-hexanol (Merck) was distilled. Water was of Super-Q-Millipore grade filtered through a 0.2 μm Millipore Millistak-GS-filter. Potassium oleate of 99.7% purity was prepared by reaction of equimolar amounts of oleic acid (Riedel) and KOH (Riedel) in hot ethanol. After evaporation of ethanol, the solution was cooled to promote crystalization. The product was washed twice with acetone and five times with diethyl ether; after filtering it was dried in vacuo. Samples with the ratios hexanol/K-oleate = 1.63 wt/wt and (hexanol plus K-oleate)/n-dodecane = 0.41 wt/wt have been studied by increasing the water content from above the CMC [7] ($c > 0.10$) to $c = 0.36$ (where c is the weight ratio, water : total).

Method

The typical Cotton-Mouton effect set-up is used for the detection of the magneto-optical effects. Linearly polarized light with its plane of polarization at $\pi/4$ with respect to the direction of the magnetic field enters the cell containing the isotropic microemulsion sample. An analyzer having the optical axis at $\pi/2$ (with respect to the polarizer optical axis) stops the light. In case the magnetic field induces birefringence, the light leaves the sample elliptically polarized and the light intensity (I) measured by the detector is given by [13]:

$$I = I_o \sin^2 (\delta/2) \tag{1}$$

where I_o is the light intensity incident on the sample and δ is the phase shift between the components, respectively parallel and perpendicular to the magnetic field direction, of the polarized light after leaving the cell. The phase shift depends on the induced birefringence (Δn) according to the relation $\delta = (2\pi/\lambda)\, l\Delta n$ with λ wavelength of the laser beam and l length of the light path within the cell. In the Cotton-Mouton effect [14] Δn is given by $\Delta n = C\lambda H^2$ with H field intensity; C is a constant which depends on the diamagnetic susceptibility and the optical polarizability anisotropies of the orientating object (molecules, molecular aggregates or microemulsion droplets, etc.).

A block diagram of the experimental set-up is shown in Figure 1. The sample is contained into a Hellma glass cell (C) of polarimeter quality and internal dimensions $1\ \text{cm}^2 \times 4\ \text{cm}$, closed by a teflon plug. The sample holder device is thermally controlled by a Haake thermostat (T). The sample temperature is $T = 23\ ^\circ\text{C}$. Thermal gradients have been measured inside the cell by means of a thermocouple; the vertical gradient is $\sim 0.02\ ^\circ\text{C/cm}$, the temperature

being higher in the middle of the cell, in correspondence with the optical windows (0.6 cm wide and 1 cm high); the horizontal temperature differences are smaller than 0.01 °C. Furthermore a 0.4 °C temperature increase was directly measured by a platinum probe placed into the cell, during an experimental run with a 1 h field application time. The static magnetic field is produced by a Varian 4005 electromagnet (EM). The field intensity is measured by a Hall probe (Siemens SV 210) previously calibrated by comparison with a gaussmeter (accuracy ± 2%). Fields up to 0.8 tesla were used, the pole-piece gap being 50 mm; field uniformity within the cell better than 1%. A He-Ne laser beam (NEC, 5 mW polarized) enters: a beam splitter (BS) for continuous recording of the beam intensity by means of the power meter PM 2; a total reflecting prism (R), a polarizer (P), the cell (C), an analyzer (A), a light spatial filter (F) (microscope objective of 8 mm focal lenght plus 25 μm pin-hol), a power meter PM 1. P and A are high quality 10 mm Glan-Laser prisms (Karl-Lambrecht); to rotate the prism a micrometric device is used (angular accuracy $\simeq 3\ 10^{-4}$ rad). A mechanical apparatus orient the cell with respect to the field and to the laser beam directions. The light spatial filter increases the signal to noise ratio of the transmitted beam [15] leading to a ratio between light intensity at crossed and parallel polarizers equal to 10^{-7} (100 times smaller than without it). The power meter PM 1, built in the laboratory using a large area silicon photodiode (Hamamatsu S 875-1010) in photovoltaic mode, is completely controlled by an internal microprocessor; it measures the photodiode electric current (range $10^{-8} \div 10^{-3}$ A, full scale; minimum measurable current is 10^{-11} A) or the light intensity (in watts) directly. The calibration was made by comparison with an EG & G (model 460) power meter. The linearity of the instrument is better than 1% up to 10^{-3} A; absolute accuracy of the current measurements on the lower scale, ± 3%, and on the other scales, ± 1%; absolute accuracy on the intensity values, ± 10%. An analog to digital converter (A/D) sends the electric signals (respectively proportional to the incident and transmitted light intensities) to a HP 9816 S desk computer (MC) for data acquisition and numerical elaboration.

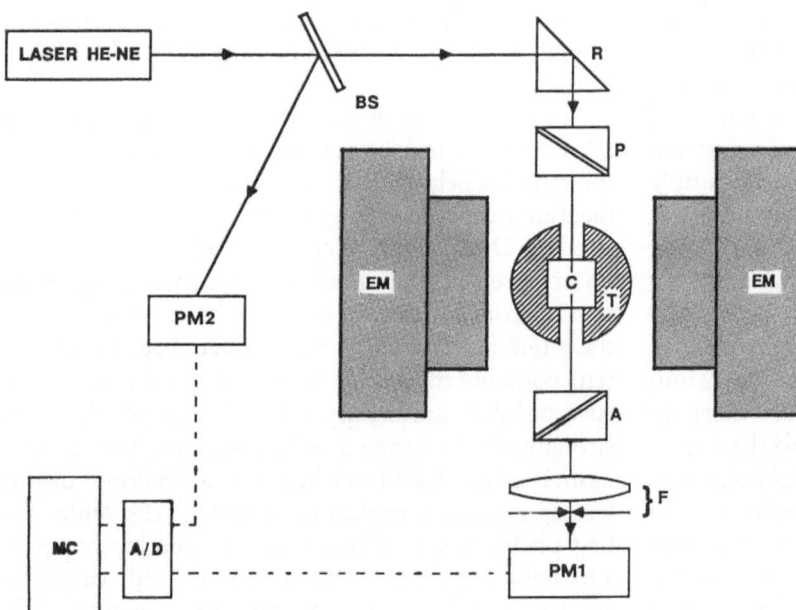

Fig. 1. Block diagram of the experimental set-up for magneto-optical measurements (see text)

An accurate matrix analysis of birefringence measurements is given in Reference [16]. The schematization of the cell as a linear retarder, δ_r, being the residual phase shift, is reasonable for polarimeter cells with thin glass windows if a small spot beam enters the cell [16], as it does in our case. Because the residual light intensity measured at crossed polarizers arises from both the apparatus background component and the cell stress-induced birefringence, the light intensity difference (ΔI), due to the sample induced birefringence, for small δ and δ_r, results consequently:

$$\Delta I \simeq I_0\, \delta/2\, (\delta/2 + \delta_r) \qquad (2)$$

at a second order approximation. Because the term linear in δ may be important also for small δ_r values, the residual light must be as small as possible. δ_r, evaluated using Equation (1), is $\sim 10^{-3}$ rad. As the relative sign of δ with respect to δ_r is not known, the value of the birefringence as low as 10^{-7} can be given with an accuracy of 10% and the minimum measurable birefringence is of the order of 10^{-8}. To test the apparatus, the nitrobenzene Cotton-Mouton constant has been measured with the maximum available magnetic field intensity; the correct order of magnitude was obtained. The alignment accuracy introduces an indetermination in the optical path length inside the cell smaller than that specified by the factory (0.01 mm). The stability of the instruments as well as the mechanical stability were tested over time intervals longer than those used during the measurements. The samples were hermetically closed and no change in the light intensity value was recorded in the absence of magnetic field.

Results and discussion

Typical curves of the light intensity transmitted at crossed polarizers (I) vs. time (t) are reported in Figures 2 and 3 (the arrows correspond to the field switching on ($\uparrow$) and off ($\downarrow$)). In Figure 2, the $I(t)$ curves for a single sample ($c = 0.26$ wt/wt) are reported at different magnetic field intensities (0.56 tesla, 0.70 tesla and 0.81 tesla, respectively, for (a), (b) and (c) curves). In Figure 3, $I(t)$ curves for samples at different concentrations belonging to the microemulsion monophasic region of the phase diagram are reported (field intensity 0.81 tesla). For all the curves, the sample temperature was $T = 23\,°C$. The main characteristics of the phenomenon are as follows: (1) Very long response time, (2) no signal saturation observed for any of the tested samples, even for a three hours field application, (3) exponential signal decay after switching off the field (time constant in the range $5 \div 10$ min for all the samples) except in some cases where an oscillatory trend appears superimposed to the curve. In any case, the light intensity with parallel polarizers showed no change because of the applied field over the whole field application time. As no instantaneous variation was observed on the $I(t)$ curves after the application of the field, the birefringence either linked

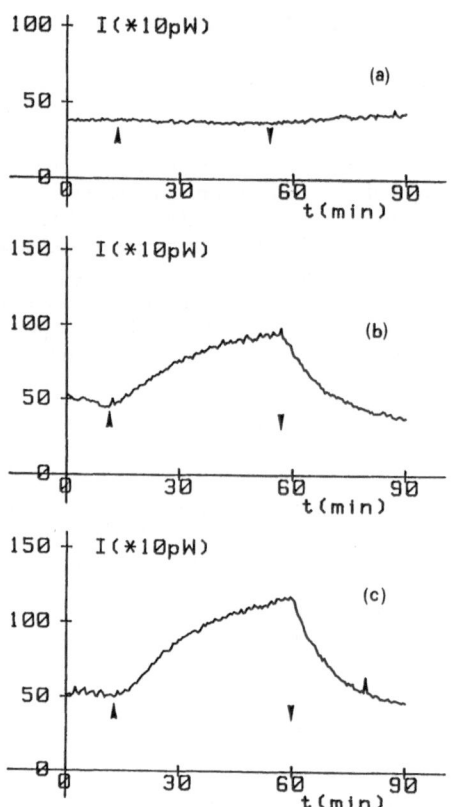

Fig. 2. $I(t)$ plots of a microemulsion sample with $c = 0.26$ wt/wt and $T = 23\,°C$ for different magnetic field intensities: (a) 0.56 tesla; (b) 0.70 tesla; (c) 0.81 telsa. Field switching on and off correspond to arrows up and down, respectively, for all the figures

to distortion or to orientation, induced by the magnetic field, must be lower than 10^{-8}, the limit of resolution of the apparatus.

We will now focus our attention on curves (b) and (c) of Figure 2 and (b), (c) and (d) of Figure 3, obtained for samples belonging to the central part of the phase diagram monophasic region and for which the trend is similar. Due to the very long response times, the observed effect cannot be ascribed either to a deformation or to an orientation of the globules and of parts of the interfacial film [8, 9], as expected, because the system contains molecules with a too low diamagnetic susceptibility anisotropy ($\Delta\chi$). We recall that field strengths in the range $2 \div 20$ tesla have been used to induce a magnetic birefringence on microemulsions having surfactant molecules with high $\Delta\chi$ value. The first problem is to distinguish between birefringence and light depolarization. In fact, in both cases, we expect an $I(t)$ increase for the given position of the

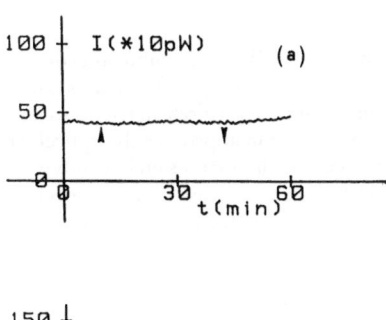

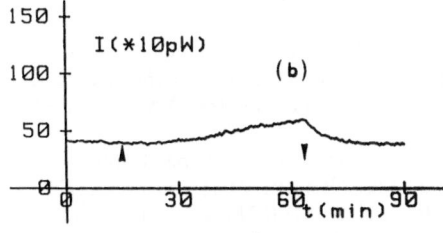

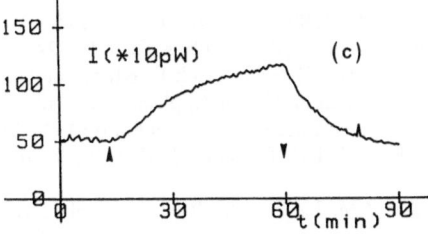

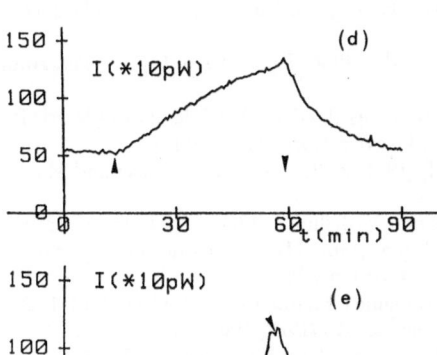

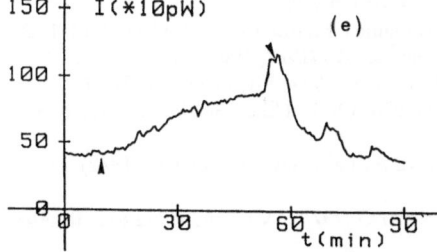

Fig. 3. $I(t)$ plots with a magnetic field intensity of 0.81 tesla at $T = 23\,°C$ for microemulsion samples at different concentrations of the phase diagram monophasic region (a) $c = 0.12$; (b) $c = 0.22$; (c) $c = 0.26$; (d) $c = 0.31$ and (e) $c = 0.36$ (proportions given in wt/wt)

polarizers optical axes. Under the hypothesis of birefringence, we expect to observe no $I(t)$ variation ($I(t)$ always being the light transmitted at crossed polarizers) when the incoming polarization plane is at $\pi/2$

with respect to the magnetic field direction; on the contrary an $I(t)$ variation similar to that under discussion was indeed detected, indicating that the observed phenomenon is mainly due to a light depolarization in crossing the sample. On the other hand, experimental misleading effects, such as slow drifts in the electronics or the mechanical set-up, as well as modification of the sample during the experiment, have been excluded after very careful controls (see Method).

The fact that the phenomenon is related to a light depolarization excludes the occurrence of any anisotropic state induced by the applied field within the sample. The small observed temperature changes inside the sample (see Method) may be responsible in producing motions within the sample which could explain the phenomenon. We recall that in a pure fluid layer of thickness a, convective instabilities develop only when heating is done from below, for a temperature difference (ΔT) such that the Rayleigh number is higher than 1708 [17]. For heptane, with $\Delta T = 0.03\,°C$ and $a = 1.4$ cm, the Rayleigh number is lower but close to the threshold value (1650). For a microemulsion, in analogy with the behaviour of mixtures, a convective instability may also occur when heating is done from above and with a critical ΔT drastically lower than for a pure fluid [18, 19]. For a sample far from a critical point and a phase transition, a 0.5 °C thermal stabilization is usually considered to ensure that the sample is stable; however, thermal diffusion and convective instabilities are likely to occur in this case, due to the measured horizontal and vertical thermal gradients. A thermal diffusion effect (Soret effect) has been experimentally observed, e. g. in macromolecular solutions [20] far from a critical point, with a vertical gradient of a few °C/cm. By measuring the vertical index of refraction gradient over times of the order of 1 h, a very rapid variation was observed due to the thermal expansion of the sample for the step-wise rise temperature gradient, followed by a very slow variation, due to the build-up of the thermally induced concentration gradients; for a thermal diffusion ratio $k_T < 0$, a convective instability was finally observed over very long periods (more than 90 min) which destroyed the concentration gradient. In order to verify whether thermal diffusion or convective motion occurs within the sample, we performed an interferometric analysis by means of a point diffraction (Smartt) interferometer (Ealing) [21]. The interferometric pattern has been observed before and during the measurements; no change of the index of refraction was detected over the whole sample window in the limit of resolution of the apparatus

($\sim 10^{-5}$). Therefore no macroscopic motion has been observed, even though we cannot exclude motions over smaller spatial scales, involving very low index of refraction variations for which a more sensitive method should be used.

Returning to Figure 2, the $I(t)$ variation increases as the field intensity increases, at constant thermostat temperature and field application time. However, this fact cannot be completely and certainly ascribed to the effect of the field, as the sample temperature shows small (max. 0.4 °C) but systematic variations with the field intensity. On the other hand, should we exclude the temperature changes effect, the only possibility of a sample-magnetic field interaction could be through the small but finite conductivity of the microemulsion sample [22, 23]. In principle we cannot exclude that, because of the thermal gradients, the ions in the microemulsion interact with the field itself; however, the magnetic field has a damping effect on the motion of a conductive medium, hence, it should have a stabilizing effect. Thus, the existence of a field threshold is probably due to the smaller heating of the electromagnet at lower field intensity.

In Figure 3 $I(t)$ curves for samples at different concentrations are reported for a constant field intensity and identical field application time. The existence of a threshold concentration, below which no effect is detectable, is clearly shown; a low variation of $I(t)$ is displayed at concentration $c = 0.22$ wt/wt (curve b). On the $c = 0.12$ wt/wt concentration (curve a), tests were repeated in different thermal gradient conditions, with no appreciable change in the results.

A peculiar behaviour is displayed by the sample with concentration value (0.36 wt/wt) close to the phase transition towards the liquid crystalline state. The occurrence of oscillations superimposed to the usual trend suggests that some kind of instability develops within the sample; however, we could not detect any macroscopic convection with the Smartt interferometer over periods of several hours. Oscillatory trends have also been observed for lower concentrations during longer periods of measurement.

In conclusion, despite being unable to exclude a sample-magnetic field interaction, the temperature seems to play the main role in the reported experiments and the thermal stabilization of 0.5 °C is not accurate enough to investigate phenomena which are clearly of hydrodynamic nature.

Acknowledgements

The authors wish to thank Prof. G. Poggi for stimulating discussions and criticisms during the course of this work. Thanks also to Dr. P. H. Guering for helpful discussions on birefringence measurements, and to Dr. G. Molesini for help in improving the optical set-up; to Dr. S. Ciliberto for very useful discussions on hydrodynamics; to Mr. M. Falorsi and Mr. B. Sarti who constructed the mechanical set-up and to Mr. G. Lauria who made the apparatus for the Hall probe controller.

References

1. de Gennes PG, Taupin C (1982) J Phys Chem 86:2294
2. Bellocq AM, Biais J, Bothorel P, Clin B, Fourche G, Lalanne P, Lemaire B, Lemanceau B, Roux D (1984) Adv Coll Interf Sci 20:167
3. Ruckenstein E (1985) Fluid Phase Equilibria 20:189
4. Senatra D, Gambi CMC, Neri AP (1981) J Coll Interf Sci 79:443
5. Senatra D, Gambi CMC (1984) Il Nuovo Cimento 3D:75
6. Senatra D, Gabrielli G, Guarini GGT (1986) Europhys Lett 2:455
7. Bellocq AM, Fourche G (1980) J Coll Interf Sci 78:275
8. Meyer CT, Poggi Y, Maret G (1982) J Phys, Paris 43:827
9. Maret G, Dransfeld K (1985) In: Herlach F (ed) Strong and Ultrastrong Magnetic Fields and Their Applications: Topics in Applied Physics, Springer Verlag, Berlin, Vol 57, pp 143–204
10. Guering P (1985) thesis, Lab de Physique de L'Ecole Normale Superieure, Paris
11. Guering P, Cazabat AM, Paillette M (1986) Europhys Lett 2:953
12. Hilfiker R, Eicke HF, Geiger S, Furler G (1985) J Coll Interf Sci 105:378
13. Born M, Wolf E (eds) (1980) Principles of Optics, Pergamon Press, New York
14. Bottcher CJF, Bordewijk P (eds) (1980) Theory of Electric Polarization, Vol 2, Elsevier North-Holland Inc
15. Siegman AE (ed) (1971) An Introduction to Laser and Masers, McGraw-Hill Inc
16. Piazza R, Degiorgio V, Bellini T (1986) Opt Comm 58:400
17. Chandrasekhar S (ed) (1961) Hydrodynamic and Hydromagnetic Stability, Oxford Univ Press
18. Schechter RS, Prigogine I, Hamm JR (1972) Phys Fluids 15:379
19. Velarde MG, Schechter RS (1972) Phys Fluids 15:1707
20. Giglio M, Vendramini A (1977) Phys Rev Lett 38:26
21. Malacara D (ed) (1978) Optical Shop Testing, J Wiley & Sons, New York
22. Shah DO, Tanjeedi A, Falco JW, Walker RD (1972) AICHE Journal 18:1116
23. Senatra D, Giubilaro G (1978) J Coll Interf Sci 67:448; 457

Received January 22, 1987;
accepted January 29, 1987

Authors' address:

Dr. C. Gambi
Department of Physics
L. E. Fermi 2
I-50125 Florence, Italy

Progress in Colloid & Polymer Science   Progr Colloid & Polymer Sci 73:161–166 (1987)

Static fluorescence quenching in the study of micellar systems

A. Malliaris

N.R.C. „Demokritos", Athens, Greece

Abstract: The effect of a number of polar and non-polar organic molecules on the micellization of sodium dodecyl sulfate at low surfactant and additive concentration, has been studied by fluorescence probing methods. The solubilization and its effect on micellar parameters and structure is examined and some general trends are discussed. Special emphasis is focused on the application of static fluorescence quenching method, since it can provide important and unique information in spite of its simplicity.

Key words: Micelle, solubilization, fluorescence, sodium dodecyl sufate.

Introduction

The ability of aqueous micellar systems to solubilize water insoluble, or sparingly soluble, compounds by incorporating the hydrophobic molecules in the aggregated phase, constitutes one of the most remarkable properties of these systems [1]. Such solubilizates have been extensively used, either to deliberately modify the micellar entity, or to probe the environment at the site of their solubilization, thus providing valuable information concerning the structure and properties of aqueous micelles [2]. Several physical parameters of the additive compound such as polarity, charge, molar volume, chain length, branching, planarity, substitution, etc., have been found to be important in determining solubilization [3].

The effect of additives on micellar parameters has been investigated in numerous cases [3]. The main interest in this respect has been focused however, on medium chain length alcohols, because of their wide use as cosurfactants in the preparation of microemulsions from surfactant solutions. Thus, parameters such as the critical micelle concentration (CMC) [4], the degree of micellar ionization (α) [4, 5], the counterion binding constant [6], the mean micelle aggregation number (N_S) [7, 8], the number of micelle-bound additive molecules (N_A) [7, 8], and other properties of

micellar solutions [9–11] have been determined with respect to both the nature and the concentration of the additive and the surfactant. Although some approximate generalizations have been proposed [12] the complexities of these systems have been recognized [12].

In this presentation we report on the solubilization in micelles of SDS, of long chain ($n = 8, 9, 10$) linear alkanols, ketones and alkanes. Our main objective was to investigate the phenomenon of solubilization from the point of view of the fluorescence probing methods, which can provide valuable information about the internal structure of these systems. SDS was preferred for this study over other common surfactants because it presents uniquely favorable conditions for the application of static fluorescence methods, as it will be shown later on in the discussion. Linear additive molecules were chosen since they are expected to disturb the micellar structure less than other bulkier solubilizates. Few of the additives studied here had been examined before [7, 8], but under different conditions. In this work we have restricted our studies to the limit of the low surfactant and additive concentration so that we can safely assume spherical or nearly spherically-shaped micelles in order to simplify calculations.

Experimental

Specially purified SDS (BDH) was recrystallized from acetone and ethanol, while the absence of traces of unreacted dodecanol was confirmed by i. r. spectroscopy. Cetylpyridinium chloride (CPyC, Serva research grade > 99 %) was recrystallized from acetone and ethylacetate. All alkanes were of the highest quality (Aldrich, gold Label or spectrophotomemtric grade) and were used as received, whereas 1-alkanols were distilled and their purity was confirmed by i. r. spectra and the value of the refractive index. Ketones (Aldrich) were purified by vacuum distillation and the formation of bisulphite addition compounds. Pyrene (Aldrich, 99 % +) was extensively zone refined. Fluorescence spectra were obtained using a previously described instrument [13].

The CMC values in the presence and absence of the additives were determined by electrical conductometry using a Metrhom Herisau E512 conductometer in conjunction with a thermostated conductivity cell capable of regulating sample temperature at 25 ± 0.01 °C. Note that alkane additives did not induce measurable changes in the CMC of SDS at the present concentration range. Triply distilled water having electrical conductivity less than 2–3 μS/cm was used throughout this work.

Micellar composition (N_S, N_A)

The micelle aggregation number N_S is one of the most important micellar parameters and several experimental techniques have been developed for its determination. Methods such as elastic [14], or quasi-elastic light scattering [15], ultracentrifugation [16], membrane osmometry [17], gel filtration [18], etc., have been used for the determination of the micelle molecular weight, radius, or mean aggregation number. However, in the case of ionic micelles, all these require extrapolation of the data to the CMC in order to eliminate intermicellar coulombic interactions. Such an extrapolation is not always possible, particularly when the micellar composition and N_S depend on the surfactant concentration (C_S), as for instance is the case in mixed micelles [19]. In these systems the micellar composition depends strongly on (C_S), and therefore extrapolation to the CMC is not possible. Fluorescence probing methods have conveniently solved the problem of the intermicellar interactions since they are not, in principle, influenced by them, and therefore they can provide information concerning N_S practically at any surfactant concentration of ionic micelles without the need of added electrolytes [19].

The static fluorescence quenching method

The static fluorescence quenching method, first developed by Turro and Yekta [20], relies on a Stern-Volmer type relationship, modified by the assumed Pois-

son distribution of fluorophor and quencher species among the micelles. It is a very simple experimental technique which gives the micellar concentration (M) directly without any complicated computations. It has been however widely critisized for underestimating N_S [21, 22]. Here we discuss the possibilities and limitations of this method with respect to the determination of the micelle aggregation number. Note however that other parameters can also be calculated, by static fluorescence quenching, under appropriate conditions [21].

According to this method, the fluorescence emitted from a micelle-bound fluorophor is quenched by the appropriate quencher molecule, which also resides in the micellar phase. The pertinent experimental information is the decrease of the fluorescence intensity as the quencher concentration increases. It is immediately evident that the fluorescence quenching depends strongly on the micelle/water distribution of the fluorophor and the quencher. When both the fluorophor and the quencher can be viewed as immobille [21], i. e. they do not exchange between the micellar and the aqueous phase during the fluorescence life time of the fluorophor, the steady-state fluorescence quenching can be described by Equation (1) [21],

$$I/I_o = K \cdot \exp\left(-[Q]/[M]\right) \sum_{i=0}^{\infty}$$

$$([Q])/[M])^i/i! \,(K + i). \tag{1}$$

Where I_o and I are the fluorescence intensities without and with the quencher, $K = K_{unq}/K_q$ is the ratio of the unquenched fluorescence rate to the rate of the fluorescence quenching, $[Q]$ and $[M]$ are the quencher and micellar concentrations respectively, and i expresses the number of quenchers inside a micelle. K_{unq} is equal to τ^{-1} where τ is the fluorescence life time of the fluorophor in the absence of quencher. Clearly, K_{unq} is the sum of the radiative and the radiationless decay of the fluorescence. Figure 1 shows plots of $\ln(I_o/I)$ vs $[Q]$ according to Equation (1) for some K values and for a typical micellar concentration $[M] = 1.3 \times 10^{-3}$ M.

In the case where $K = 0$. i. e. the ideal case of a fluorophor with very long fluorescence life time (small K_{unq}) in conjunction with a very efficient quencher (large K_q), Equation (1) becomes Equation (2)

$$\ln(I_o/I) = [Q]/[M]. \tag{2}$$

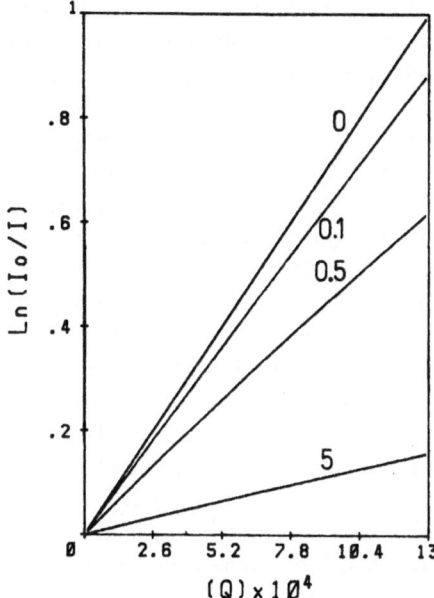

Fig. 1. Computer simulations of Equation (1) for $K = 0, 0.1, 0.5$ and 5. $[M] = 1.3 \cdot 10^{-3}$ M

Note that the slope of the line obtained by ploting $\ln(I_o/I)$ vs $[Q]$ according to Equation (2), is equal to $1/[M]$. Therefore in this limiting case ($K = 0$) one can obtain the micellar concentration $[M]$ from static fluoresence quenching. Evidently, if K is known from independent measurements, then fitting the data (I_o, I and $[Q]$) to Equation (1) allows determination of $[M]$ even in the general case with $K \neq 0$. Furthermore, if $[M]$ has been obtained through some other method, then use of Equation (1) will allow determination of K. Finally, if in addition to $[M]$, K_{unq} is also known (e. g. from τ), then the all important intramicellar quenching constant K_q can be calculated from $[M]$, K_{unq} ($= \tau^{-1}$) and Equation (1).

However, even if K is not exactly zero but it remains small, and also $[Q]/[M]$ is kept low, Figure 1 shows that N_S can be determined with acceptable accuracy. For example for $K = 0.1$, shown in Figure 1, the error in the value of N_S, compared to that for $K = 0$ is estimated to be only $6 - 10$ %, i. e. about the same as the accuracy of the time dependent fluorescence quenching method [23]. On the contrary, when $K = 0.5$, the error in Ns can become as large as 40 %. Note that since K_{unq} has a more or less fixed value (τ^{-1}) in a certain environment, it is K_q which must become large in order to obtain small K values, and consequently reliable aggregation numbers. Large K_q values are

obtained provided, (i) an efficient quencher/fluorophor pair is used, (ii) there are no quencher or fluorophor interactions with the environment which would impair their mutual approach necessary for the quenching process, and (iii) the effective quencher concentration is high. This last condition means that, in the case of micellar quenching, the micelles must be small so that the local concentration of the micelle-bound quencher is high. Indeed, the curve with $K = 0.1$, corresponds to the case of small SDS micelles (surfactant concentration ca. 0.1 M), with pyrene as the fluorophor and the surfactant cetylpyridiniumchloride (CPyC) as the quencher [24]. On the other hand, a curve with $K = 0.3$ corresponds to micelles of cetyltrimethylammonium chloride (CTAC) at $[CTAC] = 0.12$ M and the same as before fluorophor/quencher pair [25]. In the case of CTAC K is higher than in the case of SDS because CTAC micelles are larger than SDS micelles, but mainly because of the well known specific interaction between pyrene and the quaternary ammonium group [16], which reduces the magnitude of K_q compared to that in SDS micelles. This example shows that the static fluorescence quenching method is applicable in the case of small K values, i. e. in small micelles and in the absence of any specific interactions which tend to reduce K_q.

In our measurements here we have used pyrene as the fluorophor ($\tau = 350$ ns) and CPyC as the quencher. Both these molecules remain associated with the micelle for times longer than the life time of the fluorescence of pyrene [25] the former due to its hydrophobicity and the latter due to the formation of mixed micelles with the SDS [19]. Also they constitute an efficient quencher/fluorophor pair for which we have found $K_q = 6.5 \cdot 10^9$ s^{-1}, from fluorescence decay measurements in homogeneous media. Finally, note that the aggregation number measured here for pure SDS micelles at $[C_S] = 0.04$ M, $N_S = 65$, is in excellent aggrement with values reported in the literature [20].

Calculation of N_S and N_A

The micelle aggregation number N_S either in pure SDS or in SDS/ additive mixed micelles, is evaluated from the experimentally available micellar concentration $[M]$ (Eq. (2)) through Equation (3),

$$N_S = ([C_S] - [S])/[M] \qquad (3)$$

where $[S]$ stands for the concentration of the non-micellized, or free, surfactant. Since $[S]$ is not easily

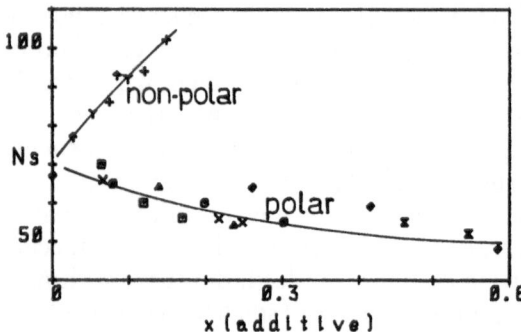

Fig. 2. Plot of additive mole fraction vs surfactant aggregation number (N_S). (◆) octanol: (□), nonanol: (▲), decanol: (✖), 2-octanone: (O) 2-nonanone: (✕), 2-decanone: (+), octane: (Υ), nonane: (↑), decane

accesible it is usually set equal to the [CMC], although it is known that the amount of free surfactant decreases as [C_S] increases [26]. This approximation, adopted in the present study, is not a very serious one since from published calculations for SDS in water [27, 28] it turns out that settingg [S] = [CMC], in this surfactant concentration range, results to underestimation of N_S less than the overall error of the experimental method involved [16]. Furthermore, in the presence of additives the [CMC] itself changes in a way dependent on both [C_S] and [C_A] [4, 29]. The values of CMC used here in the presence of additives were determined by electrical conductivity measurements as described in the Experimental Section.

For the determination of the additive molecules per micelle N_A, the partition coefficient k of the additive between the micellar and the aqueous phases must be known. However, when $k \gg 1$ N_A can be taken as equal to the ratio [C_A]$_T$/[M], where [C_A]$_T$ is the total additive concentration.

The micellar aggregation number (N_S)

It is immediately evident from Figure 2 that there is a definite differentiation in the effect which the additive has on N_S, according to its polarity. Thus, N_S decreases upon addition of polar additives to the SDS micellar solution, while it increases when the additive is non-polar. This behavior can be rationalized in terms of the different solubilization sites of these additives inside the SDS micelles.

The polar additives, which solubilize in the palisade region, lower the charge density on the micellar sur-

face and counterions are released from the Stern layer into the aqueous phase. This release of counterions increases the micelar ionization and therefore destabilizes the aggregates. Formation of smaller micelles partially compensates the excess coulombic energy, and restabilizes the system. However, when [M] becomes large enough, at low N_S, intermicellar repulsions, due to close approach of highly ionized micelles, oppose any further increase of [M], and consequently the decrease of N_S stops. Therefore the systemm comes to equilibrium, but with some N_S value lower than N_S in the absence of the additive. It should be mentioned here that at high SDS and alcohol concentration N_S reverses its trend and increases at increasing [C_A] in a way dependent on the nature of the additive [30].

A totally different behavior is exhibited by N_S when the additive is hydrocarbon. Thus, increase of [C_A], at constant [C_S], always induces increase of N_S (Fig. 2). This behavior can again be explained on the basis of the site of solubilization of the additive. In this case the additive solubilizes in the interior of the micelle away from the surface area. Since hydrocarbons do not bear a polar group which would allow some contact with water, they are protected from the aqueous phase by the surfactant ions of their host micelle. In each micelle, N_S surfactants form a spherical shell surounding N_A hydrocarbon molecules. Clearly, such shells contain more surfactant ions than the ions contained in the spherical micelle formed in the absence of hydrocarbon additive.

The number of additive molecules per micelle (N_A)

The number of additive molecules per micelle N_A, systematically increases as the additive concentration, of either polar or nonpolar molecules increases, the rise of N_A being slightly steeper when the additive is non-polar (Fig. 3). Although at low additive concentration the number of solubilized molecules per micelle is higher for non-polar than for polar additives for the same χ(*add.*) (Fig. 3), the maximum value of N_A is rather higher for polar than for non-polar additives. Indeed the solubility of polar compounds in salt-free amphiphile solutions is much higher than that of the corresponding hydrocarbons. This behavior can be understood, at least qualitatively, in terms of the free energy of the micelle/water interface. Thus, theory shows that an important part of the free energy of micellization comes from the free energy of the interface, which in turn is affected by the extent of solubilization and by the interfacial tension between solubi-

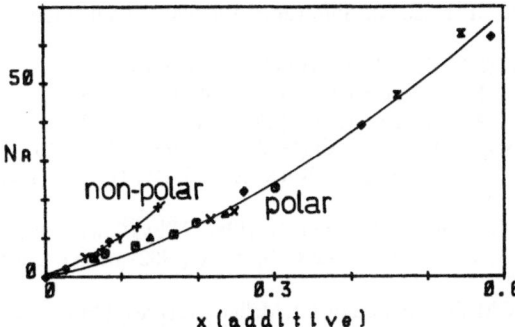

Fig. 3. Plot of additive mole fraction vs number of solubilized molecules per micelle (N_A). Symbols as in Figure 2

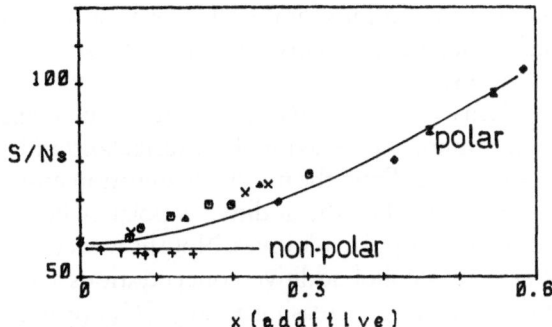

Fig. 4. Plot of additive mole fraction vs interfacial surface area per ion. (S/N_S, A^2/ion). Symbols as in Figure 2

lizate and water. Based on this reasoning the empirical Equation (4) has been proposed [10], which relates R_M to the surface tension σ,

$$R_M = 13.2 \, \{\sigma v^{2/3}/KT\}^{-2.3} \tag{4}$$

expressed in dyn/cm and the molar volume of the additive v expressed in A^3, KT is expressed in ergs. R_M here denotes the solubilization ratio, i.e. the ratio of the maximum number of solubilizate molecules to the number of surfactant molecules in a micelle. Clearly R_M is an experimentally accessible parameter, characteristic of the solubilization [10]. In view of the fact that the surface tension against water is about three times lower for polar molecules, such as alcohols or ketones, compared to the corresponding hydrocarbons [31] it is obvious from Equation (4) that polar molecules should solubilize more than non-polar ones in aqueous micelles. Note however, that at high alcohol concentration the system becomes quite complicated since the highly dynamical nature of surfactant/alcohol mixed micelles [32] introduces considerable ambiguity in the values of N_S and N_A obtained by time dependent fluorescence quenching and therefore many published results should be viewed with extreme cautiousness.

Surface charge density

Another important parameter in pure or mixed micelles is the surface charge density (ions/area unit), or its inverse, the interfacial area per surfactant ion A/N_S (A^2/ion), which relates to the coulombic energy of the interface. Here A stands for the total water/micelle interfacial area. It is characteristic that when

$\chi(add.)$ increases, the parameter A/N_S increases in the case of polar additives, but it remains nearly constant in the case of non-polar additives (Fig. 4). More interestingly, if one takes into account the accuracy of these measurements it appears that the increase of A/N_S with $\chi(add.)$ is the same for all polar additives, while it remains equal to its value in the pure SDS micelle (ca. 58–59 A^2/ion) when the additive compound is an alkane.

The importance of this parameter has been pointed out by several authors [8] and its dependence on $\chi(add.)$ and the nature of the additive is consistant with the dependence of the degree of micellar ionization α on these factors. Thus, values of α close to 1 have been found for alkyltrimethylammonium bromide micelles in water/alcohol media [33]. Such a high value of α has been interpreted in terms of the formation of surfactant/alcohol mixed micelles, built of only few detergent ions and a large number of alcohol molecules. Figures 2, 3, 4 clearly show the effect of alcoholic additives, i.e. when $\chi(add.)$ increases N_S decreases while N_A increases. Under these conditions the interfacial area per ion becomes large. On the other hand, solubilization of alkanes does not affect very much the ionization of the micelles and α remains nearly constant. In the surfactant/alkane mixed micelle the additive molecules aggregate away from the surface, while the detergent ions arrange themselves around the oily core in such a way as to optimize the interfacial energy. This results to the constancy of A/N_S with $\chi(add.)$.

Conclusions

Summarizing the findings of this work in association with other results from the literature we can draw

166 *Progress in Colloid & Polymer Science, Vol. 73 (1987)*

the following conclusions for the effect of additives on the micellization of SDS at low surfactant and additive concentrations.

The polarity of the additive appears to be the main factor determining the behavior of the surfactant/additive mixed micelle. Thus, the micelle aggregation number N_S decreases when the additive is polar, while it increases with non-polar additives. Note however that studies at high alcohol additive concentrations have shown that N_S starts increasing at large $[C_A]$ (e. g. > 0.6 M for n-pentanol). On the contrary the number of additive molecules per micelle N_A, always increases with increasing additive concentration for either polar or non-polar solubilizates. The dependence of N_S and N_A on $[C_S]$, in the case of polar additives, has the consequence that the micellar volume can first go through a minimum value before it starts increasing as $[C_A]$ increases. For non-polar additives the micellar size increases monotonously as the additive concentration increases, since both N_S and N_A increase.

The other important parameter, the surface charge density A/N_S, also demonstrates different behavior according to the polarity of the additive. Thus, with polar additives A/N_S increases with increasing $[C_A]$, whereas with non-polar ones it remains independent of the additive concentration and equal to its value in the pure SDS micelle, i. e. ca. 58–59 A^2 per surfactant ion.

Finally, as far as the static fluorescence quenching method is concerned, we should point out that it can give good and reliable results, provided some experimental conditions are fullfiled.

References

1. Mukerjee P, Bunsenges B (1978) Phys Chem 82:931
2. Fisher LR, Oakenfull DG (1977) Chem Soc Rev 6:25
3. Fendler JH, Fendler JE (eds) (1975) Catalysis in Micellar and Macromolecular Systems, Academic Press, New York
4. Zana R, Yiv S, Strazielle C, Lianos P (1981) J Coll Interf Sci 80:208
5. Manabe M, Koda M, Shirayama K (1980) J Coll Interf Sci 77:189
6. Larsen JW, Tepley LB (1974) J Coll Interf Sci 49:113
7. Almgren M, Swarup S (1983) J Coll Interf Sci 91:256
8. Almgren M, Swarup S (1982) J Phys Chem 86:4212
9. Vikingstad E, Kvammen O (1980) J Coll Interf Sci 74:16
10. Chaiko MA, Nagarajan R, Ruckenstein E (1984) J Coll Interf Sci 99:168
11. Russell JC, Wild VP, Whitten DG (1986) J Phys Chem 90:1319
12. Lianos P, Lang J, Strazielle C, Zana R (1982) J Phys Chem 86:1019
13. Paleos CM, Stassinopoulou CI, Malliaris A (1983) J Phys Chem 87:251
14. Anacker EW (1970) In: Jungermann E (ed) Cationic Surfactants, Marcel Dekker, New York
15. Mazer NA, Benedek GB, Carey MC (1976) J Phys Chem
16. Anacker EW, Rush RM, Johnson JS (1964) J Phys Chem 68:81
17. Birdi KS (1972) Koll-Z Z Polym 250:731
18. Coll H (1970) J Phys Chem 74:520
19. Malliaris A, Binana-Limbele W, Zana R (1986) J Coll Interf Sci 110:114
20. Turro NJ, Yekta A (1978) J Am Chem Soc 100:5951
21. Infelta PP (1979) Chem Phys Let 61:88
22. Lianos P, Zana R (1980) Chem Phys Lett 72:171
23. Malliaris A, LeMoigne J, Sturm J, Zana R (1985) J Phys Chem 89:2709
24. Malliaris A, unpublished communication
25. Malliaris A, Lang J, Zana R (1986) J Chem Soc Faradday Trans 1, 82:109
26. Hartley GS (1938) J Chem Soc
27. Stilbs P, Lindman B (1981) J Phys Chem 85:2587
28. Lindman B, Puyal MC, Brun B, Gunnarson G (1982) J Phys Chem 86:1702
29. Malliaris A, Binana W (1984) J Coll Interf Sci 102:305
30. Lianos P, Lang J, Zana R (1982) J Phys Chem 86:4809
31. Good RJ, Elbing E (1970) Ind Eng Chem 62:54
32. Malliaris A, Lang J, Zana R (1986) J Phys Chem 90:655
33. Lianos P, Zana R (1980) Chem Phys Lett 72:171

Received January 21, 1987;
accepted January 26, 1987

Author's address:

Dr. Angelos Malliaris
Department of Physics, N.R.C.
„Demokritos"
Aghia Paraskeri
Athens 15310, Greece

Progress in Colloid & Polymer Science   Progr Colloid & Polymer Sci 73:167–173 (1987)

Surfactant self-association in some non-aqueous systems.
A preliminary report on self-diffusion and NMR relaxation studies

K. P. Das[1]), A. Ceglie[2]), M. Monduzzi[3]), O. Söderman, and B. Lindman[4])

[1]) Department of Chemistry, Vidyasagar College, Calcutta, India
[2]) Dipartimento di Chimica, Universita deglie Studi die Bari, Bari, Italy
[3]) Universita di Cagliari, Dipartimento di Scienze Chimiche, Cagliari, Italy
[4]) Physical Chemistry 1, Chemical Center, Lund University, Sweden

Abstract: The question is examined as to whether surfactants are aggregated or not in some nonaqueous systems commonly referred to as micellar solutions and microemulsions. The surfactant used is sodium dodecyl sulphate which is investigated both in a polar solvent and in three- and four-component solutions of surfactant, polar solvent, alcohol (pentanol-octanol) and hydrocarbon. The polar solvents used are formamide, N-methylformamide and N,N'-dimethylformamide. Solution structure and molecular organisation are investigated by multi-component self-diffusion studies (Fourier transform NMR spin-echo technique) and multi-field ^{2}H NMR relaxation. From a comparison with the analogous aqueous systems it is found that the segregation into polar and nonpolar domains as well as the degree of self-association is much less (and often insignificant) in the nonaqueous systems.

Key words: Surfactant organisation in non-aqueous media, NMR self-diffusion, NMR relaxation, non-aqueous microemulsions.

Introduction

Nonaqueous solutions of surfactants have an inherent fundamental and applied interest, but investigations of surfactant aggregation in nonaqueous systems are also highly relevant for the understanding of the mechanisms involved for aqueous systems. In our group we have developed, over some years, techniques which allow a characterisation of the state of aggregation of surfactant systems on a microscopic-molecular level. In particular, we have used multi-component self-diffusion studies (reviewed in Ref. [1]) to characterize the gross features of structure in terms of geometric features of water-oil segregation and multi-field NMR relaxation (reviewed in Ref. [2]) to characterize the order and motions of the aggregated surfactant state; such studies can sensitively distinguish between aggregated and nonaggregated (molecule-disperse solutions) states of the surfactant molecules.

While these approaches have been used and thoroughly tested for a broad range of aqueous systems for quite some time, we have also recently directed our interest towards different nonaqueous systems. Generally, these systems are chosen as analogous to the aqueous ones, with another polar solvent being substituted for water. These studies are still at quite an early stage and we are far from a thorough investigation of these problems. However, a number of significant observations already appear to be emerging. In the following, a brief overview of the progress of our investigations will be presented. As full reports will be given at a later stage no experimental details will be included in this report. As regards the experimental techniques used and the principles of analysis these have previously been given in detail in connection with our studies of aqueous systems, both in original articles [3–9] and in reviews [1, 2]. All results reported below pertain to 25 °C.

A brief literature survey

Micelles in non-aqueous media

Studies of micelle formation of surfactants in non-aqueous media began as a natural extension of investigation of surfactant behaviour in aqueous medium with varying additives including water miscible liquids. A large number of studies carried out with mixed solvents were taken to imply that although the CMC's of surfactants altered with the composition of the mixed solvent, micellisation occurred even when most of the water was replaced by the non-aqueous solvent. This prompted the search for micelles in water-free solvents. Up until now numerous systems of various kinds have been reported to show micellisation in non-aqueous solvents. Reinsborough et al. [10–14] in a series of papers reported micellisation behaviour of long-chain ionic surfactants in molten pyridinium chloride employing a large number of experimental techniques. Ray reported the micellisation of cationic [15] and non-ionic [16] surfactants in solvents such as various alcohols and diols and formamide, but did not observe micellisation in N-methylformamide, methanol and ethanol. Gopal and Singh [17, 18] found similar behaviour with other surfactants. However, later studies from other groups claimed micellisation to occur not only in formamide [19] but also in N-methyl formamide [20], N,N'-dimethylformamide [20, 21], acetamide [20], dimethylsulfoxides [20, 21], glycerol [22, 23] and even in methanol [24]. Evans et al. [25–28] in a series of papers reported on micellisation in hydrazine and in molten ethylammonium nitrate. In most of these studies, micellisation was inferred through the observation of a break in physical properties such as conductivity, surface tension etc. but none of the studies proceeded to provide more direct information on such things as micelle size, aggregation number, etc. In an attempt to deduce such information it was recently revealed by Almgren et al. [29] that micellisation of sodium dodecyl sulphate (SDS) apparently does not occur in pure formamide at 25 °C, despite the usual conductance break. It was pointed out by these authors that such a break in physical properties, as has been used to find CMC's in most cases, may have a quite different origin in non-aqueous systems. According to recent work by Rico and Lattes [30] Krafft points of ionic surfactants (SDS and cetyltrimethylammonium bromide) are much higher in formamide than in water and hence micellisation in such systems below these temperatures should not occur. In view of these recent findings it is at present rather doubtful whether many of the systems in the above cited studies actually involved micellisation.

Non-aqueous liquid crystals

Lyotropic liquid crystals in a non-aqueuos medium were first observed by Friberg's group [31] in the lecithin/ethylene glycol system. Since then this group has documented liquid crystals in many other systems [32–35] mostly with lecithin. Among the non-aqueous solvents found to give liquid crystals were different chain length alkane diols [32–34], low molecular weight polyethylene glycols [35] different cellosolves [35] and amines [35]. Liquid crystals [36] were also reported in systems containing only lecithin and $C_{12}E_4$ or $C_{12}E_7$. Evans et al. [37] observed liquid crystallinity in the dipalmitoylphosphatidyl choline/fused ethylammonium nitrate salt system. Recently, detailed phase equilibria studies have revealed the presence of liquid crystalline phases in ionic surfactant systems such as SDS/glycerol/decanol [38, 39] and Aerosol OT/formamide/toluene or decane [40]. Among the physical properties studied are bilayer spacing from X-ray measurements [31–40], magnetic alignment behaviour [33], solvent mobility in the bilayer [32], repulsive forces between bilayers [41] and degree of surfactant orientation [39].

In addition it may be mentioned here that lyotropic liquid crystals with polymers in non-polar solvents have been rather well characterized (see Refs. [42–44] for reviews).

Non-aqueous microemulsions

Microemulsions are a rather new-born member of the family of non-aqueous surfactant systems and only a few systems are described at present. Friberg reported a microemulsion stability region existing in the systems ethylene glycol/decane/lecithin [45] and glycerol/p-xylene/triethanolammonium oleate and oleic acid [46]. Recent studies by this group revealed extensive isotropic regions in systems like glycerol/alcohol/surfactant/oil [47, 48] using various surfactants and alcohols. Fletcher et al. [49] reported that the system glycerol/heptane/Aerosol OT gave microemulsions. Rico et al. [50–53] have initiated a rather extensive study of microemulsions in non-aqueous solvents like formamide. They have also reported on non-aqueous perfluorinated surfactant systems [54]. Recently, work has been initiated with solvents like ethanediol, formamide and its derivatives and large isotropic

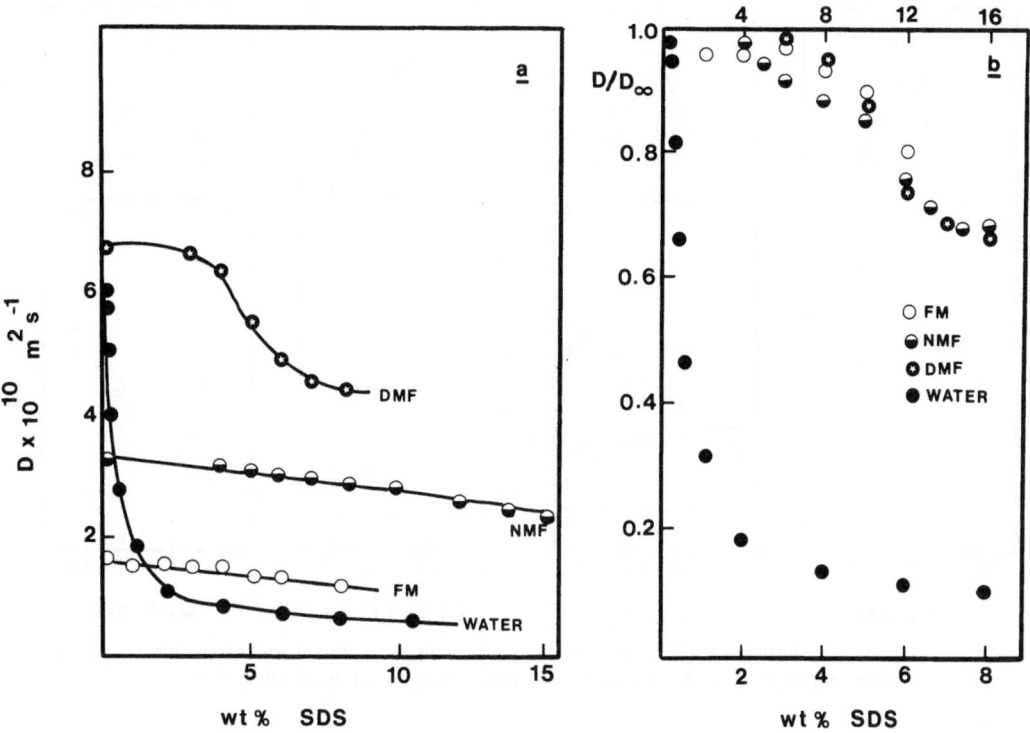

Fig. 1. a) SDS diffusion coefficients in different solvents as a function of SDS concentration. Data for water taken from Reference [56], b) Self-diffusion coefficients relative to those at infinite dilution of SDS. Upper SDS concentration scale refers to NMF only

regions have been reported [40] in systems such as Aerosol OT/toluene/polar solvent.

Although the existence of regions of microemulsions is known for many systems, practically nothing is known about the microstructures of these systems. From an analogy in phase behaviour between aqueous and non-aqueous systems, the structure are often thought to be the same [50–54]. However, our recent studies [48, 55] on some of these systems have revealed considerable differences in microstructure between aqueous and non-aqueous systems. It seems that a large number of different systems have to be investigated before any definite conclusions can be drawn.

Self-Diffusion Studies

Binary systems surfactant/polar solvent

In view of the controversy raised from the recent studies, as pointed out above, we have re-examined the question of surfactant self-association in formamide (FM), N-methylformamide (NMF) and N,N′-dimethylformamide (DMF) by the NMR self-diffusion method. Surfactant self-diffusion coefficients (D) as a function of concentration are presented in Figure 1a. Data from the aqueous system (from Ref. [56]) are also included in the figure. Figure 1b displays the data on a relative to infinite dilution scale, see Table 1. The observations can be summarised as follows.

1. SDS diffusion in these non-aqueous solvents changes very slowly with concentration and no sharp changes could be noticed over the concentration range studied (much above reported CMC's). On the contrary, a sharp fall of D_{SDS} occurs in water above the CMC.

2. The relative D value of SDS in water falls steeply to about 0.1 above the CMC, whereas in the non-aqueous solvents it lies above 0.65 up to almost the solubility limit of SDS.

3. Although it is possible to cosolubilize a hydrophobic probe, tetramethyl silane, readily in aqueous SDS solution above CMC, it was impossible to solubilise TMS in the concentrated non-aqueous solution [55].

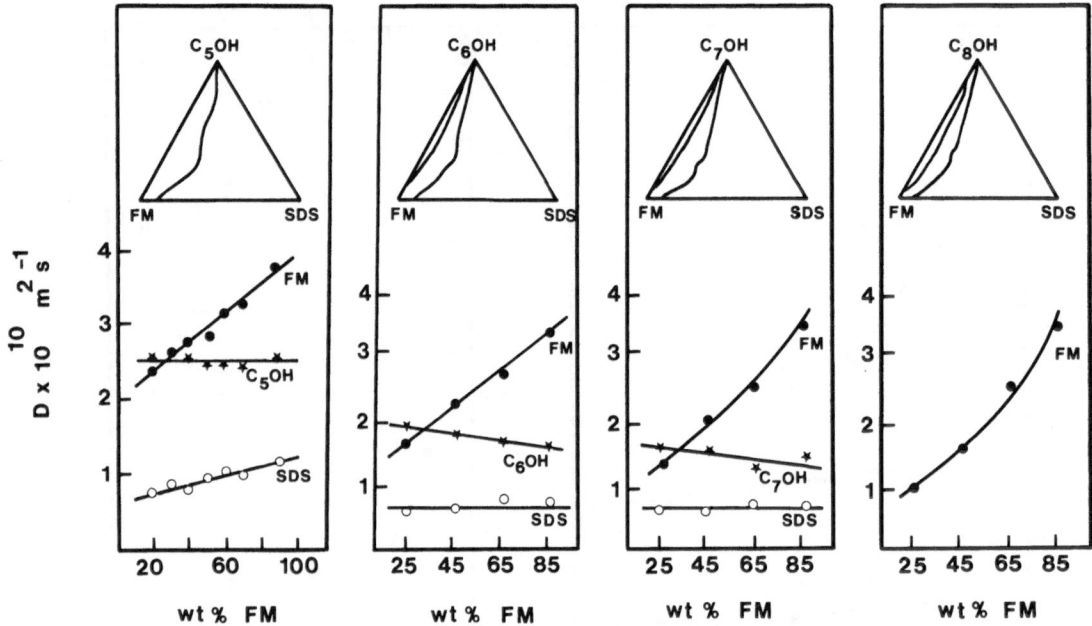

Fig. 2. Isotropic stability region and self-diffusion coefficients in three component non-aqueous microemulsions. All samples in C_5OH system corresponds to 5 wt% SDS and other systems contain 10 wt % SDS

All these results confirm beyond doubt the absence of well-defined micelles of SDS in these non-aqueous media at 25 °C.

Ternary microemulsions surfactant/alcohol/polar solvent

We have determined the isotropic stability region of systems such as formamide/SDS/$C_nH_{2n+1}OH$, with n = 5, 6, 7, 8. These are shown in Figure 2 together with the self-diffusion results. The following features are worth noting.

1. All the four systems have large isotropic solution region extending from the formamide corner to the

Table 1. Self-diffusion coefficients[a]) of some of the neat components (D^o) and of SDS at infinite dilution in polar solvents (D_∞)[a]) at 25 °C

Component	$D^o \times 10^{10} m^2 s^{-1}$	$D_\infty \times 10^{10} m^2 s^{-1}$
Formamide	5.21	1.62
N-methylformamide	8.48	3.24
N,N'-dimethylformamide	16.2	6.72
Water	22.7	6.20
1-pentanol	3.10	
1-hexanol	2.30	
1-heptanol	1.80	
1-octanol	1.40	

[a]) from Reference [55]

alcohol corner. In analogous aqueous systems [55], a similar phase behaviour was noted only with butanol, while higher alcohols gave disconnected isotropic phases.

2. Formamide diffusion increases with increasing FM content, but over most of the isotropic region it is quite rapid. Relative (to neat FM) diffusion D/D^o (see Table 1 for D^o) is > 0.5 with pentanol, > 0.3 with hexanol, > 0.25 with heptanol and > 0.2 with octanol. Thus, although relatively higher obstructions are encountered with increasing chain length of alcohol, none of these systems corresponds to significant confinement of FM in droplets for which $(D/D^o)_{FM}$ should be ≪ 0.1.

3. The alcohol diffusion coefficients are always close to their values in neat alcohol ($D/D^o > 0.8$), so alcohol molecules occur in a continuous medium in all the composition range studied.

4. The SDS diffusion coefficients in the case of n = 5, 6 and 7 were always found to be above $0.65\,10^{-10}$ m² s⁻¹ (no data of D_{SDS} were obtained for $n = 8$ due to signal overlap and low signal/noise ratio) which corresponds to a $(D/D_\infty)_{SDS}$ value > 0.40 (D_∞ is the infinite dilution value in FM). Such high values are inconsistent with extensive surfactant aggregation.

5. This behaviour is entirely different from the behaviour in analogous aqueous systems [55] (results not

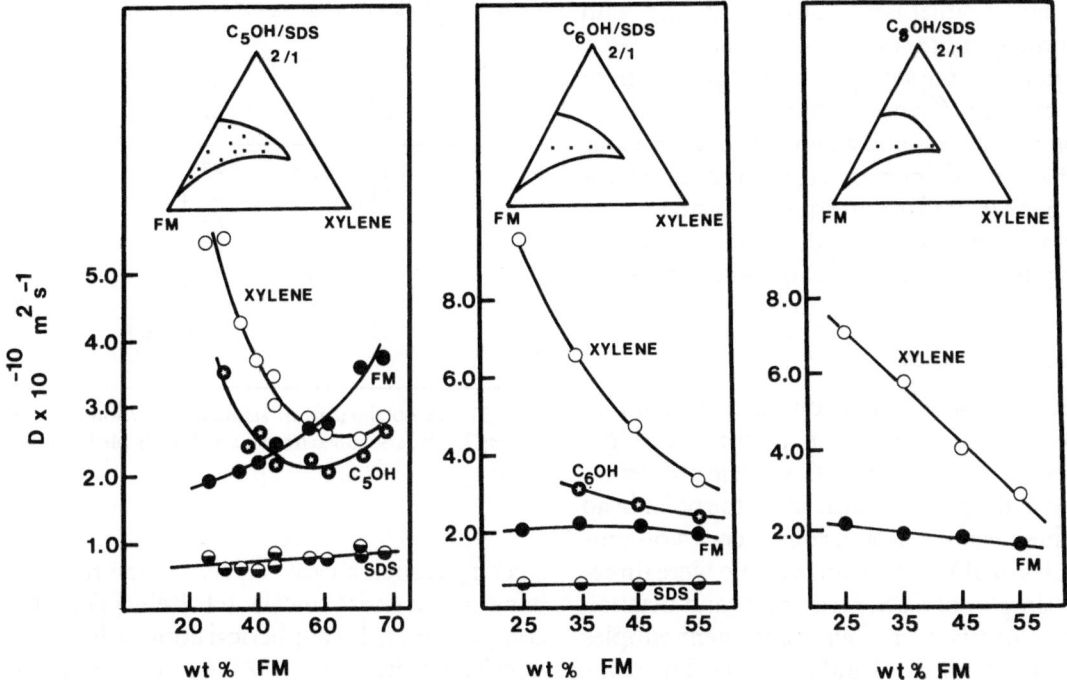

Fig. 3. Isotropic stability region and self-diffusion coefficients in four component non-aqueous microemulsions. Sample compositions are indicated ($\cdots$) in the phase diagram

presented here). Here water diffusion was found to decrease rapidly with increasing alcohol chain length. Thus $(D/D^o)_{water}$ fell below 0.04 with heptanol indicating distinct water droplet structures.

The conclusion that can be drawn from the above results is that these 3-component non-aqueous microemulsions are almost structureless and closer to simple solutions. The analogous aqueous systems are indeed very different and involve clear-cut surfactant organization.

Quaternary microemulsions surfactant/alcohol/hydrocarbon/polar solvent

Systems investigated were of the type formamide/SDS/p-xylene/$C_nH_{2n+1}OH$ with $n = 5$, 6 and 8. Figure 3 shows the self-diffusion data as well as the isotropic range in pseudo-ternary compositon diagrams. We note the following points.

1. A single isotropic region exists in each of the 4-component non-aqueous systems. This is in contrast to analogous aqueous systems [55] where two disconnected phases appear with $n > 5$.

2. The formamide diffusion coefficient was $> 2 \times 10^{-10} m^2 s^{-1}$ in all the composition range studied. Thus $(D/D^o)_{FM} > 0.40$.

3. The alcohol diffusion coefficients in the pentanol and hexanol systems were of the same order of magnitude as those of the pure alcohol. The octanol diffusion coefficient could not be determined due to signal overlap and low signal/noise ratio.

4. The SDS diffusion coefficient in the case of pentanol and hexanol was above $0.65 \times 10^{-10} m^2 s^{-1}$, i.e. $(D/D_\infty)_{SDS}$ always > 0.40.

5. The p-xylene diffusion coefficient was in all cases higher than the D values of the other components over all the composition range studied.

6. In analogous aqueous systems (with toluene instead of p-xylene), the self-diffusion behaviour was found to be very different [57]. Water diffusion was found to fall sharply with increasing n and $(D/D^o)_{water}$ became < 0.02 with $n = 8$; D_{SDS} was also found to fall appreciably with increasing n, and with $n = 6$, $(D/D_\infty)_{SDS}$ in water became 0.05, as compared to > 0.40 in the formamide system.

The conclusion for both the 3- and 4-component systems appears to be that formamide microemulsions are almost structureless at 25 °C while the analogous aqueous systems show considerable surfactant aggregation. It seems that the alcohol solubility in polar and non-polar (oil) solvents plays an important role in determining microstructure. Although higher alcohols like octanol have a rather large immiscibility gap with FM, it is definitely more soluble in FM than in water. This causes not only the structure and organization to be different but also affects the appearance of the isotropic region in aqueous and analogous non-aqueous systems. It has been demonstrated from recent measurements [55] that similar non aqueous systems with other polar solvents like N-methyl formamide and N,N'-dimethylformamide are also quite structureless.

NMR relaxation studies

To complement the self-diffusion experiments, a ^{2}H NMR relaxation study has been performed. Eight samples were prepared with SDS specifically deuterium-labelled in the position adjacent to the polar head group. Of these eight samples, four were two-component samples of SDS in formamide, two were three-component samples of SDS, formamide and octanol, and the remaining two were four-component samples of SDS, formamide, octanol and p-xylene. The compositions of the samples are given in Table 2.

^{2}H NMR relaxation experiments were performed at two magnetic field strengths and the relaxation data are given in Table 3. Starting with the two-component samples, it is clear from the data, that within the experimental error, the longitudinal relaxation time (T_1) equals the transverse relaxation time (T_2) for all compositions studied. Moreover, the most concentrated samples have identical T_1's at 6.0 and 8.5 T. This indicates that the extreme narrowing condition is fulfilled *i.e.* $\omega\tau \ll 1$, where ω is the Larmor frequency and τ_c is the characteristic time scale for the motions responsible for the NMR-relaxation. In the present context this implies that $\tau_c \ll 4$ ns. Now, if micelles were present, one would expect fairly long correlation times,

Table 2. Composition of samples for relaxation studies

Samples (% w/w) No.	SDS[a])	formamide	octanol	p-xylene
1	7.99	92.01	–	–
2	5.01	94.99	–	–
3	1.00	99.00	–	–
4	0.20	99.80	–	–
5	16.05	54.98	28.97	–
6	10.01	40.19	49.80	–
7	13.27	30.03	26.80	29.90
8	13.37	44.99	26.68	14.96

[a]) Deuterium-labelled in the α-position

Table 3. ^{2}H Relaxation data at 25 °C for samples in Table 2

Sample	T_1[a]) ms	6.0 T T_2[b]) ms	8.5 T T_1[a]) ms
1	42.9 ± 0.3	45 ± 2	46.1 ± 1.0
2	53.3 ± 0.3	62 ± 3	
3	64.3 ± 0.8	67 ± 4	
4	66 ± 3	64 ± 4	
5	30.4 ± 0.8	16 ± 1	35.9 ± 0.3
6	31 ± 1	18 ± 2	36.7 ± 0.6
7	33.5 ± 0.7	12.6 ± 0.8	38.3 ± 0.3
8	38 ± 1	15 ± 1	39.1 ± 0.3

[a]) T_1's as obtained from standard inversion-recovery experiments; [b]) T_2's as obtained from Carr-Purcell-Meiboom-Gill experiments

since τ_c in such a case is given by the rotational tumbling of the entire micelle (cf. Refs. [2] and [8]). For comparison, τ_c for a spherical lithium dodecylsulphate micelle in water is 4 ns [58]. As the viscosity of formamide is more than three times the viscosity of water, spherical micelles, if present in formamide, would be expected to have $\tau_c \approx 10$ ns. Clearly, this is not compatible with the data in Table 3. Rather, the data suggest, in accordance with the self diffusion data presented above, that SDS is present essentially as monomers in formamide. If any association of SDS occurs in formamide it is restricted to rather small complexes. Finally, we note that there is a slight concentration dependence in the relaxation data for samples 1 to 3. Between samples 3 and 4 there is no significant difference, in accordance with the fact that these samples are very dilute.

For the three- and four-component samples, however, the situation is different. There, T_1 is not equal to T_2 at 6.0 T. Moreover, the T_1's at 6.0 differ slightly but significantly (perhaps with the exception of sample 8) from those obtained at 8.5 T. Clearly, we are no longer in the extreme narrowing regime, and motions with characteristic time scales, τ_c, that fulfill the criterion $\omega\tau_c \approx 1$ are present. In line with the above reasoning this implies that at least a fraction of the SDS molecules are aggregated into some sort of interface. This is reasonable since formamide is only partly miscible with octanol and p-xylene, implying that some sort of separation into formamide-rich and alcohol/oil-rich regions exists.

In order to proceed further in the analysis of the relaxation data for samples 4–8 one needs a model for the NMR relaxation in these types of systems. However, in the present context with a rather limited data set, a more detailed analysis than the one presented above is hardly meaningful. To interpret the data further one requires relaxation data at additional field strengths (as well as data from anisotropic phases in these or analogous systems). In conclusion, the ^{2}H NMR relaxation data support a picture with no (or very minor) aggregation in the two-component SDS/formamide system, while if octanol or octanol/p-xylene is added some sort of aggregation of SDS appears to occur.

Finally, it may be worth mentioning that we have started field-dependent ^{13}C relaxation studies of these systems. The results are still preliminary and will be presented at a later date.

Acknowledgements

This work was supported by The Swedish Board of Technical Development (STU). The stays in Lund of K.P.D., A.C. and M.M. were supported by STU, NATO-CNR (National Council of Researches-Italy) and NATO, respectively.

References

1. Lindman B, Stilbs P (1987) In: Friberg SE, Bothorel P (eds) Microemulsions, CRC Press, Boca Raton Ch 5
2. Lindman B, Söderman O, Wennerström H (1987) In: Zana R (ed) Surfactant Solutions, New Methods of Investigation, Marcel Dekker, New York p 295
3. Guering P, Lindman B (1985) Langmuir 1:464
4. Olsson U, Shinoda K, Lindman B (1986) J Phys Chem 90:4083
5. Söderman O, Walderhaug H, Henriksson U, Stilbs P (1985) J Phys Chem 89:3693
6. Nery H, Söderman O, Canet D, Walderhaug H, Lindman B (1986) J Phys Chem 90:5802
7. Söderman O, Walderhaug H (1986) Langmuir 2:51
8. Söderman O, Canet D, Carnali J, Henriksson U, Nery H, Walderhaug H, Wärnheim T (1987) In: Rosano H, Clausse M (eds) Microemulsion Systems, Dekker, N. Y, in press
9. Lindman B, Ahlnäs T, Söderman O, Walderhaug H, Rapacki K, Stilbs P (1983) Faraday Disc Chem Soc 76:317
10. Bloom H, Reinsborough VC (1967) Aust J Chem 20:2583
11. Bloom H, Reinsborough VC (1968) Aust J Chem 21:1525
12. Bloom H, Reinsborough VC (1969) Aust J Chem 22:519
13. Reinsborough VC, Valleau JP (1968) Aust J Chem 21:2905
14. Reinsborough VC (1970) Aust J Chem 23:1473
15. Ray A (1969) J Am Chem Soc (1969) 91:6511
16. Ray A (1971) Nature, London 231:313
17. Gopal R, Singh JR (1973) J Phys Chem 71:554
18. Gopal R, Singh JR (1970) Kolloid Z Z Polym 239:699
19. Escoula B, Hajjaji N, Rico I, Lattes A (1984) J Chem Soc Chem Commun 1233
20. Singh HN, Saleem SM, Singh RP, Birdi KS (1980) J Phys Chem 84:2191
21. Singh HN, Singh S, Tewari KC (1975) J Am Oil Chem Soc 52:436
22. Ionescu LG, Fung DS (1981) J Chem Soc Faraday Trans 1, 77:2907
23. Ionescu LG, Fung DS (1981) Bull Chem Soc Jpn 54:2503

24. Varma RP, Bahadur P, Dayal R (1980) Rev Roum Chimi 25:201
25. Evans DF, Yamauchi A, Roman R, Casassa EZ (1982) J Coll Interf Sci 88:89
26. Evans DF, Chen SH (1981) J Am Chem Soc 103:481
27. Ramadan MS, Evans DF, Lumry R (1983) J Phys Chem 87:4538
28. Evans DF, Ninham BW (1983) J Phys Chem 87:5025
29. Almgren M, Swarup S, Löfroth JE (1985) J Phys Chem 89:4621
30. Rico I, Lattes A (1986) J Phys Chem 90:5870
31. Moucharafieh N, Friberg SE (1979) Mol Cryst Liq Cryst 49:231
32. Larsen DW, Friberg SE, Christenson H (1980) J Am Chem Soc 102:6565
33. Larsen DW, Rananavare SB, Friberg SE (1984) J Am Chem Soc 106:1848
34. Larsen DW, Rananavare SB, El-Nokaly M, Friberg SE (1982) Finn Chem Lett 6-8:96
35. El-Nokaly M, Friberg SE, Larsen DW (1984) J Coll Interf Sci 98:274
36. Ganzuo L, El-Nokaly M, Friberg SE (1982) Mol Cryst Liq Cryst 72:183
37. Evans DF, Kaler EW, Benton WJ (1983) J Phys Chem 87:533
38. Friberg SE, Liang P, Liang Y-C, Greene B, Gilder RV (1986) Colloids and Surfaces 19:249
39. Friberg SE, Ward AJI, Larsen DW, in press
40. Bergenståhl B, Jönsson A, Sjöblom J, Stenius P, Wärnheim T (1987) Progr Coll & Polym Sci, in press
41. Persson PKT, Bergenståhl B (1985) Biophys J 47:743
42. Blumenstein A (ed) (1978) Liquid crystalline structures in polymers, Academic Press, New York
43. Hines WA, Samulski ET (1973) Macromolecules 6:794
44. Blumenstein A (ed) (1978) Mesomorphic order in polymers, ACS symposium series, 74
45. Friberg SE, Podzimek M (1984) Coll & Polym Sci 262:252
46. Friberg SE, Wohn CS (1985) Coll & Polym Sci 262:156
47. Friberg SE, Personal communication
48. Das KP, Ceglie A, Lindman B, Friberg SE (1987) J Coll Interf Sci 116:390
49. Fletcher PID, Galal MF, Robinson BH (1984) J Chem Soc Faraday 1, 80:3307
50. Rico I, Lattes A (1984) Nouv J Chimie 8:429
51. Cecutti C, Rico I, Lattes A (1984) Tetrahedron Lett 25:5041
52. Samii A A-Z, Savignac A, Rico I, Lattes A (1985) Tetrahedron 41:3683
53. Gautier M, Rico I, Samii A A-Z, Savignac A, Lattes A (1986) J Coll Interf Sci 112:484
54. Rico I, Lattes A (1984) J Coll Interf Sci 102:285
55. Das KP, Ceglie A, Lindman B (1987) J Phys Chem, in press
56. Lindman B, Puyal M-C, Kamenka N, Rymden R, Stilbs P (1984) J Phys Chem 88:5048
57. Stilbs P, Rapacki K, Lindman B (1983) J Coll Interf Sci 95:583
58. Stilbs P, Söderman O, Walderhaug H (1986) J Magn Resonance 69:411

Recieved December 24, 1986;
accepted January 20, 1987

Authors' address:

Prof. Björn Lindmann
Physical Chemistry 1, Chemical Center
Lund University
Box 124
S-22100 Lund, Sweden

Progress in Colloid & Polymer Science   Progr Colloid & Polymer Sci 73:174–179 (1987)

Hexadecyltrimethyl ammonium sulphate-water system.
Phase diagram and micellization

D. Maciejewska[1]), A. Khan, and B. Lindman

Division of Physical Chemistry 1, Chemical Center, University of Lund, Lund, Sweden

Abstract: The binary phase diagram of the hexadecyltrimethylammonium sulphate-water system has been determined by ^{2}H NMR and polarized microscopy methods. At 295 K, the phase diagram comprises a large isotropic solution region (up to 40.5 % surfactant) and isotropic cubic, I_1, (41–46.5 %) and anisotropic hexagonal, E, (51–60 %) liquid crystalline phases. On increasing the temperature, the I_1 phase dissappears quickly while the E phase extends to lower water contents. There is, furthermore, a formation of a second cubic, I_2, and a lamellar phase, D. D phase extends to almost water free surfactant at high temperature. The conductivity data in the solution phase yield a CMC of 5.8×10^{-4} molal surfactant ion and a degree of SO_4^{2-} binding, $\beta = 0.83$. Both ^{1}H NMR linewidth and self-diffusion coefficient measurements of surfactant ion indicate the predominance of small spherical micelles in the entire micellar solution region. The small tendency for micelle growth and the appearance of the phase diagram are notable features which are discussed on the basis of the properties of analogous systems with monovalent counterions.

Key words: H̲exadecyltrimethylammoniumsulphate, p̲hase diagram, liquid c̲rystal, deuteron and proton N̲MR, m̲icellization.

Introduction

Recently we have studied the effect of the valency of counterions on the phase equilibria and self-association phenomena of anionic surfactant systems and demonstrated significant differences between mono- and divalent systems [1–6]. The cationic surfactant systems with divalent counterions [7, 8] have, however, not been studied in any detail, which is necessary in order to have a clear picture of the role of divalent counterions in surfactant systems. Here we present the binary phase diagram of the hexadecyltrimethylammonium sulphate (C_{16}TAS)-water system. We also report and discuss the NMR and conductivity data of aqueous C_{16}TAS solutions. C_{16}TAS has a low Krafft point, an added advantage over the analogous long-chain anionic surfactants, e. g. calcium and magnesium alkyl sulphates where the Krafft point is high.

Materials

Hexadecyltrimethylammonium bromide (C_{16}TAB) was purchased from Th. Suchardt, München, West Germany. Hexadecyltrimethylammonium sulphate (C_{16}TAS) was prepared by converting C_{16}TAB solution to the hydroxide form on a Dowex 21 K ion exchange resin (British Drug Houses). The C_{16}TOH was immediately neutralized with H_2SO_4 to pH 6. The solution was lyophilized and a white crystalline product was obtained. Chemical analysis shows that the product has a purity higher than 99 %.

Sample preparation

The samples were prepared by weighing appropriate amounts of substances into glass tubes which were sealed-off immediately. They were mixed by repeated centrifugation for several days and were kept at the appropriate temperature for a month to attain equilibrium.

Methods

The phase diagram of the surfactant system has been determined by ^{2}H NMR and polarizing microscopy techniques.

Polarizing microscopy

All samples were initially examined against a crossed-polaroid for sample homogeneity and occurence of birefringency. The texture of the liquid crystalline samples was examined by polarizing microscope first at 295 K and then as a function of temperature where the temperature was increased at a rate of 2 K per minute.

^{2}H NMR

The application of water deuteron NMR in characterizing phases and determining the phase boundaries in surfactant systems is well established [1]. ^{2}H NMR produces quadrupolar splitting (Δ) in the hexagonal and lamellar liquid crystalline phases and $\Delta_{lam} = 2\Delta_{hex}$ if the local conditions are the same [9]. Moreover, Δ-value is concentration dependent. For a multi-phase sample, where the water exchange between the phases is normally slow on NMR timescale, the ^{2}H NMR spectrum is a superposition of the spectra of the individual phases and the isotropic phase(s) always yields single resonance signals. Therefore, the analysis of ^{2}H NMR spectra provides a direct determination of the phase diagram.

The ^{2}H NMR spectra were recorded at 300 K at a resonance frequency of 15.35 MHz on a modified Varian XL-100-15 pulsed spectrometer working in the Fourier transform mode using an external proton lock.

Self-diffusion and ^{1}H NMR linewidths measurements

The self-diffusion coefficients of surfactant ion and water were measured with the Fourier transform pulsed-gradient spin-echo (FT PGSE) technique [10, 11] monitoring the ^{1}H NMR spectra on a Jeol FX-60 NMR spectrometer. The experimental conditions, data evaluation etc. were as recommended [11].

The same spectrometer was used to measure the linewidths of the $-(CH_2)_n-$ and $-N(CH_3)_3$ ^{1}H signals for the C_{16}TAS solutions by a single 90° pulse. The measurements were done at 300 K with ^{2}H$_2$O as solvent. The contribution to the ^{1}H linewidth due to the magnetic field inhomogeneity was less than 1 Hz.

Conductivity Method

The conductivity of the aqueous C_{16}TAS solutions was measured by E 382 conductometer (Metrohm A.G., Switzerland) using a standard cell EA 608 with the cell constant, $C = 0.85$ cm^{-1} at 300 K.

Phase diagram

C_{16}TAS is easily soluble in water at 295 K giving micellar solution, L_1, which extends up to 40.5 wt% surfactant (Fig. 1a). On increasing the surfactant concentration, there appears a cubic liquid crystalline phase, I_1, (41–46.5%) followed by the hexagonal liquid crystalline phase, E, (51–60%). Above 60% surfactant, E phase coexists with the surfactant crystals (hydrated).

On increasing the temperature, I_1 phase disappears at 303 K and the E phase extends to higher surfactant concentration with a maximum of 80% at 324 K. This is followed by the formation of the second cubic phase, I_2 (80–84%). Finally, at low water contents and high temperature, there forms the lamellar liquid crystalline

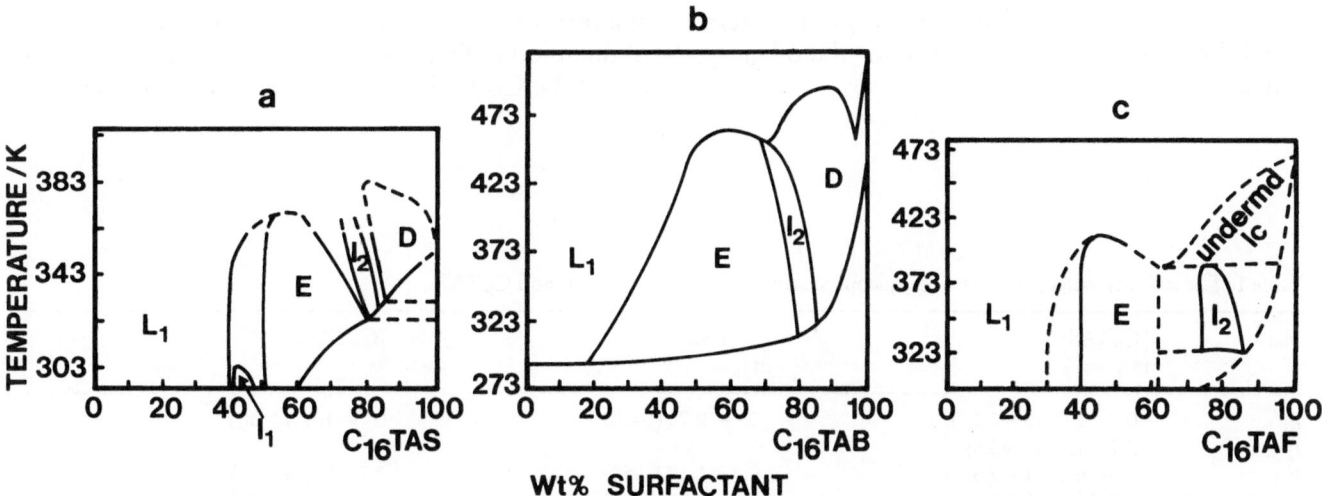

Fig. 1. Binary (temperature vs. composition) phase diagrams for the systems water-hexadecyltrimethylammonium salt with (a) sulphate (C_{16}TAS), (b) bromide (C_{16}TAB) [14] and (c) fluoride (C_{16}TAF) [15] as counterions. Phase notations: L_1, normal micellar solutions; E, hexagonal; D, lamellar, I_1 and I_2 cubic liquid crystalline phases

phase, D, which extends to almost pure surfactant at a temperature of ca. 350 K.

^{2}H NMR produced singlets for the samples in the micellar and cubic phases and for the E and D phases a single ^{2}H NMR quadrupolar splitting was obtained with $\Delta^2H_{lam} = 2\Delta^2H_{hex}$ when recalculated to identical concentration and temperature as expected from theoretical considerations [9]. The microscopic textures [12] observed in the E and D phases also confirm the NMR results. The phase diagram constructed from NMR and polarizing microscopic data is shown in Figure 1a. For comparison, the binary phase diagrams of C_{16}TAB [13,14] and C_{16}TAF [15] systems are also included (Figs. 1b and 1c).

The occurrence of phases in aqueous systems of C_{16}TA$^+$ with Br$^-$, F$^-$ and SO$_4^{2-}$ counterions does not show any dramatic differences except that I_1 phase presents only in the sulphate system and the lamellar phase is not reported for the fluoride system. However, the undetermined liquid crystal at low water contents indicated in Figure 1c for the fluoride system may be the lamellar phase. The phase diagram for the chloride system has not been published but the preliminary results show that this system does not differ much from the other systems. The thermal stability of E, I_2 and D (not for fluoride) follows the order bromide > fluoride > sulphate.

The ranges of stability of different phases of the C_{16}TAS, C_{16}TAB and C_{16}TAF systems are shown in Table 1. Important points to note are that the aqueous solubility of C_{16}TAS is higher than that of C_{16}TAF or C_{16}TAC which are again higher than that of C_{16}TAB. The stability ranges of D and I_2 are almost identical for all systems. The E phase has for the C_{16}TAB system a much broader stability range than for the other systems.

Micellization

Conductivity

The conductivity of aqueous C_{16}TAS solutions as a function of concentration at 298 K has been analyzed according to a procedure given by Mukerjee and Mysels [16]. A CMC value of 5.8×10^{-4} molal surfactant ion ($C_{16}H_{33}N(CH_3)_31/2SO_4$) is obtained from the intersection of two straight lines. The CMC obtained here agrees well with the published data [8] and is lower than that of the aqueous C_{16}TAB (9×10^{-4} m) [17] and C_{16}TAC (1.1×10^{-3} m) systems [18]. From the slopes of the two straight lines [19–21] a degree of counterions binding, $\beta = 0.83$ was obtained. This value may be compared with 0.8 with Ca^{2+} and 0.7 with Mg^{2+} in the octylsulphate systems [5], 0.71 with Br$^-$ and 0.55 with Cl$^-$ in the hexadecyltrimethylammonium systems [22]. β values with divalent counterions, though available only for a limited number of systems, are generally much higher than those with the monovalent counterions as expected from electrostatic theory.

NMR self-diffusion

The experimental self-diffusion coefficients measured at 298 K for surfactant ion, D^a, and water, D^w, are shown in Table 2. The self-diffusion data in the C_{16}TAS system follow closely those reported for the C_{16}TAC [22, 23] and C_{16}TAB [22] systems. Detailed accounts of the interpretation of self-diffusion data in terms of self-association have been given previously [22, 24].

In accordance with the previous analysis D^a is assumed to attain the value of self-diffusion coefficient of the micelle, D^m, at surfactant concentrations much above the CMC. By extrapolation of D^m values, D_o^m

Table 1. The stability ranges of phases of aqueous systems of C_{16}TAS, C_{16}TAB and C_{16}TAF

Phase	[C_{16}TAS] m^a)	[C_{16}TAB] m	[C_{16}TAF] m
L_1	0 – 2.0 (0–40.5)^b)	0 – 0.7 (0–20)	0 – 1.4 (0–30)
I_1	2.1– 2.6 (41–46.5)	—	—
E	3.0– 6.1 (51–79.5)	0.7–11.1 (20–80)	2.2– 5.0 (40–61)
I_2	12.0–15.5 (80–84)	11.1–16.9 (80.2–86)	9.4–18.7 (74–85)
D	17.0– (85–)	17.0– (86–)	not reported

^{a}) m is the molality of surfactant ion; ^{b}) within brackets is given wt% of surfactant

Table 2. Self-diffusion coefficients of surfactant ion, D^a and water, D^w of the aqueous C_{16}TAS solutions at 295 K

C_{16}TAS m	$D^a \times 10^{11}$ m^2 s^{-1}	$D^w \times 10^9$ m^2 s^{-1}
0.061	5.58	–
0.091	5.11	–
0.158	5.31	1.75
0.319	4.44	1.62
0.531	3.41	1.43
0.760	2.23	–
0.993	1.56	1.24
1.277	1.34	1.12
1.585	0.86	1.01
1.951	–	0.67

$= 6.7 \times 10^{-10}$ m^2 s^{-1} is obtained at infinite dilution of micelles. The plot of D^m/D_o^m vs. volume fraction, ϕ, is linear (Fig. 2) i. e. it follows the equation $D^m = D_o^m (1 - k\phi)$ with the slope $k = 1.5$ if the volume fraction of the micelles is taken to include the water of hydration (cf. below). If the water of hydration is not taken into account in obtaining ϕ, k has a value of 2.4. Recent theoretical studies on the self-diffusion of interacting particles as a function of concentration predict $k = 1.5$–2 for hard spheres with hydrodynamic interactions [25–29]. Therefore, the observed D^m values can be explained in terms of spherical micelles in the entire L_1 region of the C_{16}TAS system.

Assuming monodisperse spherical micelles, the D_o^m value is further used to obtain the hydrodynamic radius, $r = 2.9$ nm of the micelle from the Stokes-Einstein equation. From the calculated values of the alkyl chain length (2.174 nm), head-group (0.402 nm) and the aggregation number (94) [30], we obtain the hydration number, $n_h = 11$. n_h is defined as the average number of water moleculas per surfactant ion diffusing with the micelle as a kinetic entity.

The water self-diffusion coefficients, D^w, were also used to obtain n_h as described previously [22, 24]. In this way we obtain $n_h = 12 \pm 1$ in the concentration range of 5–39 wt% of surfactant.

^1H NMR linewidths

The micellar growth and sphere-to-rod shape transition of micelles may be shown to lead to an important change in NMR relaxation [31, 32]. Thus, for example, the sharp signals of ^{1}H NMR for the spherical micelles change to very broad signals for large non-spherical micelles [31]. This is due to the fact that as the surfactant aggregates increase in size it takes a longer time for them to rotate as well as for a surfactant molecule to diffuse around the curved micelle surface. The long correlation time for large aggregates will make the ^{1}H NMR transverse relaxation rate rapid.

The ^{1}H NMR linewidths of the $-N(CH_3)_3$ and the main methylene groups of the alkyl chain were measured as a function of C_{16}TAS concentration. A few

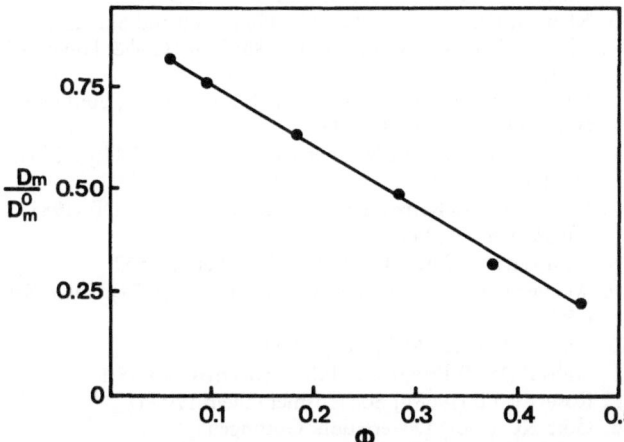

Fig. 2. A plot of D^m/D_o^m against micelle volume fraction, ϕ, where D^m/D_o^m is the ratio between observed micelle self-diffusion coefficient (D^m) and that at infinite dilution (D_o^m)

Talbe 3. ^{1}H NMR linewidths, $v\,1/2$ (Hz) measured as a function of surfactant concentration for the aqueous C_{16}TAS solutions at 298 K

[C_{16}TAS] m^a)	$v\,1/2$/Hz $-N(CH_3)_3-$	$v\,1/2$/Hz $-(CH_2)_n-$
0.03	4.3	7.7
0.06	4.6	7.8
0.09	4.3	7.7
0.16	3.7 (4)^b)	6.3 (8)
0.32	3.6 (12)	7.8 (39)
0.53	3.7 (48)	7.1 (96)
0.76	5.5	8.6
1.28	6.4 (> 2600)^c)	10.0 (> 5000)
1.79	5.0	9.3
1.95	6.6	9.2

^{a}) m is the molality of surfactant ion; ^{b}) within bracket are given in ^{1}H NMR linewidth for the aqueous C_{16}TAB system [31]; ^{c}) ^{1}H NMR linewidth in the hexagonal liquid crystalline phase for the C_{16}TAB system

Progress in Colloid & Polymer Science, Vol. 73 (1987)

typical values are shown in Table 3. The proton linewidths observed in the entire L_1 region of the C_{16}TAS system are comparable with those obtained in the aqueous C_{16}TAB system below 0.3 M surfactant concentration where spherical micelles are shown to form (Table 3). Above 0.3 M for C_{16}TAB solutions a dramatic broadening of the ^{1}H NMR signals due to the formation of large rod-shaped micelles is observed [31]. The small linewidths observed in the entire micellar solution region indicate that no substantial micellar growth is taking place in the C_{16}TAS system. The dynamic light scattering study on the dilute C_{16}TAS solutions with concentrated Na_2SO_4 also shows no micellar growth in the C_{16}TAS system [7].

Discussion

As expected from electrostatic considerations of the counterion distribution [33, 34], C_{16}TA$^+$ with a divalent counterion (SO_4^{2-}) gives a lower CMC and a higher degree of counterion binding than with a monovalent counterion (F^-, Cl^- and Br^-). Furthermore, in line with previous observations [1–6] of anionic systems with divalent counterions and with theoretical predictions [35–37] the lamellar phase of C_{16}TAS systems, when in equilibrium with a dilute aqueous solution (with an alcohol as third component), can swell much less with water than the corresponding systems with monovalent counterions (unpublished observations).

In many respects, both as regards micellization and liquid crystalline phase formation, the aqueous C_{16}TAS-water system behaves in a similar way as the monovalent cases. A striking, and at first sight perhaps unexpected, observation is the very low tendency of the C_{16}TAS micelles to deform from the spherical state and grow in size. This is seen in the micellar size studies (^{1}H NMR linewidths and self-diffusion), in the formation of an I_1 cubic liquid crystalline phase and the very high concentrations needed for the formation of the hexagonal liquid crystalline phase. In many ways the behaviour resembles that of ionic surfactants with monovalent counterions having a high degree of counterion dissociation [38–40]. From the low degree of counterion dissociation alone we would expect a larger tendency of micellar growth for C_{16}TAS than for, for example, C_{16}TAB but the actual behaviour is strikingly opposite to that. The conservation of a spherical aggregate shape up to high concentrations can in general be attributed to a large effective area in the polar region [41, 42]. This can occur both as a result of strong repulsive interactions, often of an electrostatic nature,

between headgroups or for sterical reasons. The large size of the sulphate ion in combination with a high degree of counterion association could be taken to suggest a sterical effect which gives a large effective area per polar headgroup. The phase diagram of the aqueous C_{16}TAS-water system shows indeed very close resemblance to the low temperature phase diagram of aqueous $C_{12}H_{25}(OCH_2CH_2)_8OH$ [43], a nonionic surfactant with a very bulky polar headgroup. (In such size considerations, one must, of course, take into account hydration, for both oligo (ethylene oxide) chains and counterions.) However, as shown recently [6] the details of the counterions distribution in divalent counterion systems are probably different from those in monovalent counterion systems due to image charge interactions. These cause the associated divalent counterions to be repelled away from the immediate vicinity of the micelle surface resulting in higher interheadgroup repulsions in spite of a high degree of counterion association. Too few systems have yet been studied to allow conclusive statements about the mechanism but we note that micellar growth has been found to be small also for a number of anionic surfactant systems with divalent counterions [5].

Acknowledgements

The stay of D. M. in Lund was made possible by a grant from The Swedish Board of Technical Development. Håkan Wennerström is thanked for useful comments.

References

1. Khan A, Fontell K, Lindblom G, Lindman B (1982) J Phys Chem 86:4266
2. Khan A, Fontell K, Lindman B (1984) J Coll Interf Sci 101:193
3. Khan A, Fontell K, Lindman B (1984) Coll and Surf 11:401
4. Khan A, Fontell K, Lindman B (1985) Progr Coll & Polym Sci 70:30
5. Lindström B, Khan A, Söderman O, Kamenka N, Lindman B (1985) J Phys Chem 89:5313
6. Khan A, Jönsson B, Wennerström H (1985) J Phys Chem 89:5180
7. Biresaw G, McKenzie DC, Bunton CA, Nicoli DF (1985) J Phys Chem 89:5144
8. Sepulveda L, Cortés J (1985) J Phys Chem 89:532
9. Wennerström H, Lindblom G, Lindman B (1974) Chem Scr 6:97
10. Stilbs P, Moseley ME (1980) Chem Scr 15:176
11. Stilbs P (1987) Progress in NMR Spectroscopy 19:1
12. Rosevear FB (1968) J Soc Cosmet Chem 19:581
13. Götz KG (1961) Dissertation, Göttingen
14. Wolff T, Bünau GV (1984) Ber Bunsenges Phys Chem 88:1098
15. Khan A, Fontell K, Lindblom G (1982) J Phys Chem 86:383
16. Mukerjee P, Mysels KJ (1971) Natl Stand Ref Data Ser, US, Natl Bur Stand, no 36

17. Czerniawski M (1966) Roczn Chem 40:1935
18. Ralston HV, DuBrow PL, Egenberger DN, Harwood HJ (1947) J Am Chem Soc 59:2095
19. Evans HC (1956) J Chem Soc 579
20. Hoffmann H, Ulbricht W (1977) Z Phys Chem 106:167
21. Hoffmann H, Tagesson B (1978) Z Phys Chem 110:8
22. Lindman B, Puyal MC, Kamenka N, Rymdén R, Stilbs P (1984) J Phys Chem 88:5048
23. Fabre H, Kamenka N, Khan A, Lindblom G, Lindman B (1980) J Phys Chem 84:3428
24. Lindman B, Puyal MC, Kamenka N, Brun B, Gunnarsson G (1982) J Phys Chem 86:1702
25. Lekkerkerker HNW, Dhont JKG (1984) J Chem Phys 80:5790
26. Ohtsuki T, Okano K (1982) J Chem Phys 77:1443
27. Evans GT, James CP (1983) J Chem Phys 79:5553
28. van Megen W, Snook I (1984) J Chem Soc Faraday Trans 2 80:383
29. Snok I, van Megen W, Tough RJA (1983) J Chem Phys 78:5825
30. Tanford C (1973) In: The Hydrophobic Effect, Wiley, New York
31. Ulmius J, Wennerström H (1977) J Magn Reson 28:309
32. Lindman B, Söderman O, Wennerström H (1986) In: Zana R (ed) Surfactant Solutions, Vol 22, Marcel Dekker Inc, New York, p 263
33. Gunnarsson G, Jönsson B, Wennerström H (1980) J Phys Chem 84:3114
34. Lindman B, Wennerström H (1980) Top Curr Chem 87:1
35. Wennerström H, Jönsson B, Linse P (1982) J Chem Phys 70:4665
36. Jönsson B, Linse P, Åkesson T, Wennerström H (1984) In: Mittal KL, Lindman B (eds) Surfactants in Solution, Vol 3, Plenum Press, New York, p 2023
37. Guldbrand L, Jönsson B, Wennerström H, Linse P (1984) J Chem Phys 80:2221
38. Fontell K, Mandell L, Lehtinen H, Ekwall P (1968) Acta Polytech Scand Chem Incl Met Ser 74:111
39. Balmbra RR, Clunie JS, Goodman JF (1969) Nature, London 222:1159
40. Ekwall P (1975) In: Brown GH (ed) Advances in Liquid Crystals, Vol 1, Chap 1, Academic Press, London
41. Israelachvili JN (1985) In: Intermolecular and Surface Forces, Academic Press, London New York
42. Jönsson JB, Wennerström H (1987) J Phys Chem 91:338
43. Mitchell DH, Tiddy GJT, Waring L, Bostock T, McDonald MP (1983) J Chem Soc Faraday Trans 1 79:975

Received December 30, 1986;
accepted January 20, 1987

Authors' address:

Dr. Ali Khan
Division of Physical Chemistry 1
Chemical Center
University of Lund
P. O. Box 124
S-22100 Lund, Sweden

Charge carrier induced reactions in colloidal semiconductor systems

M. Grätzel

Institut de Chimie Physique, Ecole Polytechnique Fédérale, Lausanne, Switzerland

Abstract: A survey of charge carrier induced reactions in colloidal dispersions of semiconductor particles is given. Band gap excitation of the semiconductor produces electron-hole pairs which diffuse to the surface, where they undergo redox reactions with substrates or are trapped by suitable catalysts. The dynamics of electron or hole ejection from the semiconductor across the Helmholtz layer to acceptors in solution has been analyzed, yielding quantitative information on the rate of interfacial charge transfer. The inverse process, i. e. electron injection from donors into the conduction band of the semiconductor, has also been examined; the most prominent example being the sensitization of wide band gap semiconductors. Reactions discussed include the photochemical cleavage of hydrogen sulfide and water, the photo-uptake of oxygen on colloidal oxides as a means of O_2 activation and the reduction of nitrite to ammonia.

Author's address:

Michael Grätzel, P.h.D., Professor of Chemistry,
Institut de Chimie Physique
Ecole Polytechnique Fédérale
CH-015 Lausanne, Switzerland

The double-helix of DNA as a colloidal matrix for photosensitized electrical charge separation

P. Fromherz, G. Reinbold, B. Rieger, and C. Röcker

Abteilung Biophysik der Universität Ulm, Ulm-Eselsberg, F.R.G.

Abstract: The organization of photochemical reactions in micelles of surfactants suffers from the dynamic nature of those assemblies and the ill partitioning of hydrophilic and hydrophobic regions. We consider the double-helix of DNA as some liquid-crystalline polymerized micelles with a more static and better defined core/coat structure.

We have studied the photoinduced electron transfer from intercalated ethidium to condensed methylviologen. The catalytic factor of the matrix is half a million, mainly due to the stoichiometric accumulation of methylviologen by the phosphate backbone. The time constant of the forward reaction is around 1 ns, that of the recombination around 1 ms.

We have further studied energy transfer from fluorescent donors localized in the small groove to intercalated ethidium. Most efficient one-dimensional antennas are built up as described by Perrin-Förster transfer.

The construction of sequention energy-electron transfer systems with chemical stabilization of separated charges is attempted, in defined finite synthetic DNA sequences.

Authors' address:

Prof. Dr. Peter Fromherz
Abteilung Biophysik der Universität Ulm
D-7900 Ulm-Eselsberg, F.R.G.

Kinetics of colloid particle transfer to interfaces

Z. Adamczyk

Institute of Catalysis and Surface Chemistry, Polish Academy of Sciences, Cracow, Poland

Abstract: The phenomenology of colloid and suspension particle transfer and interactions with boundary surfaces (collectors) is presented. Various forces affecting particle transport are discussed, including hydrodynamic and external ones acting over large separations and specific surface forces acting on molecular distances. The role of dynamic tangential interactions which are not predicted from the classical DLVO theory in particle adsorption and deposition phenomena is elucidated. Various theories aimed at describing quantitatively colloid particle transfer and adsorption rates are presented, including the Levich-Smoluchowski convective diffusion theory, the surface boundary-layer approach and adsorption theories. The range of validity of these approximations is estimated using results of exact numerical solutions of mass balance equations in which the forced convection, hydrodynamic wall correction effects, external and specific forces are considered [1]. It is shown that all these approximations fail for particle size larger than about 0.1 μm and the discrepancy increases for high flow rates (Reynold's number).

By contrast, our general theory [1] is applicable for the suspension size range as well, predicting a minimum deposition rate for particle size around 1 μm for all collectors immersed in stagnation point flows, e. g., spherical, cylindrical, rotating disc, etc. For larger colloid particles and suspensions, the theory predicts a very significant effect of electrical double-layer forces which can increase deposition rates many times for low ionic strength (10^{-4} M and less) when the zeta potential of particles and collector surface are of opposite signs.

These theoretical predictions are compared with experimental results obtained by the direct microscope observation method using the stagnation point-flow cell [2, 3]. Monodisperse latex suspensions of particle size ranging from 0.2 to 4 μm were used and the collector surfaces were made of glass and mica modified by an amino silane. The kinetics of particle deposition was studied systematically as a function of particle size, bulk number concentration, and flow intensity (Reynolds number). Also the geometric blocking effect appearing at surface coverage larger than a few per cent of a monolayer coverage was investigated. These and previous experiments [4] performed for mineral particles of high specific density confirm that the general convective diffusion theory [1] can adequately describe particle transfer and deposition rates for a broad range of particle size and Reynolds number for various flow and interface geometries. No adjustable empirical parameters are required to attain an agreement with experimental results in the case of the bulk transport controlled regime (barrierless deposition). On the other hand, the kinetics of surface barrier limited transport can only be properly described when using some empirical parameters. Also, in order to account for the high surface coverage deposition kinetics, one has to introduce some Langmuir-type kinetic models containing adjustable constants.

References

1. Adamczyk Z, Dąbroś T, Czarnecki J, van de Ven TGM (1983) Adv Coll Interf Sci 19:183
2. Dąbroś T, van de Ven TGM (1983) Coll & Polym Sci 261:694
3. Adamczyk Z, Zembala M, Siwek B, Czarnecki J (1986) J Coll Interf Sci 110:188
4. Adamczyk Z, Pomianowski A (1980) Powder Techn 27:125

Author's address:

Zbigniew Adamczyk
Institute of Catalysis and Surface Chemistry
Polish Academy of Sciences
30-239 ul. Niezapomiajek
Cracow, Poland

Progress in Colloid & Polymer Science   Progr Colloid & Polymer Sci 73:182 (1987)

Adsorption of surfactants on kaolinite

J. Garnes, H. Høiland[1]) and A. Skauge

Norsk Hydro Research Centre, Bergen, Norway
[1]) Department of Chemistry, Bergen-University, Bergen, Norway

Abstract: The adsorption isotherms of two anionic surfactants (sodium dodecyl sulfate, SDS and sodium dodecanoate, SDC), one cationic (dodecyl trimethylammonium bromide, DTAB) and two nonionics (nonylphenol polyoxyethylene, NPEOH, and dodecyl polyoxyethylene, DEOH) on kaolinite have been determined at 35 °C. The ionic strength was 0.05 M NaCl for all the isotherms. By selecting surfactants with similar hydrophobic structure, we can compare the effect of the hydrophilic head-group on the adsorption properties.

In addition to the static adsorption results, electrophoretic mobility of the kaolinite particles, with and without surfactants adsorbed, was measured as a function of pH by the Laser-Doppler technique.

The kaolinite particles have a net negative surface charge at neutral pH. The mobility data combined with XRD, SEM and BET surface area measurements suggest that the particles can be regarded as cylindrical disks with a positive charge on the edges, covering 20 % of the total surface area.

All the isotherms with exception of DEOH show a maximum adsorption level at CMC. For dodecyl polyoxyethylene the adsorption density increases above the cmc, probably due to reorientation of the adsorbed layer.

DTAB show a Langmuirian type of isotherm. However, there is a high positive surface charge of the kaolinite/DTAB from the electrophoretic mobility data, and together with the level of adsorption density, DTAB must exceed monolayer adsorption. The isotherm for the other nonylphenol nonionic is also a Langmuirtype, but deviates from Langmuirian at very low concentrations.

Calculation based on the plateau level adsorption show that both DTAB and NPEOH exceed monolayers. The anionic surfactants SDS and SDC both approximates monolayers, while the nonionic, DEOH, has a plateau level adsorption which is below that of monolayer adsorption.

The electrophoretic mobilities show that the nonionics preferably adsorb on the negative sites. Comparing the nonionics, the nonylphenols adsorb more strongly than dodecyl polyoxyethylenes. The higher adsorption of the nonylphenol cannot be explained by a difference in ethoxylation.

SDS adsorb primarily on the positive edge surface of the kaolinite particles.

The isotherm of sodium dodecanoate is measured at the natural pH of the carboxylates. Abvoe pH 7.8 the edge surface on the kaolinite will have a net negative charge. The high adsorption density and the electrophoretic mobility of kaolinite/dodecanoate indicate that SDC adsorb on the negatively charged surface or there must be a strong ion exchange with hydroxyl ions on the edge surface.

The level of adsorption density for anionic, cationic and nonionic surfactants having the same hydrophobic structure can be ranked by the adsorption maximum:

Cationic > Anionic > Nonionic.

The solution concentration of the surfactants at the plateau level are different, and if the adsorption is ranked as a fraction of the solution concentration, the order would be:

Nonionic > Cationic > Anionic.

In addition to the static adsorption, we find that the electrophoretic mobility data improves the possibilities of interpreting the adsorption isotherms.

Authors' address:

Jan Magne Garnes
Norsk Hydro Research Centre
N-5001 Bergen, Norway

Progress in Colloid & Polymer Science　　　　　　　　Progr Colloid & Polymer Sci 73:183 (1987)

The paramagnetic relaxation in structural and dynamic studies on lyotropic systems

T. Ahlnas[1]), C. Chachaty, A. Faure, J. P. Korb[2]), H. Néry[3]), and A. M. Tistchenko

Département de Physico-Chimie, C.E.N. de Saclay, Gif-sur-Yvette cedex, France, and
[1]) Physical Chemistry I, Sweden, and
[2]) Laboratoire de Physique de la Matière Condensée, Ecole Polytechnique, Palaiseau, France, and
[3]) Laboratoire de Méthodologie RMN, Université de Nancy I, Vandoeuvre cedex, France

Abstract: Paramagnetic probes have proved useful for obtaining structural, dynamic and conformational information on surfactants in direct micelles. We have developed the paramagnetic relaxation technique for investigating other lyotropic systems such as the reversed micellar and lamellar ones.

We report recent results on inverted micelles of sodium diethylhexyl phosphate/ water/benzene or cyclohexane and on the lamellar liquid crystalline phases of sodium diethyl-hexyl and dibutyl phosphate/water, using Mn(2+) and VO(2+) as paramagnetic probes.

To interpret the paramagnetic relaxation in the inverted micellar systems, the inter- and intramolecular contributions have been determined by a computational procedure, taking into account the statistical weight of all the surfactant conformers. The geometry of binding of the paramagnetic ions to the polar head is determined from the relaxation of the phosphorus and of the first carbons of the chains.

The population analysis of the surfactant conformers undergoing gauche-trans isomerizations gives, particularly, the chain segment density distribution from the center of the polar core of reversed micelles. Filling the surfactant layer (palisade) to constant hydrocarbon density enables an estimation of the solvent penetration. The measured paramagnetic relaxation rates of the solvent nuclei confirm the predicted extent of solvent penetration and provide, moreover, the diffusion coefficient of the solvent in the hydrocarbon layer of the micelle, which is nearly one tenth that of the neat solvent.

The order parameters of the C-H bonds in the surfactant alkyl chains were extracted from multifield carbon 13, deuteron and phosphorus relaxation experiments in diamagnetic micellar solutions. Consistence with the order parameters deduced from the paramagnetic relaxation are obtained for only one possible orientation of the local director.

A similar approach is used in the investigation of the conformations and dynamic behaviour of surfactants in the lamellar phases. Due to the bidimensionality of the system, a large intermolecular contribution to the paramagnetic relaxation is expected [1]. This was experimentally verified by the field dependence of the relaxation rates. The intermolecular contribution is related to the mean lateral diffusion of the surfactant and of the probe. In the two lamellar phases under study, a surfactant diffusion coefficient of ca. 2×10^{-7} cm^2/s is obtained in good agreement with other NMR methods [2].

References

1. Korb JP (1985) J Chem Phys 82:1061
2. Chachaty C, Quaegebeur JP, Caniparoli JP, Korb JP (1986) J Phys Chem 90:1115

Authors' address:

C. Chachaty
Département de Physico-Chimie
C.E.N. de Saclay, F-Allai
Gif-sur-Yvette Cedex, France

Progress in Colloid & Polymer Science   Progr Colloid & Polymer Sci 73:184 (1987)

Center of mass and rotational diffusion in concentrated suspensions of rodlike macroparticles

W. Hess

Fakultät für Physik, Universität Konstanz, Konstanz, F.R.G.

Abstract: For a model of hard rods, the influence of the excluded volume effects on the rotational and the center of mass diffusion coefficient are calculated. Both quantities are found to decrease monotonously as $(c\,d\,L^2)^{-1}$, where c is the concentration, d the diameter and L the length of a rod, as long as there is no structural order in the system. This result differs from the Doi-Edwards theory [1], which has so far been mostly used for the interpretation of the dynamic properties of interacting rods, but it is in good agreement with measurements of the rotational diffusion coefficient of Zero and Pecora [2] and of the c.o.m. diffusion coefficient by Maret [3], if one assumes that the effective diameter of the rods is of the order of five times the geometric diameter.

References

1. Doi M, Edwards SF (1978) J Chem Soc Faraday Trans 2, 74:560,1789
2. Zero KM, Pecora R (1982) Macromolecules 15:82
3. Maret G, to be published

Author's address:

Walter Hess
Fakultät für Physik
Universität Konstanz
D-7750 Konstanz, F.R.G.

HNC study of colloidal solutions within the primitive model — Phase separation

L. Belloni

CEA – IRDI/Départment de physico-Chimie, CEN-Saclay, Gif-Sur-Yvette cedex, France

Abstract: In the framework of the primitive model, the colloidal solutions are considered as mixtures of charged hard-spheres immersed in a continuous dielectric solvent. The small ions are treated explicitly. The equilibrium and structural properties of such multi-components systems are obtained with the help of the HNC integral equation for a large range of size and charge dissymmetry. In the strong interaction regime, the compressibility equation leads to a spinodal decomposition at low concentration. The phase diagram and the critical point are calculated for different colloidal charges. It appears that the strong coulombic attraction between polyions and counterions induces an effective "solvatation" of a pair of colloids. This phenomenom, which cannot be obtained with the classical one-component DLVO approach, should be of interest in the theory of colloidal stability since a phase transition is expected in absence of Van der Waals forces.

Author's address:

Luc Belloni
CEA-IRDI
Département de physico-Chimie
CEN-Saclay
F-91191 Gif-Sur-Yvette Cedex, France

Progress in Colloid & Polymer Science   Progr Colloid & Polymer Sci 73:185 (1987)

Evanescent wave photon correlation spectroscopy to study Brownian diffusion close to a wall

K. H. Lan, N. Ostrowsky, and D. Sornette

Laboratoire de Physique de la Matière Condensée, CNRS UA 190, Université de Nice, Nice Cedex, France

Abstract: We report the first complete light scattering study of the dynamics of free Brownian particles, in the immediate proximity of a reflecting wall. The particles were probed by an evanescent wave with variable penetration range, and their scattered light analyzed at different angles with a photon correlation technique. We have shown [1] that the measured correlation spectrum, very different from the single exponential expected in the bulk geometry, may be simply and completely interpreted in terms of the combined wall's mirror effect and evanescent wave geometry, with no adjustable parameter. The spectrum is actually the product of two correlation functions. The first ("transverse") decays with the time it takes for a particle to diffuse in a plane parallel to the wall on a length scale $q_{\parallel}^{-1}$, where $q_{\parallel}$ is the component of the scattering wave vector parallel to the wall. The second one ("longitudinal") involves two length scales, the penetration depth ξ and the inverse of the longitudinal component of the scattering wave vector, q_z^{-1}. For $q_z\xi \leq 1$, this function has a long tail ($t^{-1/2}$ behavior), interpreted as a finite size effect, which strongly modifies the spectrum if the "transverse" time is slow enough. This effect is important and should be present in all experiments attempting to measure fluctuation dynamics at interfaces; it must therefore be accounted for and distinguished from other factors affecting the dynamics close to a surface, and due to the various wall/fluid interactions generally referred to as "surface effects".

Futhermore, we expect important changes in the correlation function when the longitudinal length scale of the problem (particle's radius, wall/suspension interaction range, etc.) becomes of the order of the optical penetration length ξ. This type of study is currently being done with Brownian particles of radius $R \sim \xi$, which should enable us to recover the z dependance of the diffusion coefficient parallel and perpendicular to the wall. It may also be quite relevant for studying surface transition dynamics, which can be viewed as the diffusion of a self-similar distribution of droplets.

References

1. Lan KH, Ostrowsky N, Sornette D (1986) Phys Rev Let 57:17

Author's address:

K. H. Lan
Laboratoire de Physique de la Matière
Condensée (CNRS UA 190) Université de Nice
Parc Valrose
F-060 34 Nice Cedex, France

Diffusion of polystyrene latex spheres in polymer solutions studied using photon correlation spectroscopy

W. Brown

Institute of Physical Chemistry, University of Uppsala, Uppsala, Sweden

Abstract: Although the diffusion of latex spheres in polymer solutions has been extensively investigated, there is some confusion regarding the interpretation of the data owing to the presence of several interacting effects. These include adsorption of the polymer, charge interactions, and steric effects such as obstruction, and these are related not only to the respective concentrations but also to the relative sizes of the components as well as their composition.

The diffusional behaviour of carboxylated polystyrene latex spheres in both dilute and semidilute solutions of various high molar mass, stiff chain polymers is described. In very dilute solutions of the polymer ($\pm 10^{-3}$ %) the average latex particle mobility decreases strongly with increasing polymer concentration; the relative variance shows a concomitant increase. Multiexponential analysis of the time correlation function leads to a bimodal fit as the simplest model:

a) Fast mode: a polymer concentration-independent D_{latex} value equal to that of the isolated latex particle at infinite dilution.

b) Slow mode: a diffusion coefficient corresponding to aggregate species with apparent radii 2 to 3 times larger than that of the latex monomer. The aggregation behavior is consistent with a bridging mechanism.

In semidilute solutions of non-ionized carboxymethyl cellulose (pH 4) in the salt-free system the interactions are apparently minimal. The latex diffusion follows the functional relationship predicted by Ogston [1] and Cukier [2]: $D/D_o = \exp(-AC^{1/2})$. However, with carboxymethyl cellulose in the fully-neutralized form at pH 9, electrostatic forces lead to deviations from this law. Similarly, in the case of the non-ionizable polymer hydroxyethyl cellulose there are strong departures from the above relationship but in this case due to adsorption of the polymer.

The results are analyzed in the light of current theoretical models and the way in which adsorption/aggregate formation can be taken into account is indicated.

References

1. Ogston AG, Preston BN, Wells JD (1973) Proc Soc Lond, A 333:297
2. Cukier RI (1984) Macromolecules 17:252

Author's address:

Wyn Brown
Institute of Physical Chemistry
University of Uppsala
Box 532
S-751 21 Uppsala, Sweden

Progress in Colloid & Polymer Science   Progr Colloid & Polymer Sci 73:187–188 (1987)

Static and dynamic properties of solutions of strongly interacting micelles

L. Cantu, M. Corti[1]), and V. Degiorgio[1])

Dipartimento di Chimica e Biochimica Medica, Università di Milano, Milano, Italy
[1]) Dipartimento di Elettronica, Sezione di Fisica Applicata, Università di Pavia, Pavia, Italy

Abstract: In the last few years considerable effort has been devoted to understanding the behaviour of interacting Brownian particles [1, 3]. If we confine our attention to scattering experiments performed with globular particles, we find that the list of the systems studied so far by light or neutron scattering includes large colloidal particles (polystyrene latex spheres) [4, 5], small proteins [6, 7], micelles [8, 11] and microemulsions [12].

The interpretation scheme for static scattering data is well established. The structure factor $S(k)$, related to the radial distribution function $g(r)$, can be calculated once the pair interaction potential $V(r)$ is known. Depending on the specific system under investigation, various approximations have been used to derive $g(r)$: the dilute gas approximation [8], the mean spherical approximation (MSA) [11], the rescaled MSA [13] the hypernetted chain approximation [7, 10, 14]. The main problem in a real system is to assign precisely the potential $V(r)$. From this point of view, the simplest system to study is a dilute collection of electrically charged identical particles in a low ionic strength solution, where the dominant contribution to $V(r)$ comes from the Coulomb repulsion and short range interactions can be neglected. Solutions of ionic amphiphiles can be good candidates for such a study, provided that their critical micelle concentration is sufficiently low.

The dynamics of interacting Brownian particles represents a very difficult theoretical problem, because the calculation of the dynamic structure factor $F(k, t)$ has to take into account not only $V(r)$, but also hydrodynamic interactions. In most cases [6, 10] the measured $F(k, t)$ is an exponential with time constant $(k^2 D)^{-1}$, where D is a collective diffusion coefficient, but in the most general situation [1–3] the shape of $F(k, t)$ may be nonexponential.

In the following we report on light scattering measurements performed on ionic micelles of biological glycolipids in very low ionic strength aqueous solution [15]. Our experiment is similar to the one performed by Cannell and coworkers [7] who studied dilute solutions of a globular protein, bovine serum albumine (BSA), at very low concentrations of added salt. Our system, however, differs fundamentally with respect to solutions of macromolecules or inorganic colloids, because the micelle is a spontaneous aggregate, its electric charge is unknown a priori, and the system may present an intrinsic polydispersity. The static and dynamic light scattering data are compared with calculations based on the HNC approximation.

Gangliosides are anionic amphiphilic glycolipids occurring in neuronal plasma membranes which form micelles in water above a very small critical micelle concentration, around $10^{-6} - 10^{-9}$ M (see [16]).

The effect of NaCl addition to the ganglioside Gm1 solution was studied at fixed ganglioside concentration. We show in Figure 1 the behaviour of the scattered intensity as a function of the ionic strength for GM1 solutions at the concentration of 0.5 mM and 1mM. The reported values are normalized by the intensity scattered from the ideal solution, derived by measuring the scattered intensity at various GM1 concentrations in 30 mM NaCl solutions and extrapolating to zero micelle concentration. The micelle molecular weight M is 470000 with an aggregation number $m = 300$.

The correlation functions, obtained at the scattering angle of 90° for various NaCl concentrations, are analyzed by the standard cumulant fit, which gives the diffusion coefficient D and the relative variance v. The ratio D_o/D is reported in Figure 2 as a function of the ionic strength I for two distinct GM1 concentrations, D_o being the diffusion coefficient of the individual GM1 micelle. D_o is derived by measuring D at various GM1 concentrations and extrapolating to zero micellar concentration (micellar hydrodynamic radius $R_H = 5.9$ nm).

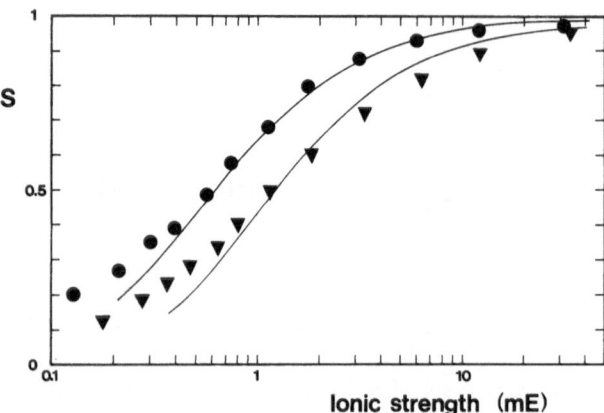

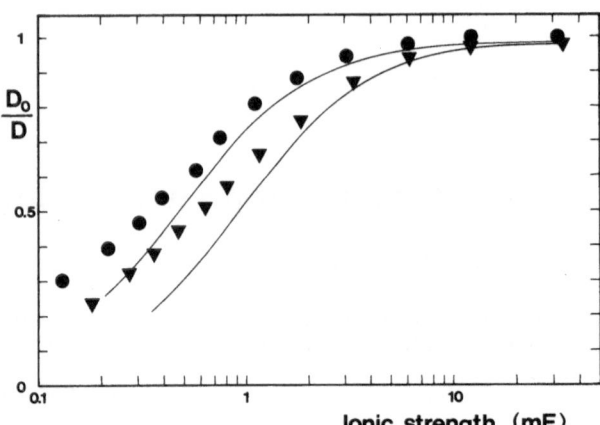

Fig. 1. Structure factor S of 0.5 mM (●—●) and 1 mM (▲—▲) GM1 solutions measured as a function of the ionic strength I at 25 °C and scattering angle of 90°. The full curves represent theoretical results

Fig. 2. The quantity D^*/D plotted versus the ionic strength for 0.5 mM (●—●) and 1 mM (▲—▲) GM1 solutions. The full curves represent theoretical results calculated as described in the text.

Since the micelle size and shape do not change with the ionic strength (see [15] and [16]), the quantity plotted in Figure 1 is the static structure factor $S(k)$ evaluated at $k = 0.023$ mm^{-1}.

We derived the theoretical structure factor $S(k)$ by using the HNC approximation for $g(r)$, with the pair interaction potential $V(r)$ consisting of a hard core repulsion plus the DLVO screened Coulomb potential.

The full curves of Figure 1 are obtained with a micellar charge $Q = 48$ electronic units.

The full curves in Figure 2 are derived by inserting the calculated $g(r)$ into the expression given by Ackerson [17] which considers hydrodynamic interactions at the Oseen level, by using for Q the best fit value derived from the static data, $Q = 48$, and by applying the correction factor proposed by Belloni and Drifford [18] to take into account the non-negligible size of small ions. As discussed in more detail in Reference [10], the trend is correct, but the agreement is not as good as for the static data.

References

1. Degiorgio V, Corti M, Giglio M (eds) (1980) Light Scattering in Liquids and Macromolecular Solutions, Plenum, New York
2. Pusey PN, Tough RJA (1985) In: Pecora R (ed) Dynamic Light Scattering and Velocimetry: Applications of Photon Correlation Sepctroscopy, Plenum, New York
3. Hess W, Klein R (1983) Adv Phys 32:173
4. Brown JC, Pusey PN, Goodwin JW, Ottewill RH (1975) Phys A8:664
5. Gruner F, Lehmann W, in Reference 1
6. Doherty P, Benedek GB (1975) J Chem Phys 61:5426
7. Cannell DS (1985) In: Degiorgio V, Corti M (eds) Physics of Amphiphiles: Micelles, Vesicles and Microemulsions, North Holland, Amsterdam, p 202; Neal DG, Purich D, Cannell DS (1984) J Chem Phys 80:3469
8. Corti M, Degiorgio V (1981) J Phys Chem 85:711
9. Nicoli DF, Dorshow RB (1985) In: Degiorgio V, Corti M (eds) Physics of Amphiphiles: Micelles, Vesicles and Microemulsions, North Holland, Amsterdam, p 429
10. Cantù L, Corti M, Degiorgio V (1986) Europhys Lett 2:673
11. Hayter JB, Penfold J (1981) J Chem Soc Faraday Trans I 77:1851; Bendedouch D, Chen SH, Koehler WC (1983) J Phys Chem 87:2621; Triolo R, Hayter JB, Magid LG Johnson JS (1983) J Chem Phys 79:1977
12. Vrij A, Nieuwenhuis AE, Fijnaut HM, Agterof WGM (1978) Faraday Disc Chem Soc 65:101; Cazabat AM, Langevin D (1981) J Chem Phys 74:3148
13. Hansen JP, Hayter JB (1982) Molec Phys 46:651
14. Schaefer DW (1977) J Chem Phys 66:3980
15. Cantù L, Corti M, Degiorgio V, For a more complete version of the work present her, Faraday Disc Chem Soc, 83, to be published
16. Cantù L, Corti M, Sonnino S, Tettamanti G (1986) Chem Phys Lipids 41
17. Ackerson BJ (1976/1978) J Chem Phys 64:242; 69:684
18. Belloni L, Drifford M (1985) J Physique Lett 46:1183

Authors' address:

Laura Cantù
Dipartimento di Chimica e Biochimica Medica
Università di Milano
I-20133 Milano, Italy

Progress in Colloid & Polymer Science   Progr Colloid & Polymer Sci 73:190 (1987)

Micelle formation of ionic amphiphiles: Thermochemical test of a thermodynamic model

I. Johnson[1]), G. Olofsson[1]) and B. Jönsson[2])

[1]) Division of Thermochemistry, and
[2]) Division of Physical Chemistry 1, Chemical Center, University of Lund, Lund, Sweden

Abstract: A thermodynamic model for the association of aqueous ionic surfactants has been developed by Jönsson and Wennerström [1, 2]. The treatment is based on a model expression for the free energy containing five contributions: a hydrophobic energy, a surface free energy of aggregation, an electrostatic free energy calculated from the Poisson-Boltzmann equation, an entropy of mixing of the micelles, and the condition that the aggregate size is limited in at least one dimension to the length of the extended amphiphile molecule. Recently, the thermodynamic model has been extended to ternary water-surfactant-alcohol systems [3].

In order to test the model, a careful study of the micellization of sodium dodecyl sulphate in pure water and in aqueous alcohol solutions has been made.

Differential enthalpies of dilution in water of 28 wt % aqueous sodium dodecyl sulphate solution have been measured using a titration microcalorimeter. In this way titration curves were determined in the concentration range 0 to 0.040 mol l^{-1} at three different temperatures. Theoretical titration curves are calculated using values derived from the thermodynamic model for concentrations of the amphiphile in the monomer and micellar state. The agreement between the calculated and experimental curves is good, which shows that the model can give quantitative predictions of the micellization process. It is observed that the decreasing monomer concentration with increasing amphiphile concentration above the CMC requires the quantitative evaluation of the composition of the reaction solutions in order to derive correct values for the enthalpy of micelle formation of ionic amphiphiles from results of calorimetric measurements.

In pentanol solutions the predicted decrease in CMC with increasing alcohol concentration is clearly seen in the titration curves. The measured enthalpy changes for the formation of the (mixed) micelles show a pronounced variation with alcohol concentration.

References

1. Gunnarsson G, Jönsson B, Wennerström H (1980) J Phys Chem 84:3114
2. Jönsson B, Wennerström H (1981) J Coll Interf Sci 80:482
3. Jönsson B, Wennerström H (1987) J Phys Chem 91:338

Authors' address:

G. Olofsson
Division of Thermochemistry
Chemical Center
University of Lund
Box 124
S-22100 Lund, Sweden

Progress in Colloid & Polymer Science   Progr Colloid & Polymer Sci 73:189 (1987)

Particular physical properties of liquid boundary layers on solid surfaces

G. Peschel

Institute of Physical and Theoretical Chemistry of the University of Essen, Essen, F.R.G.

Author's address:

Prof. Dr. G. Peschel
Institute of Physical and Theoretical
Chemistry of the Unversity of Essen
Universitätsstraße 57
D-4300 Essen, F.R.G.

Abstract: Experimental evidence has accumulated that liquid boundary layers on solid surfaces display a structure different from that of the bulk. We have devised a method to investigate the structural behavior of liquid interlayers which are created by the overlap of boundary layers. The method is based on the approach of two solid bodies, in the form of two cylindrical high-grade polished plates, in the test liquid. The one plate is flat, the other spherically formed. The flat plate is mounted on the bottom of a vessel containing the test liquid, the spherical plate is attached to a balance which can be deflected by aid of computer control. Thus, the plate separation distance can be varied down to ca. 1 nm; it can be determined by a displacement transducer connected with a strain gauge measuring bridge. Applying aqueous electrolyte solutions as test liquids a repulsion is found at separations of the order of 100 nm and smaller. This repulsion is due to the overlap of electrical double layers. At much smaller separations of the order of 7 nm a steeper repulsion becomes evident, created by the interpenetration of immobilized multimolecular hydration layers on the solid surfaces. The 'hydration forces' can be converted into the 'structural disjoining pressure'. Generally the hydration forces show a non-monotonous dependence on the eletrolyte concentration. Tests with organic liquids as alkanes and alcohols exhibit repulsive effects due to solvation forces effective in thin interlayers. Correlations exist between the decay length of the solvation forces and the bulk liquid structure.

Viscosity and quasielastic light scattering study of micellar solutions of hexadecyltrimethyl ammonium bromide in presence of 0.1 M KBr

E. Hirsch, S. J. Candau, R. Zana[1], and M. Adam[2]

Laboratoire de Spectrométrie et d'Imagerie Ultrasonores, Unité Associée au CNRS N° 851
[1] Insitut Charles Sadron (CRM-EAHP), CNRS/ULP, Strasbourg Cedex, France, and
[2] Laboratoire Léon Brillouin, C.E.N. Saclay, Gif-Sur-Yvette, France

Abstract: Micellar solutions of hexadecyltrimethyl ammonium bromide (CTAB) solutions in $H_2O - 0.1$ M KBr have been investigated by quasielastic light scattering and viscosity. In order to measure the zero shear viscosity and the longest relaxation time was used a magnetorheometer. The measurements were performed as a function of temperature (30°t 60°C) and surfactant concentration (10^{-3} M to 0.84 M). The results are consistent with a model of long flexible micelles described elsewhere. In the semi-dilute range, the elastic modulus is nearly constant, at given surfactant concentration, when the temperature is increased, whereas the viscosity and the relaxation time decrease markedly. In this range, the behavior can be qualitatively described using an analogy with an entangled network of polymer chains.

Authors' address:

E. Hirsch
Laboratoire de Spectrométrie et d'Imagerie Ultrasonores
Unité Asociée au CNRS n° 851
Université Louis Pasteur
4, rue Blaise Pascal,
F-67070 Strasbourg Cedex, France

Progress in Colloid & Polymer Science   Progr Colloid & Polymer Sci 73:191 (1987)

Migration of small hydrophobic molecules between micelles in aqueous solution

M. Almgren, and J. Alsins

The Institute of Physical Chemistry, University of Uppsala, Uppsala, Sweden

Abstract: Triplet energy transfer from 9-methylanthracene to azulene or guajazulene has been used to probe the migration of the azulenes between micelles in aqueous solution. The migration of the hydrophobic solutes between small ionic and nonionic micelles is of the temperature dependence expected for a process controlled by the diffusion through the intermicellar solution, although the rate in some cases is substantially smaller than calculated from the Smoluchowski equation. Under conditions in which the micelles grow into large, probably rod-like structures there are severe difficulties in separating the inter- and intramicellar deactivation processes. The intermicellar migration is enhanced under these conditions, in cetyltrimethylammonium surfactants on addition of chlorate ions, and in hexaethyleneglycoldodecylether at temperatures approaching the cloudpoint. The mechanism of this migration is discussed and compared with pertinent results from micelle relaxation kinetics and surfactant self-diffusion measurements.

Previous reports by others on a very rapid migration of pyrene in similar systems are shown to be probably due to a misinterpretation of intramicellar quenching processes.

References

1. Almgren M, Alsins J, Prog. Coll Poly Sci, in press

Authors' address:

M. Almgren
The Institute of Physical Chemistry
University of Uppsala
P.O. Box 532
S-75121 Uppsala, Sweden

Progress in Colloid & Polymer Science　　　　　Progr Colloid & Polymer Sci 73:192 (1987)

Liquid crystallinity in surfactant systems

A. Khan and A. Sadaghiani

Division of Physical Chemistry 1, Chemical Center, Lund University, Lund, Sweden

Abstract: The phase equilibria of surfactants in two- and three component systems were studied by water deuteron NMR, polarizing microscope and differential scanning calorimetric methods. The ternary phase diagrams of ionic surfactants of types $C_8SO_4^-$, $C_{12}SO_4^-$ with Ca^{2+}, Mg^{2+} and Ba^{2+} counterions and of hexadecyltrimethylammoniumsulphate with water and decanol; and the binary phase diagrams of calcium, magnesium and barium di-2-ethylhexylsulphosuccinate with water, reveal that the homogeneous liquid crystalline and isotropic solution phases form in these surfactant systems have many characteristics which are different from those formed in the corresponding systems with monovalent counterions. The most important difference, however is, that the lamellar liquid crystals form, with divalent counterions, have limited swelling capability and these systems often lead to the formation of reverse hexagonal liquid crystal at high decanol and low water contents. The mixed ionic surfactant system sodium and calcium di-2-ethylhexylsulphosuccinate with water forms two lamellar liquid crystals; one is stable at high water contents, and the other at low water contents, and these two liquid crystals are found to coexist. A similar observation can be made for the ternary system calcium dodecylmonooxyethylenesulphate – decanol – water. The following important observations may be made from the study of phase equilibria of uncharged surfactants hexadecyltrimethylammonium n-alkyl carboxylates ($C_{16}A^+RA^-$, where R = 1,3,7,11,13 and 15 carbons) with water: (1) all surfactant systems form micellar solutions above their respective Krafft point; (2) when R = 1, the system forms cubic and hexagonal liquid crystals and for R = 2, only a hexagonal liquid crystal is obtained; (3) when R = 7, there appears to be a hexagonal liquid crystal occupying a very narrow zone immediately after the micellar solution region and a lamellar liquid crystal also occupied a narrow area at high surfactant concentration, the two-phase region (two liquid crystals) occupies a large region and (4) when R ⪴ 11, only the lamellar liquid crystall is found to exist. Some of the experimental findings observed for the ionic and uncharged surfactant systems may be explained qualitatively by existing theories.

Author's address:

Ali Khan
Division of Physical Chemistry 1
Chemical Center
Lund University
P. O. Box 124
S-22100 Lund, Sweden

Microstructure of microemulsions

R. Strey and M. Kahlweit

Max Planck Institut für Biophysikalische Chemie, Göttingen, F.R.G.

Abstract: Investigations of phase behavior and microstructure of microemulsions are greatly facilitated if one keeps the number of components as small as possible. For this reason we have studied ternary systems consisting of water, oil and nonionic amphiphile. While the essential features of the phase behavior of such systems are well understood [1], the question of how a few amphiphilic molecules manage to solubilize simultaneously hundreds of water and oil molecules is still unanswered. In this paper we report on results of a series of experiments performed in order to supply a basis for solving this problem, applying various techniques including small angle X-ray scattering (SAXS), elelctrical conductivity and freeze fracture electron microscopy. On the basis of detailed knowledge of the phase behavior, well defined paths were chosen along which either the concentration of the amphiphile was varied at constant H_2O/oil ratio, or the latter ratio was varied at constant concentration of amphiphile. The structure of the microemulsions is shown to evolve with increasing chain length of the amphiphile. For the system H_2O-n-octane-$C_{12}E_5$ it is found to be bicontinuous over a broad range of compositions. In particular, at comparable content of water and oil freeze fracture electron micrographs show that the water- and oil-rich regions are disordered, curved and three-dimensionally interconnected. These findings are consistent with recent SAXS results on the H_2O-n-tetradecane-$C_{12}E_5$ system [2].

References

1. Kahlweit M, Strey R (1985) Angew Chem 24:654
2. Lichterfeld F, Schmeling T, Strey R (1986) J Phys Chem 90:5762

Authors' address:

R. Strey
Max-Planck-Institut für Biophysikalische Chemie,
Postfach 28 41
D-3400 Göttingen, F.R.G.

A new model for microemulsion stability

J. Biais, J. L. Trouilly, B. Clin, and P. Lalanne

Centre de Recherche Paul Pascal, CNRS, Domaine Universitaire, Talence Cedex, France

Abstract: The proposed model deals with quaternary microemulsions made of water, oil, surfactant and cosurfactant. The difficulty of a theoretical stability study is reduced by making use of the pseudophase model which allows one to consider the system as a ternary one, made of three pseudophases (water, oil and membrane).

The free enthalpy function is derived from the regular solution model. It particularly considers the water-interface and oil-interface terms as well as a contribution from the bending energy of the interface. The interaction terms are related to the experimental solubilization parameter.

Phase diagrams have been calculated in pseudophase plan as well as in constant alcohol/surfactant ratio planes. Those diagrams are in excellent agreement with experimental ones already obtained for waterdecane-pentanol-n-dodecylbetaíne.

Authors' address:

J. Biais
Centre de Recherche Paul Pascal
CNRS, Domaine Universitaire
F-33405 Talence Cedex, France

Progress in Colloid & Polymer Science
Progr Colloid & Polymer Sci 73:194 (1987)

Structural comparison between non-aqueous and aqueous microemulsions from multicomponent self-diffusion measurements

K. P. Das, A. Ceglie and B. Lindman

Department of Physical Chemistry 1, Chemical Center, University of Lund, Lund, Sweden

Authors' address:

K. P. Das
Department of Physical Chemistry 1
Chemical Center
University of Lund
Box 124
S-22100 Lund, Sweden

Abstract: Microemulsions are known to be formed also with non-aqueous solvents like glycerol, formamide and its derivatives. The isotropic region is fairly large and is often larger than for the corresponding aqueous system. Although some structural information of aqueous microemulsions is available in literature, there is currently still no information on the structure of these non-aqueous microemulsions. From a comparison of phase behaviour of these non-aqueous systems with that of the aqueous ones, the microstructures in these optically isotropic solutions are often believed to be the same. But this approach of gaining information about the structure is too indirect. We have carried out multicomponent self-diffusion measurements in these systems by Fourier Transform NMR spin-echo technique and compared the results with those of aqueous ones. Our results reveal considereable difference in micro-structural behaviour in these systems. The self-diffusion data in the aqueous system show that these systems are quite structured when a cosurfactant like hexanol or octanol is used. The self-diffusion coefficients of all the components in the non-aqueous system are quite high, so the idea of segregation into distinct domains does not seem to hold good. The non-aqueous microemulsions thus seem to be less structured than the aqueous ones and are probably closer to the structureless simple solutions.

Enzymatic transesterification of a triglyceride in microemulsions

K. Holmberg and E. Österberg

Berol Kemi AB, Stenungsund, Sweden

Authors' address:

K. Holmberg
Berol Kemi AB
Stenungsund, Sweden

Abstract: Microemulsions based on aliphatic hydrocarbon, surfactant and aqueous buffer have been used as reaction medium for the lipase catalyzed transesterification of a triglyceride and a fatty acid. Both AOT (sodium bis (2-ethylhexyl)sulfosuccinate) and certain alcohol ethoxylates could be used as surfactant to produce a triglyceride having a fatty acid composition similar to that of natural cocoa butter from a palm oil distillation fraction. The nonionic surfactant gives a higher reaction rate than AOT, presumably due to a more favourable association of water in the microemulsion. Recovery of the enzyme is easier with the former surfactant. However, the ethoxylate is found to participate in an unwanted side reaction, viz, formation of esters with free fatty acids in the solution.

Fractal cluster formation and percolation in supermolecular fluids

H. Eicke, S. Geiger, R. Hilfiker, and H. Thomas[1]

Institut für Physikalische Chemie, Universität Basel, Klingelbergstraße 80, CH-4056 Basel, Schweiz, und
[1] Institut für Physik, Universität Basel, Klingelbergstraße 82, CH-4056 Basel, Schweiz

Abstract: We have investigated fractal cluster formation near the percolation threshold of water-in-oil microemulsions by various experimental methods. The three-component system water/AOT (surfactant)/isooctane forms a well-defined super-molecular fluid consisting of monodisperse spherical water droplets of ca. 10 nm radius in oil, covered by a monomolecular layer of AOT. The system is thermodynamically stable and responds reversibly to changes of the physical parameters.

A steep conductivity increase by several orders of magnitude in a narrow temperature (or concentration) interval clearly demonstrates percolation of the nanodroplets. Dielectric relaxation measurements in the frequency range 100 kHz to 1 GHz reveal the formation of fractal clusters, as the percolation threshold is approached. Their size distribution is studied by time-resolved electro-optic Kerr-effect and by light-scattering experiments, leading to a scaling model in good agreement with experimental results [1].

An applied electric field induces structural changes resulting in a connectivity increase of the fractal cluster network. This gives rise to an anomalous current response with very slow rise times between 100 and 400 μs and strongly non-linear field dependence. Analysis of the data yields a shift of the percolation threshold with the electric field, consistent with thre results of the electrooptic experiments.

References

1. Eicke H-F, Hilfiker R, Thomas H (1985) Chem Phys Lett 120:272; (1986) 125:295; Eicke H-F, Geiger S, Sauer FA, Thomas H (1986) Ber Bunsenges Phys Chem 90:872

Authors' address

H.-F. Eicke
Institute of Physical Chemistry
Klingelbergstraße 80
CH-4056 Basel, Switzerland

An application of the optical microscopy to the determination of the curvature elasticity modulus of biological and model membranes

I. Bivas[1], P. Hanusse, P. Bothorel, J. Lalanne, O. Aguerre-Chariol

Centre de Recherche Paul Pascal (C.N.R.S.) Domaine Universitaire, Talence Cedex, France
[1] Insitute of Solid State Physics, Bulgarian Academy of Sciences, Liquid Crystal Group, 72 Lenin Blvd, Sofia 1184, Bulgaria

Resume: On présente une nouvelle méthode permettant de mesurer le module d'élasticité k_c de courbure de la membrane à partir des fluctuations thermiques de la forme d'une vésicule sphérique. Des liposomes ont été observés sous microscope travaillant au régime de contraste de l'interférence différentielle de Nomarski. Nous montrons que la valeur ainsi mesurée de k_c est celle du flip-flop libre. On discute la possibilité d'étendre la méthode à la mesure du flip-flop bloqué.

Abstract: A new method is proposed, permitting the measurement of the curvature elasticity modulus k_c of a membrane, starting from the thermally induced fluctuations of the form of a spherical vesicle. Observations of the liposomes were carried out under a microscope working in the regimen of Normarski differential interference contrast. We show that the value of k_c thus measured is the curvature elasticity modulus of free flip-flop. We discuss possibilities of modification of the method to allow the measurement of the modulus of blocked flip-flop.

Authors' address:

P. Bothorel
Centre de Recherche Paul Pascal (C.N.R.S.),
Domaine Universitaire
F-33405 Talence Cedex, France

Kinetics of the phase transition in phospholipid bilayers

A. Gnez, R. Groll, and J. F. Holzwarth

Fritz-Haber-Institut der Max-Planck-Gesellschaft, Berlin, F.R.G.

Authors' address:

J. F. Holzwarth
Fritz-Haber-Institut
der Max-Planck-Gesellschaft
Faradayweg 4–6
D-1000 Berlin 33, F.R.G.

Abstract: The Iodine-Laser Temperature-Jump (ILTJ) method has been used to measure the dynamic processes which represent the main phase transition in phospholipid bilayers. We found five well-separated relaxation phenomena between 10^{-9} s and 10^{0} s. By choosing turbidity, fluorescence of absorption as detection parameters it is possible to create a dynamic model which attempts to explain the complex molecular processes during the crystalline fluid transition.

Bilayers containing cholesterol show strong changes in the relaxation times and the co-operativity of the processes between 10^{-6} s and 10^{-1} s. The incorporation of functional polypeptides or proteins like gramicidin and bacteriorhodopsin shifts the relaxations to shorter times and reduces the co-operativity. If we combine all the kinetic results we conclude that there are certain time domains in which the interactions of phospholipids and functional units inside the bilayer favour the switching of the membrane from a passive into an active role or vice versa.

Physical gels from PVC: Dynamic properties of dilute solutions

S. J. Candau, Y. Dormoy, P.-H. Mutin[1]), and J. M. Guenet[1])

Laboratoire de Spectrométrie & d'Imagerie Ultrasonores, Unité Associée au CNRS n° 851, Université Louis Pasteur, Strasbourg Cedex, France, and
[1]) Institut Charles Sadron (CRM-EAHP), CNRS/ULP, Strasbourg Cedex, France

Authors' address:

S. J. Candau
Laboratoire de Spectrométrie &
d'Imagerie Ultrasonores
Unité Associée au CNRS n° 851
Université Louis Pasteure
4 rue Blaise Pascal
F-67070 Strasbourg Cedex, France

Abstract: In order to gain information on the gelation mechanism of PVC in diethyl malonate, investigation of dilute solutions has been achieved through the use of transient electric birefringence, dynamic light scattering and viscometry. The dynamical structure factor obtained from the cumulant analysis of the autocorrelation function of scattered field exhibits an anomalous K dependence which can be assessed to an intra molecular effect. Transient electric birefringence experiments show that the aggregates exhibit a permanent dipole which we assume to be associated with crosslinks of small syndiotactic crystallites. Measurements of intrinsic viscosity and rotational diffusion coefficient provide information on the internal structure of the aggregates formed by quenching dilute PVC solutions.

Progress in Colloid & Polymer Science   Progr Colloid & Polymer Sci 73:197 (1987)

New method in elastic light scattering from Mie-particles

O. Glatter and M. Hofer

University of Graz, Graz, Austria

Abstract: Particles in the size range $200 \leq D \leq 2000$ nm have enough information in their scattering function for a general interpretation procedure. The experimenter is faced with two basic problems. There is an essential loss of information caused by the cutoff, and the scattering theory of Mie has to be used for the description of the scattering process. It has been previously proved that the interpretation of scattering results is much easier in real space (distance distribution function $p(r)$) than in reciprocal space. It is shown to what extent it is possible to apply the corresponding techniques to light scattering data. There exists no analytical method for the transformation of Mie scattering data into real space. Therefore we have used Fourier transformation as an approximation. The results show that this approximation gives reasonable results for ideal data as well as for data with up to 10 % statistical noise. The main influence of the Mie contribution are oscillations outside the maximum particle dimension. The shape of the part up to the maximum dimension containing the important information about the structure is nearly unchanged. The principles of the method and the application to spherical particles have been published recently [J. Coll Interf Sci (1985), 105, 577–585]. We can now present results for nonspherical and inhomogeneous particles. The radial polarizability profile can be calculated by a convolution square root technique, even for particles with imperfect symmetry.

Initial results for polydisperse systems show that this method could be superior to the particle sizing methods using quasi-elastic light scattering in the size range mentioned above.

Authors' address:

O. Glatter
University of Graz
Graz, Austria

Progress in Colloid & Polymer Science

Progr Colloid & Polymer Sci 73:198 (1987)

Melittin-induced disc to vesicle transition in lipid systems

J. F. Faucon, E. J. Dufourc, J. Dufourcq, G. Fourche, J. L. Dasseux, M. Le Maire[1]), and T. Gulik Krzywicki[1])

C.R. Paul Pascal, CNRS, Domaine Universitaire, Talence, France, and
[1]) C.G.M., CNRS, F-91190 Gif-sur-Yvette

Abstract: The amphipathic peptide melittin was studied for its ability to stabilize new lipid-peptide supramolecular entities. Below the phase transition temperature T_c, and for lipid-to-protein molar ratios $Ri > 20$, the solid state 2H-NMR spectra of chain-deuterated dipalmitoylphosphatidylcholine (DPPC) exhibit a sharp line, due to very small objects undergoing fast isotropic reorientation, superimposed on a broad spectrum of lipids in their gel phase. Above T_c, and for $Ri > 20$, a single phase characteristic of large lamellae is detected by NMR and local order can be determined as a function of the bilayer depth. Higher amounts of peptide ($Ri = 4$) prevent the formation of large structures, whatever the temperature is.

Freeze-fracture electron microscopy, when performed from the gel state and at $Ri > 30$, leads to the observation of apparently unperturbed large lamellae, coexisting with very small structures. At $Ri = 15$, only disk-like particles are detected with ca. 190 Å length and ca. 40–60 Å thickness. Freezing from 50 °C, at $Ri = 30$, yields a homogeneous distribution of large vesicles of ca. 2000 Å diameter which present a rough surface.

Gel filtration on Sepharose 4B and dynamic light scattering data corroborate these findings. At high melittin content ($Ri = 4$) and $T = 20$ °C small complexes are isolated, of radii ca. 70–80 Å, with an effective lipid to peptide ratio (Rc) of ca. 20. For $Ri > 15$, the radius of praticles increases up to ca. 2000 Å at the transition temperature of the lipid, i. e., a transition from discoidal particles to large unilamellar vesicles occurs at T_c.

The above results show that melittin can promote, either by vesicularisation or fusion, the formation of new well defined supramolecular structures. This process is independent of the initial state of lipids (multilamellar or small unilamellar vesicles), but depends on the amount of melittin bound and on the physical state of the bilayer.

References

1. Dufourc EJ, Smith ICP, Dufourcq J (1986) Biochemistry, in press
2. Dufourcq J, Faucon JF, Fourche G, Dasseux JL, Le Maire M, Gulik-Krzywicki T (1986) Biochim Biophys Acta 859:33–48
3. Dufourc EJ, Faucon JF, Fourche G, Dufourcq J, Gulik-Krzywicki T, Le Maire M (1986) FEBS Lett 201:205–209

Authors' address:

J. F. Faucon
C.R. Paul Pascal
CNRS
Domaine Universitaire
F-33405 Talence, France

Progress in Colloid & Polymer Science   Progr Colloid & Polymer Sci 73:199 (1987)

Mechanical properties of swollen polyelectrolyte gels covalentyl crosslinked

R. Wehn and D. Woermann

Institut für Physikalische Chemie, Universität Köln, Köln, F.R.G.

Abstract: The static viscoelastic properties of condensation products of phenolsulfonic acid/formaldehyde gels in equilibrium with aqueous electrolyte solutions and acetonitril/water mixtures have been studied as a function of the degree of crosslinking, electrolyte concentration and counterion species, temperature and concentration of acetonitril. The experimental results can be summarized as follows:

1. The gels do not exhibit rubber elastic properties but behave like solids to a first approximation.

2. The viscoelastic properties can be described by a model composed of a Maxwell element (spring and dashpot in series) in series with several Voigt elements (spring and dashpot in parallel).

3. The viscoelastic properties of gels loaded with silver ions and potassium ions, respectively, in equilibrium with acetonitril/water mixtures differ in a characteristic way.

A model of the structure of phenolsulfonic acid/formaldehyde gels is proposed to interpret the experimental results. This model takes into account small X-ray scattering data. It can explain discrepancies between model calculations of transport coefficients of membranes formed by this type of gel and experimentally observed phenomenological transport coefficients.

Authors' address:

D. Woermann
Institut für Physikalische Chemie
Universität Köln
Luxemburger Str. 116
D-5000 Köln 41, F.R.G.

Light scattering of concentrated micellar solutions: Influence of monomers

M. Drifford, L. Belloni, M. Dubois

CEA-IRDI-DESICP-Département de Physico-Chimie, CEN. Saclay, Gif sur Yvette Cedex, France

Abstract: We have studied both the intensity and the autocorrelation function of light scattering by concentrated micellar solutions. Octyltrialkylammonium bromide series: $(C_8H_{17}N(C_mH_{2m+1})_3^+ Br^-$ with $m = 1$ to 4) have been measured at 25 °C as a function of surfactant concentration. These surfactant solutions have a high CMC (0.30–0.15 M) and the micellar aggregates are small. The influence of monomer diffusion on the dynamics of solution is analyzed: a) below the CMC, b) close to the CMC, c) Above the CMC in concentrated solutions (> 0.5 M).

Below the CMC, we consider the surfactant solution a pure electrolyte solution. From the scattered intensity we can test the non ideality correction and fit our experimental results with HNC calculations in the presence of an attractive potential. The dynamics of monomers are strongly dependent on surfactant concentration and decrease from Nernst-Hartley diffusion to an effective diffusion coefficient at the CMC.

Close to the CMC, some premicellization states are expected and we observe no noticeable effect in dynamics of solution. From the scattered intensity measured above the CMC, we can deduce the aggregation number (N), the "dry" size of micelles. A large modification of aggregation number is obtained and N varies from 20 ($m = 1$) to 13 ($m = 4$). The size is approximately constant ($R_I \simeq 12.2 \pm 1$ Å). Above the CMC a classical hard sphere behavior is derived from the volume fraction dependence of scattered intensity for $m = 1$. The increase in the polar head size induced an additional attractive potential interaction ($m = 2$ and 3). A particular behavior is detected for the propyl head group ($m = 4$). A very strong attractive potential must be used to analyze the experimental results. The large overlapping by the long head chains introduces a pseudocritical phenomena and at 70 °C we observe a two phase separation.

The dynamics of solutions are strongly perturbed by monomer diffusion. The effective diffusion decreases from the CMC, then passes by a minimum and then it increases slowly in an approximately linear way with increasing micelle concentration. Two analyse are proposed: chemical exchange between monomer and micelle and/or mixture of monomer and micelle. It is clear from our data that monomer diffusion is an important feature in the dynamics of concentrated micellar system with a high CMC.

Authors' address:

M. Drifford
CEA-IRDI-DESICP-
Département de Physico-Chimie
CEN. Saclay
Gif-sur-Yvette Cedex, France

Micellization of surfactants with divalent counterions

B. Lindström and A. Khan

Division of Physical Chemistry 1, Chemical Center, Lund University, Lund, Sweden

Authors' address:

B. Lindström
Division of Physical Chemistry 1
Chemical Centre
University of Lund
S-22100 Lund, Sweden

Abstract: Micellization of aqueous calcium octyl sulfate, sodium- and magnesium dode-cyl sulfate have been studied by a number of different techniques. Self-diffusion coeffi-cients of surfactant ions and of solubilized TMS were measured, with and without deca-nol added, over a wide concentration range by the Fourier transform NMR pulsed-grad-ient spin-echo method and those of Ca ions by the openended capillary tube method employing radio-active 45-Ca labelling. The two-site model is used to calculate concen-trations of free and micellar surfactant ions and Ca ions.

Free surfactant ion concentration is found to increase initially, having a broad maxi-mum around CMC, and then to decrease at higher total surfactant concentrations. This decrease is relatively smaller than that of surfactant systems with monovalent counter-ions. The degree of counterion binding is nearly constant over a broad concentration range. Conductivity data give a considerably higher counterion binding with Ca and Mg than with Na ions. 1-H line widths, 13-C NMR chemical shifts, and viscosity measure-ments reveal that the micelles are small and spherical over almost the entire range of the micellar solutions. However, if decanol is added, a dramatic micellar growth can be observed in the dodecyl sulfate systems. The results can be explained qualitatively by electrostatic theory developed for ionic surfactant systems.

Diffuse interface between oil/water microemulsions in the low surfactant concentration range

C.M.C. Gambi[1]), L. Léger[2]), and C. Taupin[2])

[1]) Department of Physics, University of Florence, Italy, CISM (of the M.P.I.) and GNSM (of the C.N.R.) groups, and
[2]) Laboratoire de Physique de la Matière Condensée, Collège de France, Paris Cedex, France, GRECO "Microemulsions" du C.N.R.S.

Authors' address:

C. M. C. Gambi
Departement of Physics
University of Florence
L. E. Fermi n. 2
I-50125 Florence, Italy

Abstract: The system brine (65.74 %, with NaCl salinity of 6.5 %), toluene (31.9 %), n-butanol (2.3 %) and SDS (0.04 %) displays at room temperature a new type of phases equilibrium between an upper oily transparent phase and a lower oil-in-water micro-emulsion domain; the latter exhibits two regions different in composition and structure. The oil-microemulsion interface is sharp, while the microemulsion-microemulsion inter-face is diffuse, the diffuse region being a few millimeters thick. Results of compositional and structural investigations are shown in terms of the index of refraction, gas chromato-graphy and diffusion coefficient (quasi-elastic light scattering analysis). To account for the observed results, an interpretation is proposed in terms of structural change of the o/w microemulsion (droplets coalescence).

Subject Index

Author Index